Isolated Wonder

A Scientist in the Robinson Crusoe Islands

Isolated Wonder

A Scientist in the Robinson Crusoe Islands

Tod F. Stuessy

Herbarium and Department of Evolution,
Ecology, and Organismal Biology,
The Ohio State University, Columbus, Ohio, U.S.A.,
and Department of Botany and Biodiversity Research,
University of Vienna, Austria

MISSOURI BOTANICAL GARDEN PRESS

Library of Congress Cataloging-in-Publication Data

Names: Stuessy, Tod F., author.
Title: Isolated wonder : a scientist in the Robinson Crusoe Islands / Tod F. Stuessy.
Other titles: Scientist in the Robinson Crusoe Islands
Description: St. Louis, Missouri : Missouri Botanical Garden Press, [2024] | Includes index. | Summary: "An exploration of the Robinson Crusoe Islands, a national park of Chile located more than four hundred miles west of the South American continent. The classic novel Robinson Crusoe by Daniel Defoe was based on the true adventures of Alexander Selkirk, a Scottish sailor who lived isolated in these islands from 1704 to 1709. There exist unique flowering plants and ferns that stimulate scientific questions: How and when did these plants arrive in the archipelago? What processes of evolution occurred after they colonized? Botanists on twelve expeditions over forty years from Ohio State University, University of Concepcion, University of Vienna, and other institutions have revealed secrets of this unique plant world through numerous scientific publications. Isolated Wonder explains clearly and honestly the challenges of doing fieldwork on these islands, the excitements, the disappointments, and how research can dramatically impact personal lives, and vice versa. This book provides informative and entertaining insights into what happens on botanical expeditions in an isolated region of the world."—Provided by publisher.
Identifiers: LCCN 2023048851 | ISBN 9781935641285 (paperback)
Subjects: LCSH: Botany—Fieldwork—Juan Fernández Islands. | Island plants—Juan Fernández Islands. | Juan Fernández Islands. | Robinson Crusoe Island (Juan Fernández Islands). | BISAC: BIOGRAPHY & AUTOBIOGRAPHY / Science & Technology | SCIENCE / Life Sciences / Botany
Classification: LCC QK473.J8 S87 2024 | DDC 581.983—dc23/eng/20231214
LC record available at https://lccn.loc.gov/2023048851

ISBN 978-1-935641-28-5
Library of Congress Control Number 2023943626
Managing Editor: Allison M. Brock
Editor: Lisa J. Pepper
Press Coordinator: Amanda Koehler
Cover and book design by Paula Catalano

4344 Shaw Boulevard, St. Louis, Missouri 63110, U.S.A.
www.mbgpress.org

This book is dedicated to my good friend and colleague, Daniel J. Crawford, whose consistent support, enthusiastic field participation, and laboratory analyses were major factors for success of our research in the Robinson Crusoe (Juan Fernández) Islands.

"It was bright starlight when we were clear of the bay, and the lofty island lay behind us in its still beauty, and I gave a parting look and bade farewell to the most romantic spot of earth that my eyes had ever seen. I did then, and have ever since felt an attachment for that island altogether peculiar. It was partly, no doubt, from its having been the first land that I had seen since leaving home, and still more from the associations which everyone has connected with it in his childhood from reading Robinson Crusoe. *To this I may add the height and romantic outline of its mountains, the beauty and freshness of its verdure and the extreme fertility of its soil, and its solitary position in the midst of the wide expanse of the South Pacific, as all concurring to give it its charm."*

Richard Henry Dana Jr.,
Two Years Before the Mast (1840).

Contents

Preface

There are lots of books dealing with popular science. After long careers, many scientists feel that they want to share experiences about research and their personal lives with the non-scientific public. Being a scientist is a bit different from other people, in that we are always questioning, probing, and seeking answers about the world. This can be irritating to people who do not routinely do this, and they may find it inappropriate, perhaps even rude, to have this constant search for more information. As my wife sometimes says after a series of questions about some topic: "What is this, the Inquisition?"

Scientists harbor the hope that it is worthwhile, sort of a professional duty, to explain to the general public what we do. In this age of rejection of scientific evidence by some people in favor of cult and belief systems, there is a need to place science front and center in our society. The wonderful civilization that many of us enjoy is due largely to scientific discoveries and technological advances over the past 200 years. We must seek more scientific progress—not backslide into medieval authoritarian systems. Talking to the general reader about what scientists do and why they do it just might be helpful to further a positive attitude on science. At minimum, it is worth trying.

But my motivation for writing this book is even more specific. During the past forty years of working in the Robinson Crusoe Islands, my colleagues and I have learned much about the archipelago and about the evolution of plants in oceanic islands in general. We have published more than 140 articles in scientific journals on diverse topics dealing with the evolution, biogeography, and conservation of plants of these Chilean islands (see list in Appendix 2). We brought these scientific results together in a recent book entitled *Plants of Oceanic Islands* (2018). This is a technical volume designed to help establish a new research platform for future investigations in the archipelago. In addition, I have published another book, *Environmental History of Oceanic Islands* (2020), which focuses on the historical ecology of the vegetation of the islands. This is a deeply historical volume that infers natural forces that have impacted the landscape of the islands during the past four million years, as well as those influences from humans over the past 400 years. The many historical documents from visitors to the archipelago represent a detailed case study of the impacts of humans on the ecosystem of an oceanic island, and it causes us to reflect on how our activities are now affecting the entire planet.

In addition to these journal articles and books, there is still much more to say about the Robinson Crusoe Islands. Simply put, I want to tell about my personal experiences accumulated over the past twelve expeditions from 1980 to 2011. Getting to the islands is not simple and working there is even more challenging. There have been many excitements, expectations, joys, and frustrations. I also want to explain what motivates our science—why we go so far away, why we endure such personal hardships on the trail, what we do when we are there, and what the results show. These islands, as do all oceanic islands, hold clues to the origin of new life forms, and being so close to these origins in some way gets us closer to our own.

The documentation for these expeditions derives from several sources: field collection notebooks on all expeditions from 1980 to 2011; diaries I kept in the years 1991, 1996, 1997, 1999, 2000, 2010, and 2011; a diary made on our expedition of 1986 by my participating doctoral student, Thomas Lammers; a notebook of observations and impressions in Concepción, Chile, during 1994; and file folders of letters, receipts, and memorabilia from all of the expeditions. These resources have been deposited permanently in the Library and Archives of the Herbarium of The Ohio State University.

Most of the images in this book are my own photographs from 1977 to 2011. The idea has been to show how things were in the islands when the events took place, but a modern visitor will note many changes. This is especially a result of destruction of buildings from the tsunami in February of 2010 followed by new construction. Slow modernization of the airline facilities has also occurred. Even life in the settlement of Las Casas on Alejandro Selkirk Island has changed somewhat with better communication and availability of electricity, but it is still a very isolated part of the planet. To provide historical accuracy, the year in which each photograph was taken is indicated in the Notes section at the end of the book. Photos were selected for clarity and attractiveness to illustrate points in the text, sometimes taken in years other than the time being discussed.

Much of the text is a narrative—not at all a scientific presentation. My objective has been to describe our activities in some detail, so that the reader can imagine the places, come to understand the people, and in this way enter this island world. Most deliberately included are comments on how I felt prior to and during these expeditions. How was the project first initiated? What were my impressions as I got more immersed into the project? What were my fears and anxieties on the excursions? Sometimes I digress to comment on personal or professional perspectives, including relating stories from other collecting trips or experiences. This book is not fiction; everything in it is true, to the best of my ability to remember events, places, and people. My objectives have been to inform and to entertain. I don't want anyone studying this book—I want it just to be enjoyed.

CHAPTER 1

Corrales de Molina

We were tired, dead tired. We are also scared. The eight of us stood on the slippery lava rocks beside the heaving sea, with waves crashing sporadically to our left and casting spray high in the air above us. We and our backpacks were getting wet—some members of our group had waterproof jackets and boots, but others were not so well protected. Each deep rumble of the sea announced a new wave that crashed against the rocky platform and sent vibrations under our boots and shot water high above. It was late in the day as the sun was setting, the temperature was dropping, and the sea was beginning to flex its muscles, ebbing and flowing three to five meters against the rocky platform. The white park service boat had just arrived to pick us up from the slick rocks, but the rising and falling of the sea, with swirling foam around the rocks and boat, made such a transfer extremely dangerous. The plan had been to arrive by 4:00 in the afternoon at the bottom of Corrales de Molina for calm seas and an easy jump into the waiting boat, and from there a one-and-a-half-hour routine voyage around Robinson Crusoe Island to the other side and the village of San Juan Bautista, where we were staying. But we had waited too long, and it was now after 6:00 in the evening—there had just been too much to see and do.

We had begun the day at national park headquarters in the village at 8:30 in the morning on 18 January 1996. Our guides had not yet arrived, but this was not unusual in the islands, nor is it unusual in Latin America in general, but by 9:00 in the morning they did appear. Besides myself, our group consisted of Gregory Anderson (United States), Gabriel Bernardello

(Argentina), Ulf Swenson (Sweden), Marcelo Baeza (Chile), Eduardo Ruiz (Chile), Pedro Aqueveque (Chile), and Eric Tepe (United States), plus Oscar Chamorro and Guillermo Araya, two outstanding park service guides who live on the island. The objective for this day was to collect plant specimens in a difficult-to-access area called Corrales de Molina.

We started up the trail toward Damajuana (see color insert, **1**), one of the conspicuous mountain peaks behind the village. It was a relatively gloomy day, but not raining. It was cool and the ascent difficult as usual, but nothing out of the ordinary. We rested every now and then, and then continued to climb up into the forest of *Nothomyrcia* trees, found only on this island. Since we were not collecting from this area that day (we had collected from this same route six days earlier on 12 January), we made pretty good time. Hiking up these peaks is an effort, physically challenging, but not impossible—we just took it slowly step by step. One sweats profusely, which means taking off sweaters or hoodies, but after an hour or so on the trail, the cool breeze from the ocean felt refreshing.

By about 11:00 in the morning we had reached the top ridge to the south of Damajuana Peak. I collected a few plant specimens, and the group began to consider going farther south along the ridge. We got organized and started across the narrow ridge, about a meter wide, which dropped off very sharply on the village side. We worked our way over the ridge, buffeted by gusts of wind, but by hunching down and crawling slowly I was able to manage (although frightened). Guillermo then announced that we should continue along this ridge and up the next ridge. After looking at that possibility, I realized that I just couldn't do it. I told Guillermo it wouldn't be possible for me, which I hated to say, because it was a weakness—my weakness. He said I could be roped up it, and I trusted his intentions, but I knew it would be too much—I would freeze up, my legs would then weaken, possibly leading to a fall. It was simply too steep for me.

Now, I have to confess that I suffer from considerable vertigo. This is not a good attribute when working in a volcanic island with lots of steep slopes and precipices, sometimes dropping down 500 meters or more. Some members of our group could stand at the edge of some of these drop-offs and love every minute of it, the openness, the freedom, the expansiveness of the landscape. I, on the other hand, have to stand back about three meters, looking down at the ground so as not to get frightened. This vertigo developed when I was much younger, but it became terrifyingly obvious as a teenager in the late 1950s. Our church youth group was on a tour of New York City, and among the many interesting sights we visited was the magnificent Statue of Liberty. Inside was a narrow spiral staircase that went from the base to Liberty's head (long since closed). The staircase had wire mesh all around, so it was impossible to fall away from it, but for me it was a horrible experience. Just thinking again about looking down, down, down toward the bottom makes me nervous. And the space was mostly all open—there were very few dividers internally. This was the beginning of my understanding of vertigo. Another terrible situation years later was in the Chicago Sears Tower (now called the Willis Tower), which has a glass overhang in the top observation area on the 103rd floor that allows the visitor to look all the way down to street level. I preferred to stay pressed against the wall in the center of the observation deck, right near the elevator. Or perhaps you have ridden in those glass elevators that travel inside an atrium or even on the outside of a high-rise hotel—no thanks, no way. My heart rate starts to rise, I freeze solid, and then my legs get weak—not a good syndrome when trying to traverse a narrow and dangerous basaltic ridge.

Instead of crossing this Damajuana Ridge, I offered to go down the slopes of Corrales de Molina as we had done in 1991. I could go with a guide, and we could meet the other group going across the ridge and down another slope. Guillermo said no, we all must go together, because he had only one walkie-talkie radio that was in contact with park headquarters in the village, and he also had only one rope. Thankfully, he yielded to my fears, and we all went down directly into Corrales de Molina. No big problem for me here; it was the usual steep descent through ferns and shrubs. In general, working down for me is so

Valley of the Corrales de Molina, Robinson Crusoe Island

much easier than climbing up. Guillermo had stressed that heading down would be difficult, too, because there was a steep area where a rope was needed. I felt this was certainly the lesser of two evils, and I promised to make the drop without any resistance.

We started picking our way down into Corrales de Molina, collecting plants as we went, and finally reaching the area of *Lactoris fernandeziana* at about 380 meters. This small, yellowish-green shrub with fragile stems and inconspicuous greenish-white flowers is the only species in its own family, Lactoridaceae. There are about 1000 to 2000 individuals of this species in the world,[1] and all of them occur on this one island. This is its home in the universe, but even more remarkable is that it is known from DNA analyses[2] and fossil pollen deposits[3] to be a very old family, easily ninety million years old. What is it doing, then, on a volcanic island in the Pacific Ocean that is only four million years old? I will tell you more about this later in chapter 5, but right now we were delighted to find this patch of *Lactoris*. Greg and Gabriel stayed there longer because they were interested in determining pollination mechanisms (it appears to be by wind). We took many photos of each of us with the plant—in effect a comparison of our youthful human lineage no more than 300,000 years old[4] juxtaposed against this ancient plant lineage ninety million years of age. For this short moment, our communion with this other fascinating life form was impactful. I doubt our species will have such evolutionary success as this plant lineage, because we face so many challenges—overpopulation, tensions among countries, huge quantities of armaments, pollution of air and water, global warming, loss of topsoil, and extreme weather.

Gregory Anderson examining *Lactoris fernandeziana* in Corrales de Molina

We continued downward into Corrales de Molina collecting other plant material. The day was still overcast but no rain (thank heavens!). When it rains, everything gets slippery, especially the ferns on slopes, and also the rocks. Generally no excursions are attempted when it is raining, but if showers open up when on the trail, the best strategy is to stop and wait for the rain to abate, but even then the underbrush remains soaked. These treks are dangerous enough without the complication of unsure footing. At about 1:00 in the afternoon we selected a safe lunch spot in a ferny grotto, a bit shielded from the wind under some *Nothomyrcia fernandeziana* trees, three to four meters tall, which are also only known from this island (they are said to be endemic), but they are common, being the dominant forest tree. We were already tired from the constant care needed to climb down the slopes. Any missed step could result in serious injury, a broken leg, concussion, even death. Corrales de Molina is not an easy place to visit and work and is not recommended for the casual visitor. The views from the ravines (*quebradas*) looking outward toward the sea are breathtaking. I got out my usual lunch, one can of sardines in olive oil with *pan amasado* (hard, kneaded bread, well suited for carrying in a backpack on the trail), and a small piece of chocolate with almonds. It is amazing how good just a little chocolate tastes when working in the field. We often saved some for late afternoon to have a sugar pick-me-up to keep going. My water for the day was kept in a used plastic Sprite bottle, which proved quite satisfactory.

After lunch we headed down farther onto the slopes, which in many cases consisted of switching back upward, too, zig-zagging along the steep declines. Fortunately, we had many small *Nothomyrcia* trees to hold onto, the stems of young trees being just right for a good handhold, but it was tough going. My pack was getting heavier, especially with the large bracket fungus I had collected. We had two dogs with us—one a chocolate-headed mutt and another sort of a Collie–German shepherd mix but much smaller. The latter was accumulating in its fur enormous quantities of fruits of the *trun* (*Acaena argentea*), an invasive, trailing ground cover. When the dogs got to a steep part of the trail that they couldn't handle (which didn't happen very often), they started a pitiful whining—not the best sound to be hearing when we ourselves were a bit anxious. I love dogs, and it was a comfort to have them along, like airline passengers who get permission to take their "comfort animal" into the cabin.

Climbing slowly down the steep slopes of Corrales de Molina was manageable, until we reached the drop-off where rope work was required.

Rope work during descent of Corrales de Molina. Left to right: Gregory Anderson, Gabriel Bernardello, Eric Tepe, Oscar Chamorro, Guillermo Retamal (sitting).

Guillermo and Oscar tied a rope around each of us in turn, and we climbed down with the two of them holding onto the rope above. This for me was a thousand times better than trying to climb along a high ridge, although here one can look down hundreds of meters toward the sea, but there is low vegetation, which suggested to me that I could always grab onto something to save myself, no matter if I slipped or the rope broke. As we descended this very steep drop-off, one by one, the dogs started to get nervous and began again to whine. They really couldn't make the drop as it was just too straight down, so Oscar grabbed each one, made a sling around them, and let them down slowly by the rope. The only bad aspect of this descent was waiting out in the open before dropping down over the cliff. I sat down among plants of the wiry, short shrub, *Ugni molinae*, another introduced species, to feel a bit more relaxed while I waited. My turn finally came, and it proved no difficulty to inch down the thirty-meter slope. After each person descended, the rope was released, and the next person then was roped up and helped down. At the end, the park guides coiled the rope and simply climbed down without help. They are exceptionally good climbers, very sure of foot, and without specialized climbing gear.

We continued our descent, all the while making plant collections but proceeding slowly and carefully. Each step took concentration. At last, we arrived to the lower reaches of the ravine at about 280 meters. It felt good to rest a few minutes here with the group, but Guillermo then called out that we needed to hurry down to the boat. I wanted to spend additional time doing general collecting in this area, but the best we could do was to make quick grabs of plants on the way down, being careful of footing and listening to Guillermo's instructions and encouragements.

The final and lowest part of Corrales de Molina was easy enough, except that we were walking through much trun (*Acaena argentea*) with its intermeshed surface runners that caught at our feet—one had to be careful not to twist an ankle. Further, and worse, were the numerous loose rocks under the trun, so that each step required extra concentration. At this point, my backpack and those of the others weighed perhaps twenty kilograms (around forty-five pounds) each. Guillermo was ahead of us, and he kept calling for us to come faster, which was really impossible considering the unstable substrate. We jumped across several small creeks (*saltos*) at the bottom of the quebrada, but nothing dangerous. Nonetheless, we were tired, so these crossings required special attention. I am happy to relate that in all our trips to the islands, the worst calamities have resulted in only broken fingers—nothing more serious. There were occasional stomach problems, sprained muscles, bacterial skin infections, but the injuries could have been much—very much—worse.

We assumed that when we finally arrived near the ocean, it would be easy going down the sea cliffs to the boat. This proved to be a false impression. Climbing down the sea cliffs with sharp, dark brown lava rocks was not super difficult, but we had to descend with care and not slip. At one point, I was inching along a narrow lava ledge, and suddenly something was pushing my feet—it was a dog wanting to forge ahead! We crept along and down the rocky ledges, descending them one by one, until finally coming out near the sea. There was a final small drop down onto a rocky platform that was slimy with algae, which should have been a clue to be more alert, but after this short drop, I felt relieved that the adventure was nearly over. This turned out to be completely wrong. As I walked confidently across this slippery lava exposure, suddenly a huge wave came crashing over it. I grabbed a lava dike on my right to hold on, but the wave covered me nearly to the waist, the lower part of my body now being soaked. Guillermo yelled at me to get moving off that spot, to which I responded quickly, imagining that another wave might be imminent. Most of the other men were already at the boat landing area with the other guide, Oscar; the Corporación Nacional Forestal (CONAF) boat was coming to rendezvous with us a short distance away.

Pick-up spot at the bottom of Corrales de Molina, on the southern side of Robinson Crusoe Island

It was now after 6:00 in the evening, two hours later than originally scheduled. Bernardo López and Jorge Angulo in the boat had already picked up two women phycologists who were collecting marine algae at the bottom of Villagra, another location also on the southern side of the island. Hence, the boat already contained these two women, plus one of their students, and a schoolteacher from the island, with the guides operating the boat. Our group amounted to ten persons, eight scientists plus our two guides (Guillermo and Oscar), all of which totaled sixteen—and the boat had a listed capacity of only fourteen! This was not a good calculation for ensuring a safe escape from the rocks and return to the village.

The sea was rough, very turbulent, and the boat was bobbing as much as three to five meters relative to the shore. There were stepping rocks to the sea, but they were all very slick, covered with algae. The park service boat had been anchored behind for stability, so that it could be kept from smashing forward against the lava rocks. First Marcelo tried to get into the boat. He still had his backpack on (bad idea) and stepped down to the slippery ledge near the sea. Bernardo let the boat out a little toward the rocks, and Oscar and Guillermo pulled it in close with the rope extending

Park service boat coming to pick us up from the tidal pool lava rocks at the bottom of Corrales de Molina

from the boat to the shore. Marcelo got ready for the jump. Because the boat was moving up and down, a jump had to be calculated when the wooden prow was on the upswing close to the level of the rock. Before Marcelo could make the leap, a huge wave swept up and covered him completely. He first grabbed the rocks below him, but then let loose and grabbed the air above. Luckily, his hand found the rope to the boat onto which he clung to keep from being swept over into the sea.

The situation was obviously more serious than anyone had thought. Jorge shouted to take off the backpacks to lessen weight for our jumps into the boat. At Greg's suggestion, we had also taken off our jackets from under our rain gear, and put these in the backpacks along with cameras so they would already be safe and dry in the boat. These were put on another smaller rope and run down the line into the boat. Some of the backpacks got wet when waves came crashing in, but the transfers went reasonably well. Marcelo was now totally soaked and cold. There was a strong wind, and occasionally it would rain and then stop. The life jackets from the boat were then sent over to us by rope in a sack, especially for the persons who could not swim. This was not a very pleasant situation—it was quite tense for all of us. The guides later admitted that this was the worst boat landing they had ever experienced.

Gabriel Bernardello was the next to try to leap into the boat. The guides recommended that Gabriel tie a rope from the boat around his waist, so that if he slipped he could be pulled out of the sea (this proved to be a good idea for several people). Gabriel waited for the right moment, the sea heaving up and down, the boat rising and falling, and then he leaped—and like a spider he landed perfectly, extended on the prow.

One on board safely! The other persons then tried their luck, and the sequence was Eric, Marcelo, Pedro, Greg, Ulf, me, Eduardo, Oscar, and Guillermo.

Eric had on just a T-shirt, and he stepped down to the ledge to calculate the jump. Just then a wave came and swept him off the ledge and into the water below the boat! This was extremely dangerous, especially because he was not roped up! He briefly disappeared under water and then came up again. It was a miracle that he wasn't squashed between boat and rocks. The guides pulled him off to the side and finally up into the boat—he was totally soaked and shaken (but alive and with no broken bones).

Everyone had some level of problem, although I was luckier than most. I was confident that I could calculate the jump as long as the waves didn't knock me off. When my turn came, I positioned myself, and when the boat came in and rose upward, the guides shouted "*Ahora*!" ("Now") and out I went—extended flat onto the prow—no problem. After each person landed safely everyone clapped and gave the thumbs-up sign. Eduardo got ready and jumped but missed the boat a bit, and as he disappeared into the water, he grabbed the side of the boat and hung on. The two guides and I pulled him in. He was wet but safe. Oscar, who professed not to swim, put the rope around himself, and just jumped into the water alongside the boat. We fished him out—his arms were like steel. Guillermo then made the last jump, and we were all safely aboard.

I mentioned Oscar's arms. These island guides, who have lived here all their lives, grew up climbing up and down these mountains and valleys. They are all lean and wiry, not an obese person among them. And they are really strong—very tough, tenacious, and powerful. They may not look like weightlifters or bodybuilders, but ounce for ounce, I would bet on the strength of these island men over anyone on earth. They are also extremely sure-footed and, obviously, do not suffer from vertigo.

Our sense of security of being in the boat was short-lived, because Bernardo tried to start the small Yamaha outboard motor, and it wouldn't engage! Finally, after several attempts it revved up blessedly, and we headed out to sea. But the problems were still not over. The sea was very rough with three- to five-meter waves. The boat slid down into and rose up out of water troughs, with occasional waves breaking over us (and into Eric's face—he took a real bath several times). I was facing the other direction, so the spray hit only my back. Greg and I discussed what we could do if the boat capsized, but we could agree only that it would be grim. I had no life preserver on at this point, because it had been passed to another person for jumping into the boat.

We finally made the open sea, but the waves continued high and rough, with occasional troughs into which we fell with a bump. Bernardo did a great job handling the boat. I felt a little queasy but kept thinking about the research project to keep my mind off the movement of the boat. Everyone was cold and unsmiling. As a good omen, the guides kept chatting and joking—never telling us how they might really have felt. The rough sea continued until we rounded the other side of the island near Puerto Francés. Waves then settled down, which was very welcome, but we were cold—Oscar and Eric were really shivering. As we neared the dock of the village of San Juan Bautista (St. John the Baptist), I began to shiver too, and I couldn't stop. I had given my woolen sweater to Marcelo to keep him warm, and the red and blue cotton hoodie under my yellow rainsuit was not adequate.

Upon finally arriving at the cement dock, we walked unsteadily up the stairs and shook hands and embraced all around—it was a good feeling to have arrived safely. Ramón Schiller, the acting chief of the island national park, came down with the service jeep, and we excitedly told him of our adventure—it was really too much—too stressful and really life threatening. We dragged ourselves up the slope to our house and got out the whiskey bottle, Johnnie Walker Red Label, which we were saving for the final party. The contents disappeared quickly. We drank hot coffee, too, and Eric and Marcelo took hot showers to warm up. There were stories all around and much chatter; we toasted being alive and safe again. The old house we were staying in really felt good. It is funny how relative everything in life is. We lived in a very rustic house in the village in 1991, but after a few days it seemed quite comfortable (see chapter 7). In my tent that night, which I hung inside my bunk bed to keep the mosquitos away (I just can't sleep with mosquitos buzzing around my ears), I was glad to be safe, thinking of family, friends, and home back in Columbus, Ohio.

But, what is all this about? Why am I telling you about these adventures? What I want to do in this book is to take you through my experiences in the Robinson Crusoe (Juan Fernández) Islands accumulated over the past twelve expeditions from 1980 to 2011. Getting to the archipelago is not so simple. I want to take you across 667 kilometers of vast, deep, blue ocean in a boat or a five- to nine-seat propeller airplane, as a small seed might be carried by the wind, or as a bird might errantly fly westward seeking the next available land. There is much to share with you—excitements, expectations, joys, frustrations. I also want to tell you about our science—why we go so far away, why we endure such personal hardships on the trail, what we do when we are there, and a bit of the results and their significance. Let me tell you how it all began.

CHAPTER 2

Meeting Mario

Many a time in winter I have relaxed at home in a comfortable wingback chair, feet up on the divan, warm coffee at the three-legged side table, with a fire crackling in the fireplace, while it is snowing outside—large, wet flakes that so often fall in the relatively mild Columbus, Ohio, winter. On such a beautiful evening I might leaf through pages of a book old or new, but all new to me, of some distant land. The photos and drawings are at first so inaccessible, but they slowly begin to assume reality. I get up and go to the atlas of maps to understand further where these territories might be, what their relationships are with other areas I might already know something about, their elevations, and so on. Back to the chair, another sip of coffee, and once more into this new land.

And so I was one particular evening in 1975 deep into Sherwin Carlquist's engaging book *Island Life: A Natural History of the Islands of the World*.[1] Such fascinating islands: Hawaii, the Galápagos Islands, which proved so stimulating to Charles Darwin on his visit there in 1835, the Canaries, St. Helena, Mallorca, and the Robinson Crusoe Islands. Further into the pages I went, and it was astonishing to see the bizarre plants and animals which, in their isolation from continental source areas, assume unusual sizes and shapes. Most incredible to me were the sunflowers, relatives of plants that produce "seeds" (actually dry, single-seeded fruits) sold in supermarkets, distant relatives of lettuce, endive, chicory, or the globe artichoke. In islands these were often trees many meters tall, with thick trunks and a bizarre series of leaf and floral forms. I continued to travel to islands that evening, flying in and out of the landscapes, over sun-

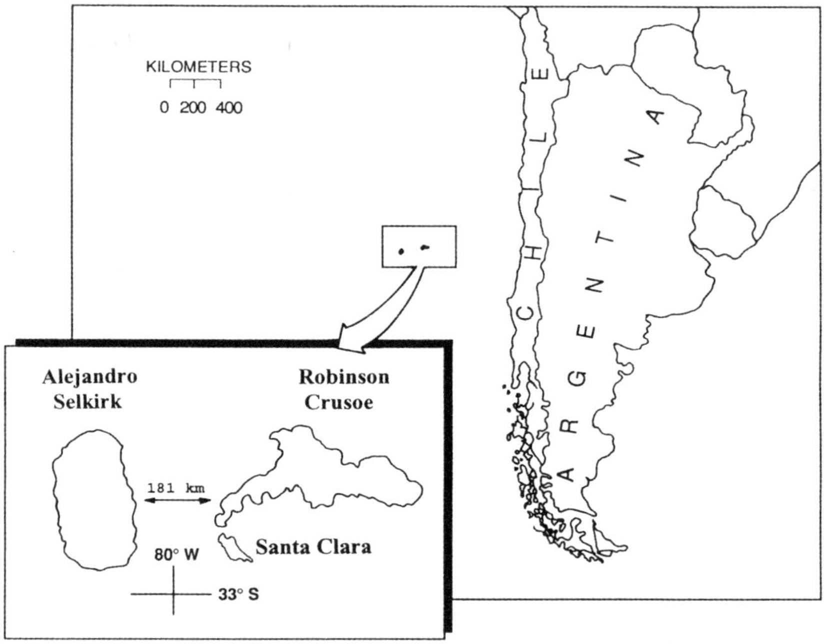

Location of the Robinson Crusoe (Juan Fernández) Archipelago

baked exposed lavas, around frigid, snow-capped peaks, down into lush, moist valleys, and along the tops of plants that really had no business being trees. I was about as close to visiting another planet as I ever dreamed possible. But honestly that evening I never for a moment believed I would actually visit some of those islands, certainly not the Robinson Crusoe Archipelago,[2] so far out in the Pacific Ocean off the coast of Chile in South America. But I did visit them, and sooner than I thought.

Later in 1975 I wrote to the Organization of American States in Washington, D.C., asking if they might have research funds to support my work on systematics of South American members of the sunflower family (Asteraceae).[3] They responded that such funds were awarded competitively only to Latin American scientists, but that there were three professors with whom I might possibly collaborate, one each in Guatemala, Ecuador, and Chile. I wrote to all three of them, and I received replies from the two in South America. The letter from Chile was especially encouraging. Dr. Mario Silva, from the Department of Botany of the University of Concepción, wrote back quickly that he was interested in developing a joint chemical-systematic project, perhaps in the sunflower family with which he already had a strong publishing record. Trained at King's College, London, in the laboratory of Sir Derek Barton, who later was awarded the Nobel Prize in Chemistry, Mario was obviously a perceptive scientist and highly motivated researcher. With his strong background, he developed one of the most highly respected natural plant products laboratories of Latin America, and particularly within Chile.

In March of 1977 I was looking out the airplane window on the Andes Mountains, armed with a small development grant from the U.S. National Science Foundation, en route to Chile to meet Mario for the first time and to talk about possible collaborative research projects. I was crossing the Andes from Buenos Aires on the Atlantic side of South America to Santiago de Chile on the western side of the continent on the last leg of the air route from Miami. I couldn't get enough of the views of these mountains—they are so broad, so many valleys, so many snowy, jagged peaks. At 35,000 feet they look almost ruggedly inviting—adventurous places to visit and hike into—but in reality they can be deadly isolating if a plane plunges haplessly downward.[4] Here in March it was late summer in the Southern Hemisphere, and the snow had receded to its maximum extent. From this perspective with the white glistening Andes stretching as far as the eye could see, I sensed the strength of South America—the powerful plate tectonic forces that resulted in the Andean uplift. I also began to realize that I was really in South America! Strange as this may seem, whenever I visit a new place, especially a place I have dreamed of visiting, it takes me a while to internalize that I am actually living out that dream and adding it to my life story. To think of this even now brings a smile to my face and a warm recollection of that first flight over the backbone of South America.

View of the Andes Mountains on my flight from Buenos Aires to Santiago, Chile

Suddenly the plane began to descend, rather too quickly I thought, as we were still quite evidently over the Andes, as far as I could see from my little porthole on the world. Lower we went, even more quickly, although the flight attendants and experienced travelers continued about their usual business of attending to cabin needs, reading the in-flight magazine—not totally consoling to me, however. Not until we began to cross the lower, drier, scrub-covered slopes of the Andes and swoop down into the agricultural fields of the central valley of Chile did I relax. We had crossed the Andes in less than half an hour! I imagined the difficulties more than 200 years ago as people struggled through the low passes in summer and were entirely shut out in winter.

Chile is isolated, like the islands I had read about during that snowy Columbus winter evening. The tall Andes Mountains isolate it on the east from Bolivia and Argentina, and obviously the Pacific Ocean isolates it on the west. To the north is the vast Atacama Desert, one of the driest places on earth. And to the south is the frozen continent of Antarctica, of which Chile claims a pie-shaped portion extending to the South Pole. Politically Chile is also isolated due to interactions between Peru, Bolivia, and Argentina. In the War of the Pacific (1879–1884),[5] Chile won thanks to the superior naval tactics of Admiral Arturo Prat and a more disciplined military force that Chile has always maintained, and Chile parted company with its neighbors to the north. And with Argentina, there has been a constant friction due to boundary disputes in the southern Beagle Channel over petroleum fields that both countries have wanted. This reached a climax in 1978, and war was averted only through the intervention of Pope John Paul II, first on the military government of Argentina and then on that of Chile. Although relationships among all countries of the Southern Cone of South America are now cordial, these background tensions continue to keep Chile isolated psychologically, especially among the older generation. It must be remembered that Chile is a small country with only nineteen million people (about the same population size as New York), but is stretched out geographically over more than 4,000 kilometers of territory (not including Antarctica), and barely seventy kilometers wide on average.

The life of a country, its rhythms and customs, are in part determined by its climate and the vegetation that it has. And so with Chile, the northern one third is warm desert, the middle zone a Mediterranean-type climate with hot and dry summers and cool, wet winters, and the southern

Map of Chile and other countries of the Southern Cone

third with the moist Magellanic forests. More specifically, the northernmost part of the country is the Norte Grande, a true desert of the driest proportions imaginable. There are areas that receive no rainfall—yes, *no* rainfall for years, and in which there is *no* visible plant life—nothing but rocks and sand. This region up into the cordillera contains economically significant deposits of copper, manganese, cobalt, and the semi-precious gemstone, lapis lazuli, with its lovely dark blue color. Farther to the south is the Norte Chico, a semi-desert zone, with cacti and drought-resistant shrubs of various plant families. The Central zone is a true Mediterranean-type climate with spiny shrubs and grassland; this is where most of the agriculture of the country occurs, and it is therefore the most ecologically disturbed region. Up into the cordillera east from the central valley is the high Andean zone, reaching to alpine meadows above timberline; it is here that the well-known ski resorts of Portillo and La Parva wait for those Northern Hemisphere aficionados who ski all year round. Farther south is the Lake Region (Región de Los Lagos) with Valdivian forests and their vast timber resources, now much planted with lodgepole pine (*Pinus radiata*) from the United States, and eucalyptus (*Eucalyptus globulus*) from Australia. This lumbering industry has created strong pressure on the native trees and flora in general. Finally, to the very south is the Extremo Sur with the much cooler and wetter Magellanic forests, similar to forests of the Olympic Peninsula of Washington in the United States and adjacent Canada. This zone includes the major permanent ice fields, which are the fourth largest in the world (after Antarctica, the Arctic, and Greenland). Most of the population of Chile is concentrated in the central zone, with 35% in the metropolitan region of Santiago alone.

Some people like Santiago; others don't. I love it. But I have so many mixed feelings about the city, that it's hard to separate them. I courted my wife here in restaurants, nightclubs, and museums; I celebrated our engagement here, beginning with rings over breakfast in the lobby of the Hotel Irazú; we were married here; and I have enjoyed countless emotionally meaningful moments with my new family. I have also spent pleasurable hours in dusty used book shops; I have savored the *barros jarpa* (hot ham and cheese sandwich) and beer in inviting sidewalk cafes; I have valued the warm summer days working in the herbarium and library of the

Museo Nacional de Historia Natural in Santiago, Chile

natural history museum in Santiago (Museo Nacional de Historia Natural) on historical collections from the past century of the Robinson Crusoe Islands; I have relished an empanada and a fresh *ensalada chilena* (a delightful salad of sliced tomatoes and blanched onions, flavored with cilantro); and it is also the first Chilean city I ever knew.

On that first day in March of 1977 my acquaintance with Santiago began as our plane finally swept down from the Andes into the Chilean central valley and landed at the Arturo Merino Benítez International Airport northwest of the city. As there were no jet ports then, I descended the portable stairway and entered a world of sun and warm breezes, all backdropped by the majestic snow-capped Andes as far as the eye could see. The airport was like virtually all others on the planet—with its customs officials, porters, and baggage carts. By Latin American standards it was organized and orderly; people stood patiently in line for the immigration and for baggage—there was no pushing or shoving common in other countries.

Renting a car at the airport and getting a map to the city proved easy, thanks to prior reservations. Navigating a large city of more than four million people (now there are more than seven million) and driving there for the first time wasn't the horror that I had imagined, based on previous experiences in Mexico, Colombia, and Ecuador. Thanks to reasonably disciplined Chilean drivers and a detailed street map, I found the road out of town toward Talca and Concepción some 500 kilometers south on the Pan-American Highway. As I tell my wife—give me a detailed street map and I can drive to the moon! (This usually uttered when she is driving, and we are lost in some suburb of Columbus.) Santiago is laid out, as most cities, following geographical features, in this case along the River Mapocho and in and around several hills, such as Cerro San Francisco and Cerro Santa Lucía. The main avenues are broad, with six to eight traffic lanes on each side, and back then with numerous buses (*micros*) of all shapes, sizes, and colors. The bus lines were mostly private, little regulated, and the buses clogged the two lanes adjacent the curb, jockeying for the best position to pick up and let off passengers. Contamination of the air was (and still is) substantial in Santiago, in part due to these buses that generally were not as well maintained as individual cars and taxis. On the other hand, buses helped to keep down the number of cars on the road, which was more ecologically sound. The real jewel of transportation in the city was the polished and gleaming metro (subway), which I did not

Santiago de Chile

see on this first visit to the city but subsequently came to love and respect. The subway is super clean, always on time, with excellent stainless-steel cars, and a modest fare. The residents of Santiago are proud of their subway, and they keep it tidy and polished as a matter of civic pride.[6]

If I had to describe Santiago to you, I would compare it to a European city, perhaps a mixture of Madrid and Paris. In the early years of the twentieth century, Santiago was very prosperous, and the elegance of the older buildings attests to this—tall, columned facades, elaborate scrollwork on balconies, many shuttered windows, all too costly to duplicate today. The city reflects its past; when international investment ran high, the mines poured forth mineral riches, and the city grew in impressive and elegant response to this prosperity. Numerous yellow-topped and black-bottomed taxicabs dart here and there in Santiago, always alert for the smallest wave of the hand, even a raised finger, to come screeching to a stop by the curbside. Perhaps you prefer to sit and watch it all happen along the avenues at an open-air café underneath colorful umbrellas.

There I was, in my dark gray Peugeot with deep red upholstery, motoring south at one hundred kilometers/hour on the Pan-American Highway (*Ruta* 5) into the scrubland of thorny legumes, grasses, and cacti. It's pretty hard to get lost in Chile, with the tall Andes always to the east, but I found one disconcerting element: the sun in the Southern Hemisphere rises and sets toward the northern part of the sky, not in the south as occurs in the Northern Hemisphere. Intuitively I always orient toward the shadows in Ohio and elsewhere—the shadow always points a bit north, especially near midday. Not so in the southern half of the world, where the shadows point south! Despite numerous visits to Chile since this first one in 1977, I still find this situation counterintuitive and have to rethink it through every time. But here south of Santiago, there was no confusion—my job was to keep on the highway straight south to Concepción. Mario Silva was expecting me.

Because I arrived in the morning in Santiago, and after arranging the rental car, I was only able to travel half the distance to Concepción by evening, which took me to the city of Talca. This is a medium-sized city of about 200,000 people, centered in agricultural land, but also with a strong manufacturing base. My only objective was to enter the city, find a modest hotel, get a good dinner and rest, and then head back on the highway toward Concepción the next morning. I followed the main street into Talca to the central plaza, and just off the center about a block was a small hotel painted light green. The hotel was narrow, sandwiched between two residential buildings, nothing special, but it looked clean and tidy. More importantly, they had a room available at a reasonable price, secure parking, and I was ready to relax and rest, especially after the long flight from the States and the drive from Santiago.

After getting settled into the hotel, I strolled to the center of town and right there on the plaza was a convenient restaurant that afforded a good view of the life of the city. Being unfamiliar with all the different foods on the menu, I ordered a thin steak (*bistek*) with French fries and a beer. When the beer arrived, I began to look out at the plaza at the many people scurrying toward home after work with packages, couples walking slowly with arms around each other's waists, mothers holding hands of children. This could be anywhere in the world, but it was Talca, in Chile, and I was there, too—my first evening in a new country.

The next morning after a quick breakfast I was on the road once more, now passing fruit stands full of cantaloupes, tomatoes, watermelons, corn, papaya, and other fruits of the season. After a few hours of driving, I took time to stop and buy a ripe cantaloupe and eat it all on the spot on the roadside—so sweet and juicy—and such an enjoyable rest from the droning monotony of the highway. Traffic was not too congested, but with this being the main commercial artery of the country, with heavy trucks and only a two-lane road, caution was needed. But I was once again amazed that here I was, driving south in the warm Southern Hemisphere, in Chile, toward an uncertain meeting with a person unknown to me. Somehow this never bothered me. I knew Mario would be there, and that it would lead to something very interesting and productive, but I never dreamed how fascinating it would all become.

After having passed and been passed by numerous buses and trucks, having taken in panoramas of snow-capped peaks for hours on end, and having crossed over numerous silt-laden mountain streams rushing toward the sea, I finally arrived in the city of Concepción. It's much smaller than Santiago, similar to the size of Talca, now about 250,000 people, which if combined with the surrounding towns, such as Talcahuano on the seacoast (and a naval base), the total population of the area might reach about 450,000. It's one tenth the size of Santiago, and it felt that way, making it relatively easy for me to find the Plaza de Los Tribunales de Justicia on which my City Hotel was located. I was relieved as I pulled up in the Peugeot, grabbed my small blue bag and briefcase, and walked up the stairs to the lobby to see if I, indeed, had a reservation that Mario had made for me. The bespectacled clerk was quite polite, and yes, the reservation was secure. Once in my room, I unpacked the essentials, brushed my teeth, and picked up the phone book. I had no phone number for Mario, but I assumed he was in the telephone *Guia*—and sure enough "Silva O., M." was there on Chacabuco Street, to which I immediately called, but whoever answered (I learned later it was the maid) let me know he was not at home, but would contact me at the hotel when he returned.

A word about Latin American names may be in order here. In the United States and England, we have a given first name, usually a second (middle) name, which might be a first name of a relative, or less commonly a family last name of a grandfather, uncle, aunt, or even the maiden name of the mother. In my case, I do have my mother's maiden name as my middle name, Falor, and as the years have rolled along and my mother has passed away, I am pleased to carry it on for her. For Chileans, there are usually several given names, to help differentiate people with similar family names, such as González or Ramírez, so common in the country. A set of given names might be something like Juan Antonio José followed by Pérez, and finally at the very end goes the mother's maiden name, say Toribio. So we might have Juan Antonio José Pérez Toribio. But usually not all the given names are used for informal purposes, so our example might be abbreviated to Juan Pérez Toribio, or shorter still to Juan Pérez. Use of the initial after the family name is correct, such as we saw with Mario Silva Osorio, or Mario Silva O. He sometimes publishes just as Mario Silva, because having studied in England, he understands the English and American naming systems and adjusts accordingly. Women in Chile always keep their maiden names. My wife, for example, is Patricia Antonieta Ghivarello Pesce, Ghivarello being the family name (of her father, of obvious Italian heritage), Patricia Antonieta her two given names, and Pesce the family name of her mother. When a Chilean woman marries, she keeps her own name but for social purposes may add "*de*" ("of") followed by the family name of the husband, although such a formality is now less common with the younger generation. Husband and wife get introduced at parties with two last names: "Please meet Tod Stuessy and Patricia Ghivarello," for example, husband and wife.

Speaking of names and Mario, he called the hotel room at 8:00 in the evening, and after making mutual phone introductions, we agreed to meet in his office/laboratory the next day at 8:30 in the morning. Mario assured me that the clerk at the desk could give me directions to the University of Concepción (he was right, as usual), and his office would be on the *tercer piso* (literally on the third floor), but for us in the United States, it is the fourth floor. When we begin counting floors, we start with the bottom floor as one, but most Latin Americans and Europeans call the bottom floor just that, the *planta baja* and begin counting with the next floor as number one. Since we would begin our discussions next morning, I took advantage of the evening to explore the pedestrian walkway that connects the Plaza de los Tribunales de Justicia with the Plaza de la Independencia, past the department store Falabella, a small supermarket, a camera store, a shirt store, and to the Nuria, the restaurant that I entered for dinner.

The Nuria was a popular downtown institution in Concepción, where we were to have many meals with university and expedition colleagues. It wasn't the best food, nor the cleanest place in town, nor the best decorated, but it had very fast service and more importantly, there were always people there. The Nuria lived at all hours of the day or evening. I loved the rows of liquor bottles behind the slicked-back hair of the Italian-Chilean cashier, who never smiled. The attentive waitresses always wore red coats with white collars as their uniforms. They never took notes, just gliding back to the kitchen and always returning with the right order somehow. A favorite dish was the *hot dog completo* with mayonnaise (Chileans love mayonnaise), ketchup, *palta* (avocado), mustard, sauerkraut, and cheese. It looked appetizing to me that first evening in Concepción, but I instead chose a *lomo* (steak), *papas puré* (mashed potatoes), and an ensalada chilena, accompanied by an Escudo beer. All came quickly, with *pan amasado* (hard, round bread), and *aji chileno* (Chilean hot sauce). There are numerous types of hot sauces throughout Latin America, all derived from the genus *Capsicum* of the tomato family (Solanaceae), and each one with a special flavor—more sweet, more sour, and with different degrees of potency. In general, Chilean food is not very spicy, the cuisine mirroring that of the many European immigrants who have colonized the country from Spain, Germany, Switzerland, and Italy. Hot sauce is provided separately, but it is rarely incorporated into the food directly. So with *lomo*, *papas puré*, and ensalada chilena, I enjoyed my first meal in Concepción thinking about how it would go with Mario in the morning. In fact, the meal was so good, and the environment so pleasant, I ordered another Escudo.

I was up at 7:00 in the morning for a continental breakfast at the City Hotel. As in parts of Europe, most Chileans eat dinner so late, at about 9:00 to 10:00 in the evening, that after going to bed shortly after the meal, one simply doesn't have much appetite in the morning. So coffee with a liberal supply of milk (*café con leche*), bread, butter, marmalade, and a small cake was the breakfast this morning. Quite enough for me; besides, I was anxious to get to Mario's laboratory. I put on a tie with my sport coat because I wanted to appear presentable and respectful.

The hotel gave me directions to the university, across the Plaza de los Tribunales de Justicia, then along the Diagonal (Avenida Diagonal Pedro Aguirre Cerda), to arrive after five blocks at the Plaza Peru, which served as a meeting place for university students. Built in the 1950s, this plaza has a lovely fountain in the center with a semicircular design, surrounded on three sides by connected buildings with many small shops on the ground floor. On the final side is Chacabuco Street, and on the other side of the street the University of Concepción. I walked to the impressive entrance of the university, the Arco de Medicina (originally the Medical Faculty, but now housing part of the Faculty of Biological Sciences), flanked by two well-established Canary Island palms (*Phoenix canariensis*). I was pretty excited walking under this arch and toward the bell tower down the crunching, firm gravel of the broad entryway. Down the walkway about a block on the left, just before reaching the clock tower, is the building of the Faculty of Natural and Oceanographic Sciences that houses the Department of Botany, from where we have planned and staged all the Robinson Crusoe expeditions.

Arco de Medicina, the main entrance to the campus of the University of Concepción

I entered the wooden double doors of the building with the polished brass handles, my footsteps echoing on the diagonally oversized black-and-white square tiles, and I started directly up the stairs, one, two, three, four flights, to the door labeled *Laboratorio de Fitoquímica* (Phytochemistry Laboratory)—Mario's domain. In I went to a world of flash evaporators, solvent odors, stacks of boxes along the wall with previously collected plant materials from throughout Chile, and people in white lab coats sequestered here and there behind their equipment. From a small inner office comes a thin man with straight black hair, presenting a warm smile with gold teeth fillings, and speaking good English, "Welcome—you must be Dr. Stuessy," he beamed with an outstretched hand. It was a relief to finally be here. "Dr. Silva, it's a pleasure to meet you at last," I responded. He added, "Please come in and have a seat, take your coat off, and please call me Mario. Would you like a coffee?" And so it began with Mario, the Mario I have come to admire, appreciate, and respect. He called for Don Gume, the trusty, middle-aged, slightly paunchy, and extremely good-natured helper who could fashion any wooden item needed for the lab, later to be of great help to us in preparing for the first expedition to the Islands. "Thanks, Mario," I replied and added, "Please call me Tod." Thus, a relationship started that led eventually to more than 140 scientific journal articles, books, and book chapters on the plants of the Robinson Crusoe Islands.

Mario Silva and his family. Left to right: Mario, Kati, Igor, Derek, and Iván. Derek holds the dog that ate a hole in the sleeve of my sports jacket.

Mario is a nervous person, which perhaps allows him to remain thin, and he is constantly in motion. Sitting and talking to him is sometimes distracting because he sits down, then gets up, then looks at some papers, then sits down again, and so on. Mario is a happy person with a supportive family of three sons (now all successful physicians and dentists) and a lovely wife, Kati. This photograph has another story attached to it. Days later, Mario invited me to dinner at his home, at which time his dog, a white terrier, chewed a hole in the sleeve of my sport coat, which I had hung on the back of a chair in the living room. My greatest challenge at that point was getting out of Mario's house before he saw the saliva-stained hole of about an inch diameter on the lower right sleeve and insisted on buying me a new coat! I was successful in hiding the damage, only years later sharing the story with them, long after the coat had been discarded. Mario radiates confidence that comes from the security of this strong family, which no doubt has contributed to his success. Were it not for Mario, we would never have gone to the Robinson Crusoe Islands, and this book would not now be in your hands. Further, I wouldn't be married to Patricia, formerly from Santiago, whom I met in the unlikely setting on the boat from Valparaíso to Robinson Crusoe Island (see chapter 9), nor would we have two beautiful sons, Daniel and Martin. More about this later, but my debt to Mario is profound. As one reflects on the strange events that form our life—the seemingly random events that result in successes, failures, joys, and sadnesses—we often focus on key people who have been particularly influential. For me Mario is clearly one of these. He will deny his importance with a few

Mario Silva with assistants and students in his phytochemistry laboratory, University of Concepción. Left to right: Roberto Rozas, Julio Vargas, Magalis Bittner, José Becerra, Zenón Rozas, Patricia Pacheco, and Mario Silva.

quick sentences as he walks nervously about the room, but I know better. I owe him a lot—all members of our research team owe him a lot.

As we talked, Mario sat down again, got up, showed me a recent publication in the *Boletin de la Sociedad Chilena de Química* (Bulletin of the Chilean Chemical Society), which our library at Ohio State didn't receive, and I had not seen. Don Gume brought the typical thick, sugary Nescafé in small University of Concepción cups and saucers, and we talked more. "Let me show you around," Mario beamed, and we began a tour of the laboratory, meeting more than a dozen students, assistants, and coworkers. This was an impressive lab operation by any standard and extraordinarily impressive for South America. The positive and proud manner by which Mario introduced each person left a very favorable impression on me—obviously Mario had people skills, and over the years we depended on these talents to solve problems that inevitably confronted us. I met Hugo Valdebenito and Patricia Pacheco, students who later came to Ohio State to complete their doctorates with me and my good friend and colleague, Daniel Crawford.

We finally returned to the office and began to discuss our collaborative research interests. To seek development funds and pay for my trip, we

Chaetanthera glabrata

had drafted a proposal to do a combined chemical and systematic (that is, a chemosystematic) study of the species of *Chaetanthera*, a group of some thirty species[7] in the sunflower family restricted to the southern Andes. Some of them are beautiful and striking, such as *C. glabrata*. Our objective would be to investigate evolutionary relationships and mechanisms using different types of information such as external features (morphology), chromosomal data, ecological assessments, and of course, secondary products chemistry, Mario's specialty. We hoped this would be a model study of analyzing plant evolution in the southern Andes; no study of the depth we planned had ever been done. Yet despite our excitement about the project, there was a very serious problem: for the techniques that Mario was using to extract and characterize the secondary plant compounds, he needed at that time at least one kilogram of dried plant material from each population. It was doubtful that we could collect such a quantity from all species. Some of them were simply too small. We had a problem now that had to be solved: we needed another avenue for cooperative research.

To get some inspiration and straighten out my thoughts, I excused myself from Mario and dropped down to the second floor (*primer piso*) to the herbarium and library of the Department of Botany. I had collected a few plants on the roadside on the way down from Santiago, and so armed with these, I went to see the Director of the Herbarium, Professor Clodomiro Marticorena. A herbarium is a collection of dried and pressed plant materials, mounted on sheets of stiff paper about 12″ × 16″, with the name of the species and data on who collected it, where, and when, often in considerable detail. When arranged in a known sequence and stored in a convenient way, these materials make up a library of plant resources from different parts of our planet, such as Chile. Professor Marticorena, a red-faced, gold-tooth-filled botanist and authority on the plants of Chile, greeted me cordially but somewhat coolly, I thought. (I learned later that I was a gringo, that is, a foreigner, to be regarded with some suspicion; it's really not surprising, due to a century of American scientific imperialism, which has only diminished in recent decades.) We chatted about our botanical interests, why I was there, and what he was working on. One of his projects was a bibliography of the Chilean flora[8] as a prelude eventually to a new flora of Chile, the first completed in 1846 by Claudio Gay, which at that time was the first complete flora for any country of South America. Marticorena wanted to repeat the effort, this time in a modern and comprehensive way, describing the more than 5000 species of ferns and flowering plants within the country. We further chatted about the plants I had collected, mostly European weeds, as Chile has had numerous waves of immigrants, especially from England, Italy, Spain, Switzerland, Yugoslavia, and Germany. He showed me the herbarium, and I was interested in the arrangement of open shelves on tracks, so that an aisle is made where one wants, rather than having aisles open all the time. An additional advantage was that in case of an earthquake (there are many in Chile), the shelves slide on the tracks and do not fall over, keeping them much safer from damage. These space-saving arrangements, called "compactors," are now used in herbaria and libraries, or in other types of storage facilities such as archives, but in 1977 they were not at all common. As far as I am aware, this was the first herbarium compactor in the New World. I mentioned the *Chaetanthera* project to Professor Marticorena, and he agreed that it was not workable for us.

Herbarium specimens stored in compactors in the Department of Botany, University of Concepción. With Prof. Clodomiro Marticorena.

Later that day I gave a seminar in Spanish to the assembled department on work I had completed on systematics of a group in the sunflower family in Mexico and Central America (the genus *Melampodium*). Here I met other faculty members such as Dr. Oscar Matthei, expert on grasses and weeds of Chile, Dr. Oscar Parra, phycologist, Lic. Roberto Rodríguez, expert on the ferns of Chile, and Lic. Eduardo Ugarte, plant community ecologist. The *Licenciado* (Lic.) is a title resulting from a five-year career program, in this case Biology, and similar to our four-year B.S. degree plus an additional year of work. I wasn't certain how well these colleagues liked my informal talk, which I judged fit the circumstances, but it was hard for me to read the audience. Were they pleased I was there? Did they view me as just another pesky gringo? Did they respect my work? I had the general feeling that I was being tolerated more than welcomed, but then again, I was worthy of healthy suspicion.

After the talk I retreated to the botanical library, which had excellent botanical holdings in general, and particularly so regarding the flora of Chile. It, too, was compacted on steel rails on the floor on which the shelves glided smoothly by a bit of pulling and shoving. I enjoyed examining the books and journals to get an idea of their literature resources, but foremost on my mind was the basic problem—what to do about *Chaetanthera*? We needed a more realistic, but equally significant, project. As I gazed out the window onto the gravel walkway below, I continued to struggle with this issue. Students were walking with arms draped around each other, touching, talking, showing much more affection than we are accustomed to offer in public in the States. Basically, I like that—we are too emotionally inhibited, perhaps due to our puritanical roots—too conservative. Couples kiss at the table in restaurants, on park benches, essentially everywhere. This for me is healthier, stronger, and somehow more natural.

And then it hit me. Juan Fernández—Robinson Crusoe Islands! These are *Chilean* islands! These fantasy islands, which had so intrigued me on that cold Columbus evening were right here off the coast 667 kilometers, and they are Chilean territory! This would be a fantastic opportunity to examine the evolution of the plants of these Chilean islands! Of course! I got up and headed up the stairs to find Mario. As everyone works till 6:30 in the evening here, I hoped I could still find him—I couldn't stand to wait to the next day. (The Chilean workday usually starts at 8:30 in the morning, ends for lunch at 12:30, starts again at 3:00 and ends once more at 6:30 in the evening—seven and a half hours spread out differently, but allowing time to visit the family for lunch, now less common in the larger cities.) Mario *was* still there in the lab, and he received me warmly, courteously thanking me for my earlier talk. I carefully broached the possibility of the Robinson Crusoe Islands with him, and he responded very positively. "Excellent idea," he enthused, "and perhaps the botanists on the first floor would like to participate in the expedition—I don't believe they have ever collected there. Why don't you see what they think about it?" And so, back downstairs I went to find Professor Marticorena, who, upon locating him and mentioning the possibility, seemed quite interested. He took me to the library, pulled open the compactor, and found three thick volumes that he brought out and put on the table for me to examine. They were Carl Skottsberg's *The Natural History of Juan Fernandez and Easter Island*.[9] "Look these over," he said, "and we can talk more tomorrow," and then he left whistling somewhat aimlessly to himself down the hallway.

I pored over the volumes. The section on the flowering plants really caught my attention.[10] Now I recalled what Carlquist had said in his book[1]—that there are tree sunflowers here in these islands that are found nowhere else on earth, in fact some twenty-five or more species. Studying the evolutionary aspects of this group would be a great project—to examine the patterns and processes of evolution of the tree sunflowers of the Robinson Crusoe Islands. Yes, that was it! I couldn't wait to tell Mario in the morning.

At 8:30 the next morning I was in Mario's office discussing the ups and downs and ins and outs of the Robinson Crusoe Islands project, with special emphasis on the tree sunflowers (family Compositae). Mario felt comfortable with the focus, because these were large plants, with easy-to-collect leaves that wouldn't hurt the whole plant. A little judicious pruning and he would have sufficient material for analysis. Further, it was clear that there were somewhere near one hundred species of flowering plants that grew in the islands and nowhere else (that is, they are *endemic* to the archipelago). Other groups that had speciated in the archipelago might also yield evolutionary information. Also there was the rare plant family, Lactoridaceae, restricted to the Islands—one of only two such families endemic to oceanic islands in the world (the other is Degeneriaceae of Fiji[11]). I thought that Mario also liked the logistic challenge the project represented—getting to the islands wasn't going to be all that easy. Mario was a master of such arrangements, and I suspected he relished the idea of helping set it all up.

Having gotten Mario's approval, I quickly descended the steps to the herbarium to talk to Professor Marticorena. There he was, immersed in his card file of the bibliography of the works on the Chilean flora. "Professor Marticorena," I began, "as you know, I have been talking to Dr. Silva about the possibility of working in the Juan Fernández [Robinson Crusoe] Islands with a focus on evolution of the tree Compositae, and last evening I read more about these groups. Would you and the other botanists here have any interest in this project, perhaps to participate in a collaborative joint expedition?" Marticorena responded quickly (as he almost always did), "Yes, that would be of great interest to us, as we are working on the Chilean flora but none of us has visited and collected in the islands. Yes, that is a very good suggestion. But, of course, we will need to find the money to be able to do it."

Yes, the money. Money is always a problem in getting a research program started, but that never must be an insurmountable hurdle at the outset. The project must come first, after which the funding must follow. So we now had a project. The remainder of my week in Concepción was spent outlining the project in more detail. We spread out literature on the table in the library and learned more about previous work in the islands, especially the earlier floristic studies of 1896 by Federico Johow,[12] a naturalized Chilean who contributed much to understanding the plants and vegetation of the islands. We read the works of Carl Skottsberg, the Swedish botanist who visited the islands on three occasions, first in 1908 and last in 1954 to 1955. And I worked over the specimens in the herbarium collected from previous botanists, comparing and contrasting the features of each of these fascinating plants. There were a sizable number of specimens for the islands already in the herbarium at Concepción due to earlier collectors such as Skottsberg, Sparre, Bock, and Behn. Each day our interest and excitement about the project grew—this was going to be an outstanding collaboration, and one that we hoped would result in a model study of the evolution of plants of isolated oceanic islands, a natural laboratory[13] waiting for us to explore the secrets of evolution that had accumulated over the past millions of years.

CHAPTER 3

Expedition Preparations

I had never before organized an expedition to an isolated island. My experiences had been on numerous field trips to Mexico and Central America in some sort of station wagon or pickup, and with one or at most two companions. In these cases, we kept passing through towns, which had food, medicines, car parts, or clothes. Sometimes a particular need could not be met in a particular village, but eventually it appeared farther down the road. As long as the need was not completely urgent, these sorts of problems created delays but not abortion of the entire trip. This is not the case in an isolated island. If you do not bring what you need, it will not be there for you to use. This means that much fact-finding about the location must take place before the trip starts. Even simple items normally assumed to be obtainable may not be so in the island. And one has to be cautious, because sometimes such information can be faulty, or misleading, or the items may be available but not for sale. The guiding principle is if there is any doubt, take your own supply.

Our expeditions to the islands have each included three to ten persons, usually in the upper range. This meant considerable preparation to be certain all materials were available during the entire time of the expedition. This is not a new challenge: armies have been dealing with issues of provisions and supply chains for millennia, but it was a new challenge for me and our colleagues in Concepción. They were also very experienced in fieldwork within Chile, using vehicles such as jeeps and vans, and for them it was always possible to find items on the road. Now we had to plan carefully so that no lack of materials might interfere with the success of our trips, or even worse, forgetting something that might threaten the health and safety of our participants.

Planning for each expedition took place in the same way. A preliminary list had to be drafted of what would be required. On the first trip, when I was co-organizer with Mario and Clodomiro, we had to compare notes by letter, as in late 1979, phones and fax were not available to us. It was theoretically possible to call internationally, but it rarely functioned well between the two universities due to having to go through several local and international operators in the United States and Chile. This meant we had to get started at least six months before the expedition, which usually took place in January and February (the principal flowering season). After a few letters back and forth, we agreed on what was needed and exactly what was required from our side.

The availability of items in Chile changed considerably over the decades as international commerce picked up and many products that previously were only available in the United States now were also available in Chile. Very important were Ziploc bags of different sizes. These are indispensable because they close tight and water cannot get in or out (if sealed properly). They have numerous uses such as keeping food away from flies, holding silica gel for drying leaf material for DNA studies, and packing vials with liquid-preserved specimens at the end of the trip for travel home in airplane luggage. We always double-bagged everything that we wished to really preserve. On our early trips we had to bring thousands of these Ziplocs of different sizes, paying airline overweight charges, but into the 1990s they could be obtained in Concepción at the large supermarkets, which made things much easier. This was true, in fact, of nearly all items. Silica gel, for example, was initially difficult to find, but later it could be purchased in a store in Concepción at the Plaza Peru near the university. Some items might be much cheaper in the United States than in Chile, which required calculation as to whether it was more cost effective to buy them in the States and pay overweight charges or pay more in Concepción. Our colleagues, initially Mario, would obtain most of the needed chemicals in Chile, usually having them sent from companies in Santiago, to be ready for our preparations in Concepción.

From Columbus, the first challenge was to get all luggage and equipment packages shipped from the airport directly to Santiago. On some occasions, this was possible, especially if we were continuing on the same airline for the entire route, but sometimes we had to retrieve everything in Miami and make the transfer of bags ourselves even on the same airline. Further, because we were traveling on government grant funds, we always sought the least expensive tickets, and this meant often changing airline companies in Miami. In Columbus, there could be a hassle depending upon the person at the ticket window, how good they felt, how much they had slept that night. Our flights were usually very early in the morning, so this helped with some personnel but not with others. We always knew the weight limits for any bag or box, but sometimes rules had changed and then we had difficulties. It was impossible to leave any bag behind, as all items were needed for the expedition. On one occasion, there was quite a scene with the ticket agent who refused to let us take some of the boxes because they were too large. We finally asked if he could just ship everything only to Miami, forgetting about the longer portion of our trip. As this was only a domestic run, this apparently avoided some international rule that was causing the problem. In Miami, we checked in the same bags and boxes on the same airline and they went through without difficulty. That is the way it always was—great uncertainty upon arriving at the Columbus airport and a great relief after all was checked and done. But for security, even if we had succeeded in getting baggage checked all the way through to Santiago, we always went to the baggage carousel in Miami to see if any of our materials had for some reason been offloaded by mistake. This was probably an unnecessary move, but I just didn't want to have some slip-up resulting in us getting to Chile without valuable research materials. Better to take precautions and be safe. The items eventually would have been shipped to Santiago, but with a time delay we could not afford.

After we and our boxes and bags arrived in Santiago, everything had to be sent by train or bus to Concepción. We frequently used Turbus, following Mario's recommendation. Not until the 1990s was it possible to fly directly to Concepción. All flights to Santiago from Miami arrived in the early morning, and the bus rides were best taken during the night. There were bus runs to Concepción during the day, but because we were traveling in summertime, a bus with lots of glass windows can really become hot and uncomfortable. Now most buses have air-conditioning, but back in 1980 this was not the case. It made sense to take the bus ride during the night, which saved time and also offered us opportunities for sightseeing in Santiago before departure. Our bags and luggage could be conveniently stored in the *custodia* (baggage check) at the Turbus terminal, ready to be retrieved toward evening.

In Concepción, our friends usually met us early in the morning at the main bus station. As we arrived and gathered our baggage, many people were hopping on and off buses to start their workday. After commandeering a few taxis to hold all equipment and people, we traveled directly to the university to integrate our materials with those of our Chilean colleagues. After the usual greetings and storing of boxes in a secure room, we normally took breakfast in the small café by the bookstore under the bell tower, ordering *huevos revueltos con jamón* (scrambled eggs and ham), or *barros jarpa* (hot ham and cheese sandwich), plus coffee. The café had outside tables with plastic, garden-type chairs, which were suitable during the summer, but inside it was pleasant with a white tile floor, light blue tile walls, and brown-red beams across the ceiling. We always had a strict schedule for when we were to get ready, coordinating with bus, plane, or ship reservations. Because of the importance of getting all materials procured and packaged properly, we always allowed sufficient time to accomplish this without undue rush. After all, if we got everything ready a bit early, fine, then we had a day for relaxation and discussions before starting the trip.

Having gotten refreshed and energized, we returned to the laboratory where work began in earnest. This preparation and sorting room was on the ground floor of the building that housed the Department of Botany, just to the right of the entry, and it was large, designed for a full class of undergraduate students. The benches were utilitarian gray, wooden, but on top painted a bright red. The gray lab stools rested upon a wooden floor, which sometimes creaked when people walked around. It was a cheery room and spacious, which was the most important point. Toward the side was a long-tiled counter, deep enough to allow for equipment sorting, but more importantly, the tile made a good surface for preparing chemical solutions for the expeditions. Any spillage could easily be wiped away. Very useful was a blackboard in the front of the room, which allowed tasks to be listed. This was the staging ground for all our expeditions to the archipelago.

Three basic equipment needs exist for any expedition: personal items, general items for living in the islands, and research items, that is, the scientific supplies. Personal items are pretty much self-explanatory, with each person taking care of his or her own needs. We organized rain gear (rubberized yellow jackets and pants) for all participants. We usually brought these from the United States; they were often very inexpensive at discount stores. We had to guess a bit on sizes, not knowing with precision the physiognomy of all participants, always better to purchase something too large rather than too small. Because there can be cold weather in the islands, the rain gear had to fit over sweaters, hoodies, or vests. Rubberized pants were normally not needed, but when in a lobster boat going to some place on the island, there could be considerable water splash, as well as spotty rain showers. I always carried mine in the bottom of the pack, just in case, along with a thin woolen sweater. My usual outerwear was a red and blue cotton hoodie, conspicuous from a distance. Hats were essential, due to the sun and heat, but each person was responsible for taking care of this need. Tough hiking boots were also required, especially because much walking is over lava, which can cut weak shoes. We would give money to the Chilean students on each trip, and they would buy in Concepción what seemed appropriate—this way getting boots that fit right. Everyone was to bring whatever work clothes they wanted, mostly jeans and work shirts. I always wore long-sleeved shirts in the field to protect against insects, bushes, and the sun, but others had their own preferences. Backpacks were essential, and most persons preferred daypacks rather than larger ones with aluminum frames. We always brought some of both, but when fighting through thick shrub cover, or working through native montane forests, having a large pack frame was cumbersome and tended to get caught on branches. On overnight treks, these larger packs were handy to support more weight, but most participants simply tied additional items to the top of their daypacks, as did I.

For relaxation in the evening in the islands after hard fieldwork, there had to be music. Music can soothe the soul, as it certainly does for me. We always purchased a cassette tape player, battery-powered, with enough batteries to last the entire trip. At the end, we would donate this to island friends. But a cassette recorder is not very useful if there are no tapes, so a few of us would head to a music store in Concepción and purchase a series of artists and rhythms, some classical, some cumbias, some popular, and I—as always—selected lots of Frankie. Listening to Frank Sinatra in a cabin on an isolated island while sipping some Chilean wine is an immensely satisfying human experience. I am talking about Frankie from the 1960s and 1970s, not before, and not after. In this period he was a real saloon singer, singing right at you and right at your heart, and that was what I, at least, needed in a tent at 1,100 meters on the highest part of Alejandro Selkirk Island, far away from anything. Yes, Frankie's character left much to be desired,[1] a mafia boss wannabe, but in those Robinson Crusoe moments, songs like "I've Got You Under My Skin" (especially when arranged by Nelson Riddle), just go right into the soul. Fortunately, at least to some extent, my colleagues tolerated this personal obsession, and, of course we listened to the other music too, usually earlier in the evening, but as we approached the "wee, small hours," it was Frankie's turn—mournful, corporeal, honest.

General items we needed for our expeditions included smaller tents for sleeping, larger tents for work rooms, and mosquito netting. We needed kitchen utensils for food preparation and also table service, all plastic plates and cups, but steel knives, forks, and spoons. We brought only essentials; much cooking in the islands involves innovation, and the señoras who cooked for us were very skilled at making do, having lived their whole lives on the islands where nothing was ever abundant. We needed a colander, pots and pans, a few plastic pitchers, and so on. Water bottles were required for the field, but we learned that ordinary plastic Coke or Fanta bottles were very resistant in the field and sealed very tight, so no special containers were needed. Occasionally someone brought a more sophisticated trekking bottle, and that was fine. We brought lots of wooden matches for lighting the propane cooking stoves, the kerosene stoves for drying the plants, and the candles. I always double-bagged these in Ziplocs for security against water damage. Lots of candles were purchased, too, as we would often be without electricity.

We purchased a large first aid kit, supplemented by all sorts of other medical supplies, rubbing alcohol, antibiotic creams, sunscreen, insect repellants, vitamins, and pills for diarrhea. Close to the University of Concepción was a good pharmacy, and we always went there, told them the number of persons in our group and the duration of the trip, and they put together what, in their view, would give us the best protection. When going to Alejandro Selkirk Island, our medical crew member also brought along a specialized bag, with syringes and other prescription medicines in case of need. This is essential in the far island because there is simply no medical attention at all. We also brought small vials containing iodine crystals. The idea was to keep the vials filled with some water, always, such that the iodine dissolved slightly forming a brown solution. Adding one capful of this solution into one liter of water was reported to disinfect it well,[2] although we never did any tests. We used this technique especially on the early trips until we became more confident of the water supply, which never caused us serious stomach distress.

Now we come to the really important part of the preparations—food and drink. On Robinson Crusoe Island there now are quite a few smaller markets, and today it would be rare to not obtain basic items, such as flour, sugar, and salt. But in the 1980s, this was not the case. A boat brought supplies from the continent once a month, and as the weeks wore on, some things just were no longer available—hence great anticipation for arrival of the next ship. Our needs had to be calculated carefully, especially for a large team. After decisions were made, we tasked Chilean expedition members to jump into cabs, head downtown, and purchase all needed items on our list. We brought lots of the basics—rice, beans, pasta, durable canned goods, especially canned fruit, as fresh fruit was only rarely available. And then there were the snacks, cookies, and always many bars of Sahne-Nuss chocolate embedded with almonds. These bars are not cheap, but we found them to be essential for that pick-me-up snack in the afternoon to get motivated for the return to camp. A few of these hefty squares can go a very long way for needed additional energy. As for beverages, we always brought along huge quantities of Coca-Cola, Fanta, Sprite, Bilz,[3] and an assortment of other soft drinks, including lots of canned juices. Also useful was the powdered Zuko from Mexico, which when mixed with water made a good-tasting drink, and it was available in different flavors. In the 1980s there were no large supermarkets in Concepción, but now they are as large and modern as those in the United States. In fact, on our last trip in 2011, my postdoctoral associate in Vienna, Patricio López-Sepúlveda, originally from Concepción, ordered and paid for all food and beverages online from Vienna with a large supermarket in Valparaíso. They packed the items, shipped them to Robinson Crusoe Island by boat, and they were waiting for us when we arrived by plane later on. Things have really changed.

And most essential, the alcohol, extremely important when working hard in the field. First the wine, red and white. We learned over the years that the best way to transport wine was in individual bottles, 0.75 liter. Cheaper, much cheaper, large carafes of wine were available, and we did try them out, but once these 5- or 10-gallon jugs were open, one had to consume the contents in a few days, and this might not happen before it

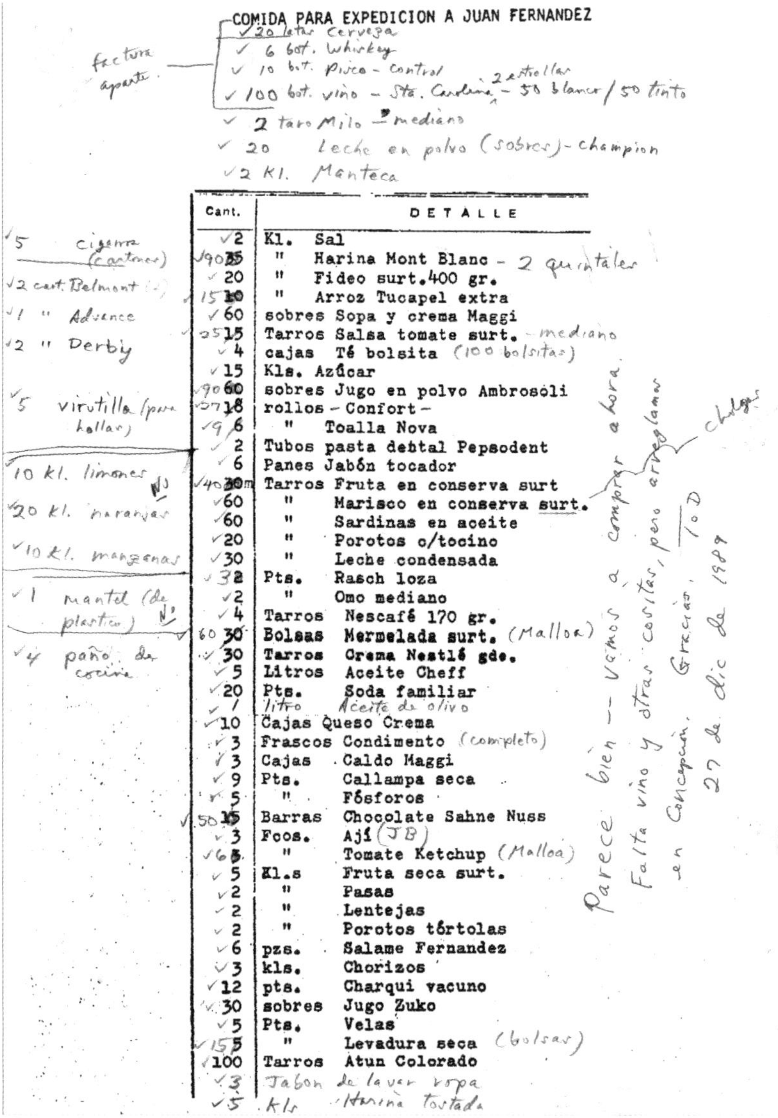

COMIDA PARA EXPEDICION A JUAN FERNANDEZ

✓ 20 latas Cerveza
✓ 6 bot. Whiskey
✓ 10 bot. Pisco – control
✓ 100 bot. vino – Sta. Carolina – 2 estrellas – 50 blanco / 50 tinto
✓ 2 taro Milo – mediano
✓ 20 Leche en polvo (sobres) – Champion
✓ 2 Kl. Manteca

factura aparte

Cant.	DETALLE
2	Kl. Sal
35	" Harina Mont Blanc – 2 quintales
20	" Fideo surt. 400 gr.
10	" Arroz Tucapel extra
60	sobres Sopa y crema Maggi
15	Tarros Salsa tomate surt. – mediano
4	cajas Té bolsita (100 bolsitas)
15	Kls. Azúcar
60	sobres Jugo en polvo Ambrosoli
18	rollos – Confort –
6	" Toalla Nova
2	Tubos pasta dental Pepsodent
6	Panes Jabón tocador
30	Tarros Fruta en conserva surt
60	" Marisco en conserva surt.
60	" Sardinas en aceite
20	" Porotos c/tocino
30	" Leche condensada
2	Pts. Rasch loza
2	" Omo mediano
4	Tarros Nescafé 170 gr.
30	Bolsas Mermelada surt. (Malloa)
30	Tarros Crema Nestlé gde.
5	Litros Aceite Cheff
20	Pts. Soda familiar
1	litro Aceite de olivo
10	Cajas Queso Crema
3	Frascos Condimento (completo)
3	Cajas Caldo Maggi
9	Pts. Callampa seca
5	" Fósforos
15	Barras Chocolate Sahne Nuss
3	Fcos. Ají (JB)
6	" Tomate Ketchup (Malloa)
5	Kl.s Fruta seca surt.
2	" Pasas
2	" Lentejas
2	" Porotos tórtolas
6	pzs. Salame Fernandez
3	kls. Chorizos
12	pts. Charqui vacuno
30	sobres Jugo Zuko
5	Pts. Velas
5	" Levadura seca (bolsas)
100	Tarros Atun Colorado
3	Jabon de lavar ropa
5	Kls Harina tostada

✓5 cigarros (cartones)
✓2 cart. Belmont
✓1 " Advance
✓2 " Derby
✓5 virutilla (para hollas)
10 kl. limones
✓20 kl. naranjas
✓10 kl. manzanas
✓1 mantel (de plastico)
✓4 paño de cocina

Parece bien — vamos a comprar ahora
Falta vino y otras cositas, pero arreglamos
en Concepción. Gracias, TOD
27 de dic de 1989

Typical working list of food items for our expeditions to the Robinson Crusoe Islands

went bad. Likewise, we experimented on one trip with paper cartons of wine, again, cheaper and easier to pack, but toward the midpoint of one of the trips, somehow many of the cartons had apparently developed small leaks such that air had gotten inside and turned the wine to vinegar. Well, it was drinkable, but just barely. We always calculated a half bottle of wine per person per day. So, if we had a group of eight persons for three weeks, this meant 84 bottles. To this must be added when we had guests for dinner and then the fiestas, at least two of them, one in the middle of the trip and one at the end. To be safe, we doubled the number of bottles to 168. If any bottles were left, which was not often, we gave away the rest to persons who had been helping us during the trip. In addition to wine, we also brought beer, and more importantly whiskey, Johnnie Walker Red Label. I am not a scotch connoisseur, but this brand seemed to go down well with everyone, and so it became our standard purchase. We also always brought a few extra bottles of scotch to be given to CONAF personnel and of course to be used for the fiestas. Not to be forgotten was a large store of pisco. Very popular in Chile and Peru, this is a drink distilled from grapes, with a taste somewhat similar to tequila (which comes from the *Agave* plant, usually produced in Mexico). There are some high-quality piscos that can be drunk straight, but these are expensive and were not within our budget. Furthermore, we enjoyed the taste of pisco mixed with Coke (making a *piscola*), Fanta, Zuko, or any other fruit juice. This was our usual drink in the evening after dinner, relaxing the muscles, and letting Frankie croon a tale of lost love. But I tell you sincerely, with all the physical exertion day after day, one just needed a drink at the end of the day to relax and fall asleep. I know, I know, you may be clucking about lack of discipline, but if you had been there with us, I think you would have accepted a piscola.

Scientific equipment also had to be accumulated in Concepción and readied for shipment to the islands. This was, after all, the core of the expedition. There have been expeditions that have ended in failure across the globe, some in very harsh circumstances, but my approach has always been to fanatically anticipate problems that might arise before they happen. The old adage "an ounce of prevention is worth a pound of cure"[4] is truly a good rule in approaching expedition planning and execution.

A fundamental tool of field research is the field notebook. Whenever the group stops to make a collection, out comes the bound notebook from the backpack and data are recorded, which means indicating date, specific location, collectors, aspects of the environment, altitude, and any information about the plants that may be pertinent to the objectives of the research. This may involve counting (estimating) sizes of populations, flower color, and what specific samples were taken (e.g., leaves in silica gel for DNA, floral buds in Carnoy's solution for chromosome analysis, wood in

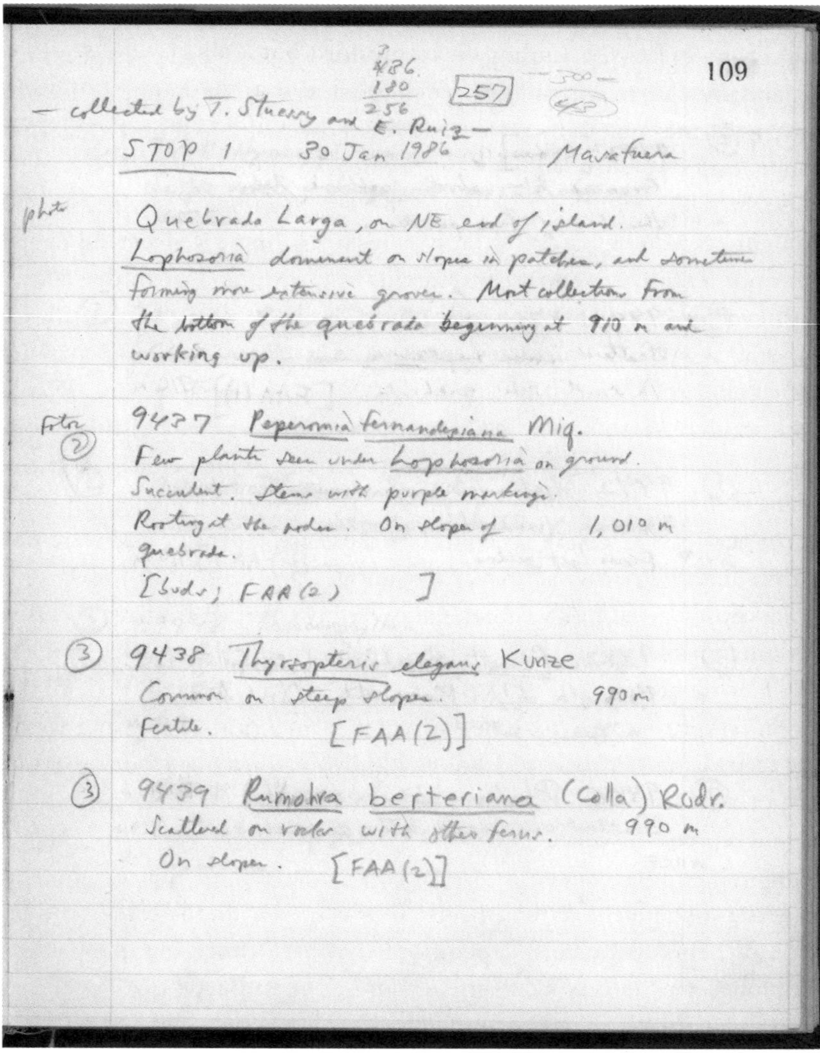

Page from a field notebook during our Robinson Crusoe expeditions (this one from 1986)

formalin-acetic acid [FAA] for anatomical study). Location is now made more precise via GPS, but we always also placed a dot on a topographic map in the field for security. Sometimes GPS, for whatever reason, can give erroneous coordinates, which are usually not detected until back in the laboratory. This was also true for elevation; I always used a pocket altimeter (Thommen) to check GPS readings, which sometimes were way off, especially when under a dense tree canopy. The altimeter varies fifty meters up or down depending on air pressure, but it always works.

For recording data in a field notebook, I have always used a simple #2 pencil,[5] kept sharp with a pocketknife, and written out the notes by hand. Pencil is the only way to do this, as inks can be capricious, and if soaked with seawater, they can run. I have used pencil to take field notes for more than fifty years, and I would continue for another fifty if that were possible. Even now, with notes that can be taken digitally, I would not trust that system in remote regions of the world. Get it down on paper with pencil and you have it captured. Even when we routinely record field data digitally in the future, followed by immediate satellite transfer to a home laboratory, I would probably still want to organize field notes by pencil for reflection later in the cabin, add more observations, pick up ideas from colleagues, or sketch out outlines for possible research papers.

Collections made in the field are each given a number, which usually follows the collecting numbers of the principal investigator of the project, such as *Stuessy 10678*. If there are two principal researchers, as on our first expedition in 1980, this issue must be sorted out before the trip begins. We elected to have two series, one for Ohio State (Stuessy) and the other for Concepción (Marticorena), which worked perfectly fine. There is a small amount of ego attendant to these collection numbers, because in publications the plants are cited, listed, and often discussed by collector and number (e.g., *Marticorena 6991*).

As a collecting trip continues and data accumulate, the notebook begins to assume an incredible value. If this gets lost, it is very difficult to use the collections, which now only have numbers on them, but without specific data. A testament to the importance of copying the field notebook at the end of an expedition comes from a real example. After I retired from the University of Vienna in 2012, our family in 2013 moved back to Columbus, Ohio, and I to the herbarium at Ohio State, where the Director, Dr. John Freudenstein, provided an office to allow me to continue research and writing. About a year after I had returned, I received two emails in the same month from investigators in Europe who had independently read two of our publications and wanted more data on plants we had cited from collections in Ecuador in 1991. I checked my set of field collection books, and guess what—this was the only one missing! Somehow during the move from Vienna to Columbus this field book had been lost. This was terribly embarrassing, and I had to write these colleagues and explain the situation. Having done that, I remembered that I had made a copy of this field notebook at the end of the Ecuadorean expedition and left it in the herbarium of the Natural History Museum in Quito. Through several emails back and forth with different persons, and with some file searching in the herbarium in Quito, the copy was found! What a relief! I asked for a duplicate copy, and within a week I had a scanned version in my office. This was printed out and now sits in chronological order with all my other field notebooks, including those from the Juan Fernández expeditions. Anticipating such a possible loss of valuable data at the end of each Juan Fernández expedition, I always made a copy of our collecting book and left it in the herbarium in Concepción before flying back to Columbus. Once that copy was handed over to Chilean colleagues, a burden was lifted from my shoulders.

Another fundamental part of field research is the plant press. This piece of equipment consists of a set of channeled cardboard pieces held between two slatted wooden frames. One objective of the press is to flatten the plants inside, by constructing a large "sandwich" held tight with two ropes or belts. Once specimens are flattened, they can be more conveniently studied and stored when back in the laboratory. During fieldwork, plants are initially placed in plastic bags (with numbers) and then pressed in the evening in camp. They are sized to fit within ordinary newspaper sheets (folded), with each one inserted between two corrugated cardboard pieces. The newspaper must always carry the field collection number, which corresponds to the data in the collection book. This also meant that we had to bring newspapers with us, because there was no certainty they would be available in the islands, and normally they were not.

Plant presses drying in the sun at CONAF headquarters, Robinson Crusoe Island

The second objective of the plant press is to dry the specimens to prevent mold from covering and destroying valuable floral structures. Once plant presses are loaded up with wet plants in newspapers separated by corrugates, the challenge is to dry them as quickly as possible, particularly if more plant material is coming in the next days. If you do not have sufficient presses in the field or you cannot dry them fast enough, life becomes very difficult. In that case, the only solution is to change newspapers every evening, substituting dry sheets for wet ones and rewriting collection numbers, which always results in human errors. Even worse—in the evening this brutally cuts into pisco and Frankie time—totally unacceptable!

Due to lack of electricity during most hours of the day and night in the islands on the early trips, we could not rely on any electrical means of drying the plants, such as a small space heater below the presses. With hot air blowing through the corrugates, plants can dry quite fast, except for succulent species that require more time. Our Chilean colleagues on field excursions in the continent had used simple kerosene stoves. These were imported from eastern Europe and were as simple a stove as one could imagine: a metal tray, and eight upright tubes with wicks extending through a welded ring above. Kerosene was put in the tray, and by capillary action the wicks took up the kerosene and burned slowly with a light blue flame, enough to last the night. Morning and evening the carbonized tips of the wicks had to be trimmed. We always constructed a wooden frame to position the plant presses above the stoves, plus a cloth skirt around the outside. The hot air rises and goes through the plant presses placed on their side with the channels of the cardboard corrugates oriented vertically. You may be asking, "What happens if the stoves malfunction, flames shoot up, and the press catches on fire?" Well, luckily this only happened once, and it occurred when we were in camp in the morning when everyone was present. During the day, one person was always stationed in camp to keep an eye on the stoves. At night, we kept our fingers crossed and hoped for the best. The stoves and presses were always secluded in a separate building away from people, equipment, and supplies. An open flame is a danger, but at least kerosene is a relatively slow-burning fuel. After consistent electricity was established on both islands, on later trips we used the innovative method of placing presses in front of the large village generators, which blew hot air right at the presses. This served us much better.

Plant-drying methods used on expeditions in the Robinson Crusoe Islands. Top, presses mounted on wooden frames, below which are square kerosene stoves; bottom, presses in front of hot air blowing from a large diesel electrical generator.

After the specimens arrive back at the laboratory, they must be processed and filed for permanent storage. All incoming materials must be fumigated to eliminate insect pests, now routinely done in low-temperature freezers at −38°F for several days. This provides an acceptable sanitization method, and it avoids using dangerous solvents or gases. The next step in the process is for the plants to be mounted with glue or tape (even sewn) onto stiff sheets of high-quality cotton rag cardstock, 12 × 16 inches. To

Specimen of *Robinsonia macrocephala* collected on Robinson Crusoe Island by Carl Skottsberg and his wife Inga in 1916, now on deposit in the U.S. National Herbarium in Washington, D.C.

these herbarium sheets are glued labels with the printed collection data (taken from the field notebook), and finally they are filed in cabinets according to the system of the institution (sometimes alphabetical by family, genus, and species, or sometimes following an indexed evolutionary sequence). It is important that these collections be deposited in reputable public institutions, so that future investigators can examine them and check on any aspect of their identification, features, or field data. There is nothing complicated about these techniques; specimens properly preserved have lasted up to 500 years,[6] certainly adequate for the length of our research activities! Attention to preparation of herbarium specimens is extremely important for successful research in systematic and evolutionary biology.[7]

Added to the plant presses were other items of equipment. Always required were tools for simple repairs, such as a hammer, pliers, wood and metal saws and files, and an adjustable wrench. Two pairs of work gloves were important. We always brought string and cord of different sizes and rope for packaging, plus a climbing rope for scaling down or up cliff areas. We always brought bound notebooks and plenty of pencils for each participant, as crew members often kept notes or diaries. We needed clippers, one pair per person, essential for making collections from woody plants. Everyone had pocketknives, but usually each person brought a personal favorite from home. We packed thousands of Ziploc bags of different sizes, plus hundreds of large, white trash bags for daily collecting. Plants were each placed in separate bags with a number, which had already been written up in the collecting book. With populational sampling, it was necessary to tag and label each individual plant, which always took more time. It is important to stress again the need for accurate and detailed field notes. Some data can be added later in the evening, but getting the basics of location and environment in the field is required for acceptable scientific work. Never trust important data to memory—get it down on paper immediately.

Also essential were chemicals for proper preservation of plant materials for anatomical and cytological study. We used a mixture of chloroform:95% alcohol:acetic acid (4:3:1) for preservation of floral buds for chromosome studies of meiosis (reductional cell division) in the laboratory. Delicate flowering buds are picked from the plant and popped immediately into a vial along with the collection number written in pencil on cardstock (not newsprint). For anatomical and embryological studies, we carried small vials of FAA, but most of these samples were prepared back in camp, even the next day, because these structures were more resilient. These solutions were kept in small vials with tight plastic caps, which kept them safe on the trail and allowed for secure packing in Ziploc bags for the return trip. Collecting for DNA studies required specific attention, but these preservations could be done the following day in camp. On two trips in 1990 and 1991, we brought liquid nitrogen in special carboys for DNA preservation (see chapter 13), which was more complicated, due mainly to concerns for safety by agents during transport to and from the islands. As laboratory techniques changed, on our later trips we used silica gel for DNA tissue preservation, which was much more convenient. We had to bring many hundreds of kilos of silica gel with us, packaged into smaller bags and boxes.

Combustibles were also on our list of needed items. To be taken in the small lobster boat of the park service (CONAF) to different locations on the islands, we had to provide them with gasoline (*benzina*) in large drums. We also needed smaller propane tanks for cooking. As mentioned above, on the early trips we brought drying stoves that ran on kerosene (*parafina*). It was not possible to obtain these items from other persons in the islands, because they had no way of replenishing their own store, except by ordering them again from the continent, which could take weeks or even months. The bottom line was: bring what you need—don't expect to find anything in the islands.

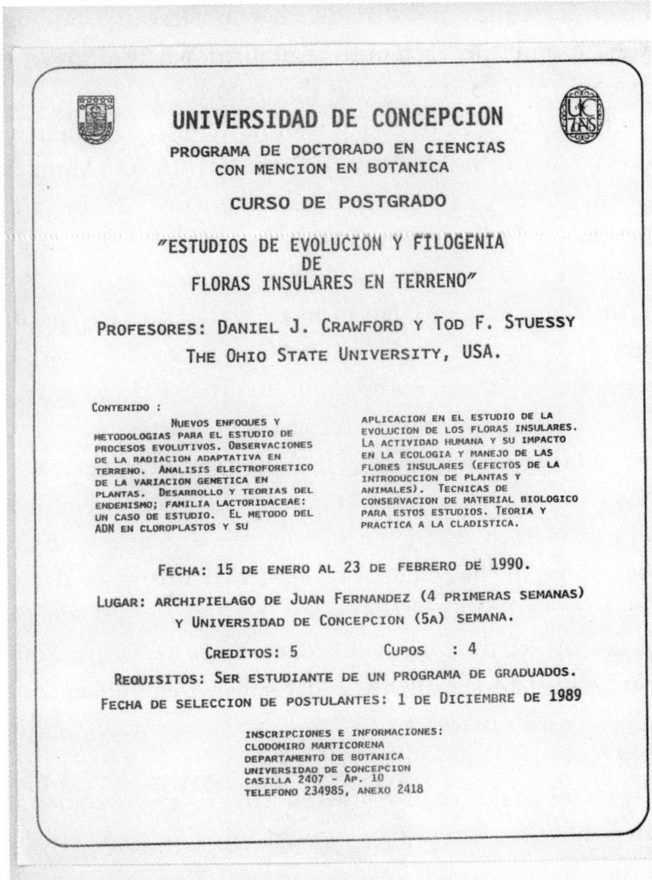

UNIVERSIDAD DE CONCEPCION

PROGRAMA DE DOCTORADO EN CIENCIAS
CON MENCION EN BOTANICA

CURSO DE POSTGRADO

"ESTUDIOS DE EVOLUCION Y FILOGENIA
DE
FLORAS INSULARES EN TERRENO"

PROFESORES: DANIEL J. CRAWFORD Y TOD F. STUESSY
THE OHIO STATE UNIVERSITY, USA.

CONTENIDO :

NUEVOS ENFOQUES Y METODOLOGIAS PARA EL ESTUDIO DE PROCESOS EVOLUTIVOS. OBSERVACIONES DE LA RADIACION ADAPTATIVA EN TERRENO. ANALISIS ELECTROFORETICO DE LA VARIACION GENETICA EN PLANTAS. DESARROLLO Y TEORIAS DEL ENDEMISMO; FAMILIA LACTORIDACEAE: UN CASO DE ESTUDIO. EL METODO DEL ADN EN CLOROPLASTOS Y SU APLICACION EN EL ESTUDIO DE LA EVOLUCION DE LOS FLORAS INSULARES. LA ACTIVIDAD HUMANA Y SU IMPACTO EN LA ECOLOGIA Y MANEJO DE LAS FLORES INSULARES (EFECTOS DE LA INTRODUCCION DE PLANTAS Y ANIMALES). TECNICAS DE CONSERVACION DE MATERIAL BIOLOGICO PARA ESTOS ESTUDIOS. TEORIA Y PRACTICA A LA CLADISTICA.

FECHA: 15 DE ENERO AL 23 DE FEBRERO DE 1990.

LUGAR: ARCHIPIELAGO DE JUAN FERNANDEZ (4 PRIMERAS SEMANAS)
Y UNIVERSIDAD DE CONCEPCION (5A) SEMANA.

CREDITOS: 5 CUPOS : 4

REQUISITOS: SER ESTUDIANTE DE UN PROGRAMA DE GRADUADOS.

FECHA DE SELECCION DE POSTULANTES: 1 DE DICIEMBRE DE 1989

INSCRIPCIONES E INFORMACIONES:
CLODOMIRO MARTICORENA
DEPARTAMENTO DE BOTANICA
UNIVERSIDAD DE CONCEPCION
CASILLA 2407 - AP. 10
TELEFONO 234985, ANEXO 2418

Announcement for a graduate student course in plant evolution and phylogeny at the University of Concepción taught on Robinson Crusoe Island in 1990

The last item to be packed up for the expedition was literature for the field library, nothing elaborate, but books and articles that would be needed for plant identification or for research discussion during the trip. We always brought along keys to species of native and endemic flowering plants that we had prepared, and Roberto Rodríguez had done the same for ferns. We always had one copy of Carl Skottsberg's studies on the plants of the archipelago[8] (see chapter 4), plus his investigations on the vegetation and reproductive biology.[9] There were times during bad weather when, confined to camp, participants wanted to learn more about the flora and vegetation of the islands, and this literature was essential. Some trips had classwork built into them, and then more study material was brought for the students.

One difficulty arose during preparations for our expedition to Robinson Crusoe Island in 1990. One of the participants, Ana María Humaña, a graduate student from the Universidad Austral de Chile in Valdivia, was to assist Professor Delbert Wiens with reproductive biological and genetic investigations of the rare *Lactoris fernandeziana*. Professor Wiens was from the University of Utah (United States) and had petitioned to accompany us on our trip, to which we happily agreed. Ana María had experience with these sorts of studies, so the arrangement was very appropriate. As we were getting materials together in the laboratory, she came to me and said that last evening she had slipped and taken a nasty fall, in the process putting a lot of weight on her right hand as she hit the floor. Her hand had swollen, and it hurt badly. Because of the physical challenges of working in the island, I realized that this was a problem that could not be ignored. She said that it was just a sprain and would go away, but I wasn't convinced. If she had broken the wrist, there would be no effective treatment in the island, and further, she would be unable to climb or help with observations. I told her that she needed to get an X-ray. If the hand were only sprained, then she could get a brace of some kind and come on the expedition. If it were broken, then unfortunately, she would have to stay back and get proper treatment. Later in the day she came back to the laboratory and presented me with an X-ray, which revealed no fracture. We were both pleased that this crisis had been resolved.

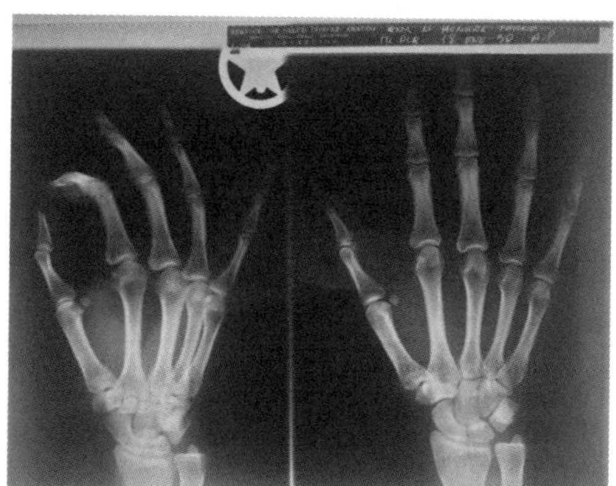

X-ray photographs prior to the expedition of 1990 of the right hand of Ana María Humaña, showing no fracture

After all items were ready in the laboratory in Concepción, they had to be packed into boxes or crates, everything double-bagged, and each finished parcel given a sequential number, with a complete list of these packages in my notebook. This way at each transfer of luggage, such as from the bottom of the bus into taxis, a numerical count of all items could be done to ensure no loss of equipment or supplies. Yes, a bit tedious but necessary. After all materials had been packed up in the laboratory, the room was locked, and we were on our own until departure time, usually that evening or the following day. The Chileans went home to spend last hours with their families. For me and the other international travelers, we usually ate dinner together and chatted about the trip ahead.

My favorite dinner spot in Concepción was always the Rincón del Pancho (Pancho's Corner). Hidden back on a short dead-end street, just off the Plaza de Independencia, the Rincón was a very small place, without any advertising sign. The front of the restaurant was nearly invisible—a small, white sign with blue letters subtly announced its existence, and curtains covered the windows, so you could not see in, and customers could not look out. Walking through the single door, inside was one rectangular room, with walls of vertical brown wooden strips, and a white ceiling, from which hung one simple glass chandelier. On the walls were four rectangular tapestries, apparently Ecuadorean, which added color to an otherwise soft, wood atmosphere. In the room were white tablecloths draped over fifteen tables, each surrounded by four metal chairs with red-brown backs.

To let you savor the Rincón, let me relate a particular evening in January of 1980 when I dined there with my postdoctoral fellow, Roger Sanders. On this occasion, classical music was being played from a sound system, piano, and there was only one couple already there, at 8:00 in the evening, early for a dinner on the town. The two waiters wearing textured reddish vests came quickly to greet us, shaking our hands warmly. I preferred a table near the wall, as this was a good position to observe who was coming and going. The waiter asked if we wanted a drink, Roger requesting a mineral water, as he did not drink alcohol, but my preference was a beer, Royal Guard, a light, quality beer. The very small bar was located toward the back of the restaurant, just to the left of the curtain leading to the kitchen. A man always stood there, somewhat balding and with slicked-back hair, well dressed in dark coat and tie, presumably the owner or manager, and he prepared drinks and also handled money coming from the waiters at the end of the meals.

My beer arrived in a small, green bottle, with black label and gold border, the waiter ceremoniously pouring it into the glass and then retiring while we examined the menu more carefully. The other waiter was now whistling along with the piano solo, probably Chopin, full of emotion. The music changed now to a popular tune, "Summer Breeze," and the other waiter switched on the TV to the national soccer tournament match, but after catching the score, he turned it off again. This restaurant is extremely quiet, a great place to talk without distracting boom-boom rhythms of some types of popular music. That sort of louder music has its own place in the world, but not in the Rincón del Pancho. The music changed again, "Send in the Clowns," one of my favorites, simple, direct, emotional. I didn't share these personal feelings with Roger, who is also a sensitive person but with his own set of emotional proclivities. I glanced at the brass and glass sconces, three on each side of the room, and now the pianist pushed deep into the keys of "As Time Goes By."

Suddenly the electrical circuit broke on the espresso coffee machine, as well as in the kitchen. I very much dislike open kitchens, where all preparations can be observed. It is a matter of trust—I presume that all is well or I wouldn't be coming to the restaurant. Especially in Latin America, space is at a premium and standards of cleanliness are not always the same as in developed countries, so better not to see this, and just enjoy the fruits of improving one's immune system by ingesting some different, but not toxic, bacterial additions to the meal. The circuit was restored and "Over the Rainbow" started, a tired tune in a way, but a pretty melody in the hands of this pianist. I could imagine an older gentleman, wearing a coat and tie, sitting at the piano, somewhat hunched over, using his own chord patterns and playing totally from the heart. These are not pyrotechnic tunes, just simple emotional infusions on a keyboard.

When the waiter came to take our order, I asked for a *filet a lo pobre*, a steak smothered with fried onions and topped by a fried egg, surrounded by French fried potatoes, but Roger preferred an *escalopa Kaiser*, a veal cutlet married with cheese and ham, and then deep fried, and accompanied by *papas duquesas* (Chilean potato puffs). I enquired about who the pianist was, and the waiter answered: "Roberto Bravo, and the album is called *Para Mis Amigas*." I learned later that Roberto Bravo, born in the same year as I (1943), was a classically trained pianist but with a huge repertoire of classical and popular music. He pushed his emotions down through his fingers into the keyboard, through the strings of the piano and out to the ears and heart of the listener. The waiter asked if I would like red wine with the filet, and of course, in Chile there are excellent wines produced in the Mediterranean region of the country. I ordered a half bottle of Cabernet Sauvignon, one of my favorites. I totally respected Roger's decision to avoid alcohol, but I couldn't help feeling that he was missing out on part of life's enjoyment. I am not anywhere as extreme as Frank Sinatra who just couldn't understand how anyone could live without drinking alcohol,[1] but I do think it has a useful role to play in human affairs.

As other diners began to filter into the restaurant, we took our time eating, as I usually do even when alone, because this is the time of day to really relax, review the events of the day, and make plans on how to structure the time for tomorrow (or later in the week). At the end of the meal I asked for an espresso, which came in a very small ceramic cup, sugary, a soothing combination of the dried and roasted seeds of *Coffea arabica* and the crushed and boiled stems of *Saccharum* spp., both plant products of immense importance for human culture. Roger preferred an herbal tea, of which there is a very good selection in Chile. At the end of the meal, the waiter came to ask if I would enjoy a small glass of pisco to settle the stomach at the end of the meal. As I slowly sipped it down, the music of Roberto Bravo continued to swirl around, and we reflected on the challenges of the upcoming expedition.

CHAPTER 4

Previous Visitors

Is history really worth anything? I wonder about this question at odd moments, even though I, of all scientists, really enjoy history, particularly history of science. It is fun to pick up an old book, perhaps several centuries old, and go back in time with that author, imagining sitting at his side while the book is being written with quill pen and by candlelight in the late evening after the rest of the family is asleep. I find it fascinating to learn how ideas develop and about the persons who dreamed them up. Because ideas rarely develop in a vacuum, it is instructive to learn the influences that have led to their development. And, of course, all this occurs in context of the ups and downs of people's lives—the hard times and soft times and all times in between. Someone might deride this fascination as being only intellectual gossip, and I yield somewhat to that criticism, but we are, after all, humans, and hence with a natural interest in members of our own species.

If we agree that history deals with past events, any past events, including attempts to understand the origin and evolution of the universe, solar system, planet Earth, and all life now found here, then I would argue that history is essential to the human condition. But if we limit our focus to history only of our own species, and that from historical records, cultural evidences of our existence, especially from written documents, then I look at the context differently. People justify the study of history by saying that "those who cannot remember the past are condemned to repeat it."[1] The only problem here is that the political and military leaders of countries, who in large measure determine the course of human events, are usually more steeped in historical knowledge than the rest of the population. And these are the very same persons who continue to repeat history, often with disastrous results!

I think it boils down simply to this: we find history fascinating because it is often entertaining, moving, and perhaps at its best, inspiring. It does chronicle the events of individuals and their achievements, and particularly in the face of hardship. We can relate to these trials and accomplishments, and they really do help us in ordering our own priorities.

This is why I want to describe persons who previously went to the Robinson Crusoe Islands. These people have had interesting histories, including a tie to Defoe's famous novel of 1719, *Robinson Crusoe*,[2] regarded by some as the first English novel,[3] that was read by many of us in high school. The tale is of a shipwrecked sailor who through personal determination and strong faith, keeps his life together and in fact becomes a stronger person through the experiences. After more than 300 years, this book is still available in bookshops today, which is more than can be said for many modern books published only twenty years ago, including some of mine! Here is the story of how the novel came about.

In 1704, the Scottish sailor, Alexander Selkirk (originally Selcraig),[4] was on the ship *Cinque Ports* as sailing master to Lieutenant Thomas Stradling, who had taken charge of the ship after the unexpected death of Captain Charles Pickering near Rio de Janeiro. They sailed under the general leadership of William Dampier, who commanded the *St. George*. The two ships continued around Cape Horn at the tip of South America and anchored successfully in Bahía Cumberland at Robinson Crusoe Island on 10 February. With the approach of French ships, Dampier and Stradling slipped anchor and gave chase, but without success of capture. They continued onward to the Peruvian coast, hoping to capture prizes as they went. At this point they split up, with Dampier following the South American coast northward toward Panama, and Stradling sailing southward toward Chile. Stradling later returned to the Juan Fernández Islands, but by this time the relationship between him and Selkirk had become strained, to such an extent that the latter demanded to be let off onto the island. Stradling at this point was happy to oblige, and Selkirk was deposited on the shore with some provisions, including clothes and bedding, a musket with powder and shot, a hatchet, one knife, a kettle, and a Bible.[5]

Selkirk stayed on Robinson Crusoe Island, at that time called Más a Tierra ("Closer to the Land"), for four years and four months. He adapted

Cave at the mouth of Puerto Inglés, Robinson Crusoe Island, which may have been used by Alexander Selkirk. With Mauricio Rondanelli and Miguel García.

relatively well, due to the availability of feral goats, abundant fish and lobster, and fruit trees and garden herbs that had survived from earlier visitors and colonists. At this time he was the only denizen of the island, and there never was a man Friday. Many ships must have come into the bay during these years, but Selkirk was fearful of making contact due to the many European wars in progress, and in isolation he was unsure what countries were allied with England or which were in conflict. A cave at the mouth of Puerto Inglés exists that may have been used by Selkirk, and at least it is so regarded today for touristic purposes. After more than four years, on 31 January 1709 Captain Woodes Rogers of the *Duke* put into Bahía Cumberland, presumably flying the British flag, and Selkirk made himself known. As luck would have it, William Dampier was also on board, and he vouched for Selkirk, allowing him to be taken along as mate on the ship. Selkirk had presented quite a wild sight, clothed in goat skins and matching hat, with long hair, and barefooted.

Selkirk returned to London on 14 October 1711, and by that time word had gotten back to England about his unusual story. He was interviewed by several journalists, and he became a person of considerable celebrity. Daniel Defoe, who had written numerous pamphlets and articles,[6] apparently never personally met Selkirk, but he was aware of information regarding this sailor who had been isolated for so many years. The book is set in the Caribbean, far from the Juan Fernández Archipelago off the

Robinson Crusoe and His Man Friday, a lithograph by Currier and Ives, 1874

coast of Chile, and Defoe introduced man "Friday" as an additional story element. The success of this book encouraged Defoe to write others, including the reasonably well-known *Moll Flanders* in 1722, but he died poor in 1731. The Chilean government, wishing in part to stimulate tourism to the islands, officially renamed the archipelago in 1962 as the Robinson Crusoe Archipielago, and the two islands Isla Marinero Alejandro Selkirk and Isla Piloto Robinson Crusoe.

But talking about Robinson Crusoe gets us quite a bit ahead of our story. The islands were discovered by the Spanish navigator Juan Fernández in 1574.[7] Prior to that time, no humans had ever lived on, nor presumably had ever visited, the archipelago. This is a very important point, because it means that no human disturbance to the ecosystem of the islands had occurred prior to European discovery; many natural alterations to the landscape had taken place over the four million years of existence,[8] but not from people. Historical documentation of human activities during the past 400+ years, therefore, provides a detailed chronicle of what, exactly, has been the human impact. Chapter 12 pulls this information together as a useful case study. Returning to the navigator Juan Fernández, he sailed from Callao, the port of Lima, Peru, but instead of taking the usual very slow route south along the coast, he sailed first westward and then turned south. The Humboldt current flows from Antarctica northward along the South American coastline and makes it difficult for ships to navigate southward with any speed. In fact, it could take three months or more to travel from Callao to Valparaíso, Chile. Because of his westward strategy, Juan Fernández made very good time, and in the process, passed by Robinson Crusoe Island, but without going ashore. He continued his voyage and arrived to Chile in record time (about three weeks), so fast, in fact, that the Spanish authorities, including the Inquisition, argued that only with the help of the Devil could anyone have done the trip so quickly. It was only after other ships had accomplished the same feat that Juan Fernández was removed from suspicion. The Spanish government kept knowledge of these islands quiet, because their strategic location for the Spanish colonies was obvious, but upon entering the seventeenth century the word started getting back to other European countries.[9]

The trip westward from Europe to the Orient was not an easy one. The most dangerous part was passing through the Straits of Magellan, which is cold and has very unpredictable weather, often fatal for the small ships of this period. As more and more ships made the passage, there was a real need for a place to let crews rest and recover from hardships, and to repair boats before continuing into the Pacific, and Robinson Crusoe Island provided a solution. It also proved to be an excellent staging location for privateers wishing to attack Spanish shipping and coastal colonies. A voyage that brought attention in Europe to the archipelago was that of Willem C. Schouten and Jacob Le Maire from the Netherlands, who arrived on 1 March 1616. Schouten published in 1619 his commentaries from this voyage,[10] including a map showing clearly the location of Robinson Crusoe Island, which informed other Europeans of its existence and potentials.

Another Dutch navigator, Jacques l'Hermite, funded by the Dutch East India Company, arrived at Robinson Crusoe Island on 5 April 1624. In addition to written observations on the landscape and vegetation, l'Hermite prepared[11] a remarkable drawing of the eastern side of Robinson Crusoe Island.

Sketch of the northeastern side of Robinson Crusoe Island, from l'Hermite in Gottfried and Merian, published in 1631

This is an extremely accurate drawing, showing the small sizes of the ships in the foreground, as well as several more anchored in Bahía Cumberland. Whoever drafted this drawing from the ship had a very good eye, because the principal peaks and ridges are clearly identifiable. This information was specific enough to provide encouragement for other European ships to use these islands as a stopping point. There was little dispute that they belonged to Spain, but nothing had been done to restrict ships of other countries. Privateers from England, France, and the Netherlands were continually using Robinson Crusoe Island as a lair to harass Spanish interests.

The Spanish finally decided to take control of Robinson Crusoe Island, and in 1750 they established a permanent colony with a fort supplied with cannons to destroy unwanted ships in the harbor. The village was called San Juan Bautista (St. John the Baptist) and the fort named Santa Bárbara. Troops arrived with rifles, muskets, powder, eighteen cannons, and more than 7,000 cannon balls, sufficient to make a good start on protecting the new village,[12] plus colonists and domesticated animals and plants. These colonization efforts were successful, and from this time forward, Spain had control of this island, but it never had strong control over Alejandro Selkirk Island, due to its further isolation, its lack of a protected bay, and the absence of a permanent settlement.

This lack of control over the far island had economic implications at the end of the eighteenth century and beginning of the nineteenth century. At this time, literally millions of fur seals (*Arctocephalus philippii*) lived and reproduced on the lower shore areas of Alejandro Selkirk Island. The market for fur pelts from these animals in Canton was strong, netting on average one dollar each, which for the epoch was a powerful economic motivation. Many ships avoided visiting Robinson Crusoe Island and elected to sail directly to Alejandro Selkirk Island, where crew members spent weeks and months killing seals, skinning them, stretching the pelts, and drying them for transport to the Orient.[13] Because the seals had no fear of humans, it was an easy matter to simply club one on the head and start the processing. More about this in chapter 12, but the killing was so substantial that by 1810, it was no longer economically worthwhile to continue the efforts due to the near extinction of populations on this island. The same had occurred earlier on Robinson Crusoe Island in the seventeenth and eighteenth centuries, but not so intensively in such a short period of time.

The village of San Juan Bautista continued to ebb and flow, with the colony at some periods gaining people, and in other periods reducing to nearly zero. Furthermore, the island was often used as a penal colony, usually for common criminals, but sometimes with political prisoners, especially during the struggle for Chilean independence from 1811 to 1818. Understandably, attempting to maintain a tranquil village with families along with hardened criminals and military troops was not a formula destined for lasting success.

The most interesting historical aspects of the archipelago for our purposes deal with the botanical activities. For convenience, these can be divided into the following stages[14]: Early Botanical period (1820–1875), Floristic period (1876–1906), Skottsberg (1907–1917), Touristic period (1918–1959), and Modern period (1960–present). In this last phase, there was an important U.S.–Chilean expedition in 1965, and then our own expeditions began in 1980.

The Early Botanical period started with the visit of Maria Graham (also known as Lady Callcott) in January of 1823. There had been one botanical collector previously, George Handisyd, who made a small collection of plants in 1690, of which two specimens survive, now residing in the herbarium of the British Museum in London.[15] Maria Graham had been recently married to Thomas Graham, a captain in the British Navy. She accompanied him on a voyage to South America, and with unbelievably bad luck, on the passage around Cape Horn he took ill and died, leaving her as a widow upon arriving in Chile. During the time, Commodore Thomas Cochrane from England had been contracted by Chile to develop the Chilean Navy and also to help liberate Peru from Spain.[16] He was subsequently offered a position as commander of the fledgling Brazilian Navy, but

Early botanical visitors to the Robinson Crusoe Archipelago. Top left, Mary Graham; top right, Carlo Bertero; bottom left, Claudio Gay; bottom right, Rodulfo Philippi.

prior to departure for that assignment, he sailed on the *Colonel Allen* to Robinson Crusoe Island, taking Maria Graham along as passenger. Cochrane and other officers collected plants in the hills adjacent Bahía Cumberland, and brought them to her for observations and pressing into specimens. She was very taken with the scenery and apparently also the company of Commodore Cochrane, but for our interests, the important point is that her collections represent the first of the early botanical collections from the archipelago during the nineteenth century. (You might wish to know more about their personal relationship, but there apparently is no more information—you will have to use your own imagination![17])

A very enthusiastic early naturalist to visit the Juan Fernández Archipelago was the Italian physician and botanist Carlo Bertero. He moved permanently to Chile in 1827 and had the opportunity to visit Robinson Crusoe Island in April through June of 1830, during which time he amassed approximately 2000 specimens. Most of these were eventually sent to the French botanist Jean Guillemin in Paris, and from there distributed to various natural history museums on the continent. Bertero sailed to Tahiti for more collecting, but upon attempting a return to Chile, he and his ship vanished without a trace.[18] Bertero had thankfully published in 1830 a report of his observations on the flora and vegetation of Robinson Crusoe Island,[19] and other botanists provided study and analysis of his valuable collections in the years that followed.

Another of the early botanical collectors was Claudio Gay from Paris. He arrived to Chile in 1828 as a professor in a new secondary school, the Colegio de Santiago. He was a very enthusiastic collector of all sorts of natural history specimens throughout Chile, and these helped form the collections of the newly founded Museo Nacional de Historia Natural.[20] He had an opportunity to visit Robinson Crusoe Island from 2 to 21 February of 1832, and he took full advantage to make significant collections, returning to Paris in March of that year with many of his specimens. He returned to Chile in 1834 and stayed collecting and working until 1842, when he sailed back to Paris for the remainder of his life. From there he worked on the massive *Historia Física y Política de Chile*, eight volumes of which comprised the *Flora Chilena*, still one of the building blocks for understanding plants of the country.[21] In addition to these new specimens from the islands, Gay published observations on the vegetation and plants,[22] including those that had been introduced from the mainland (and from Europe).

Drawing of the tsunami that struck San Juan Bautista on 25 February 1835. From Thomas Sutcliffe, 1839.

While these scientific initiatives were being accomplished, the village of San Juan Bautista was developing steadily, although always at small scale. An important event was the massive tsunami that destroyed the village on 25 February 1835. The governor of the island at that time was Thomas Sutcliffe, who was an Englishman serving in the Chilean military. The destructive wave[23] allowed him to rebuild the village, and he proved to be an efficient leader, but he pushed the islanders too hard with this project and a rebellion ensued. Sutcliffe spent many years trying to clear his name regarding the events,[24] but the whole affair was illustrative of the problems that the little community had in keeping peace, especially in the midst of so many common and political prisoners. Provisions from the mainland were also only sporadically provided to the little colony, which exacerbated tensions.

The final contributor worthy of mention during this Early Botanical period was Rodulfo Amando Philippi. He was another of the European-born naturalists who chose to come to Chile and work there on the natural history of the country, and in Philippi's case, he stayed permanently. He was born in Charlottenburg near Berlin, Germany, in 1808, but he emigrated to Chile in 1851, arriving first to the principal port of Valparaíso. He was a skilled naturalist of considerable reputation, and hence he soon was appointed Director of the Museo de Historia Natural in Santiago in 1853.[25] One of his famous expeditions was to the Atacama Desert in northern Chile, which resulted in a well-received book[26] with many attractive plates, now having an approximate value of $5,000. As Director of the Museo, Philippi sent the subdirector, Filiberto Germain, to the

Juan Fernández Archipelago at the end of October 1854, and he visited both islands. From these collections, plus observations made by Germain, Philippi published in 1856 a detailed article on the vegetation and flora.[27] Philippi himself did have the opportunity to visit Robinson Crusoe Island in 1864, but only for four days, nonetheless sufficient to publish an article on his observations.[28] The main contributions of Philippi were to provide a more comprehensive list of species (139 plants), comments on introduced species, and analysis of the island flora in comparison to that of continental Chile.

The next wave of botanical visitors to the Islands occurred during what has been called the Floristic Period (1876–1906). One of the major scientific expeditions during this period was that of the *Challenger*, sailing from Portsmouth, England, on 21 December 1872. This was an idea developed by Professor Charles Wyville Thomson of Edinburgh, Scotland, with the goal of surveying the world's oceans and making collections both in the sea as well as on land. It was a well-outfitted expedition, with about 240 persons including Thomson and five additional scientists.[29] From the botanical side, Henry Nottidge Moseley from the University of Oxford was to make observations and collections of plants. The ship stopped at Robinson Crusoe Island 13–15 November 1875, and in this short time Moseley made a collection that contained 105 different species, mostly vascular plants with a few cryptogams (mosses, lichens, algae).[30] When the expedition ended in May of 1876, the botanical collections were turned over to William Botting Hemsley at the Royal Botanical Gardens, Kew, for analysis and eventual publication. Hemsley did an outstanding job of investigating these new collections and synthesizing information from previous reports (Bertero, Gay, Philippi). The analysis included comments on biogeographic origins of the flora, in-

Important botanists who worked in the Robinson Crusoe Archipelago. Top left, Henry Moseley; top right, William Hemsley; bottom left, Federico Johow; bottom right, Carl Skottsberg.

troduced species, and percentages of indigenous species.[31] This combination of an efficient collector and a precise investigator resulted in the most detailed and careful portrayal of the flora and vegetation at that time. It also served as a template that was elaborated upon by all subsequent workers in the archipelago.

Another major contributor to understanding the flora and vegetation of the Juan Fernández Archipelago was Federico Adalberto Johow, another European naturalist who decided to make a new home in Chile. Johow was born in 1859 in Komar, Prussia, and was educated at the University of Bonn, Germany, receiving his doctorate in 1880. After a very stimulating collecting experience in the Antilles and Venezuela in 1883, he was promoted in Bonn as a professor in 1886. During this time, the Chilean government wanted to contract German professors to move to Chile and participate in the development of the new Instituto Pedagógico (teacher's college), and Johow accepted the challenge, arriving in 1889.[32] He was an effective teacher and administrator, such that the government commissioned him in 1891 to carry out an exploration of the Juan Fernández Archipelago, including both islands. He, plus three other scientists, arrived at Robinson Crusoe Island on 24 December 1891 and stayed until 3 February 1892, slightly more than one month. Johow took his commission very seriously, and the result was a book on the botanical resources of both islands, under the title *Estudios sobre la Flora de las Islas de Juan Fernández*, published in 1896. He built his treatise on the outline of analyses done earlier by Hemsley, but his mandate was broader, including recommendations on economic dimensions and improvements for the village. As for the plants, he described a number of new species and offered analyses of origins of the flora, plus a lengthy discussion on the introduced plants, including those that were threatening the native flora.[33]

At the end of the nineteenth century, a good basic inventory of the vascular flora of the archipelago was now in hand. But there was more to come, and the person who really made the difference was the Swedish botanist Carl Johan Fredrik Skottsberg. Skottsberg was born in 1880 in Karlshamn, Sweden, and showed interest early in natural history. Upon graduating from the University of Uppsala in 1901, he was invited to join the Swedish Antarctic Expedition of 1901–1903 under the leadership of Otto Nordenskjöld.[34] Although this exploring venture did not stop at the Juan Fernández Islands, the experience convinced Skottsberg that another trip to the Southern Hemisphere, especially Chile, would be warranted. Toward that end, he and two associates (Percy D. Quensel, also from the University of Uppsala, and Thore G. Halle from Stockholm) organized their own trip; in the meantime, Skottsberg completed his Ph.D. degree from the University of Uppsala and graduated in 1907.[35] Later that same year on 10 September, their small expedition left Sweden on the ship *Princess Ingeborg*, sailed along the eastern coast of South America to Brazil, the Malvinas (Falkland Islands), Tierra del Fuego, and finally up the western coast to Valparaíso. On this trip, Skottsberg and colleagues spent only eight days total in the Juan Fernández Archipelago, with four days working on Robinson Crusoe Island and two on Alejandro Selkirk Island. They returned to Sweden on 21 June 1909 with no mishap, collections and observations intact. Skottsberg published reports on their findings for the archipelago,[36] which indicated to him that much more time would be needed to provide a more definitive analysis of the flora and vegetation.

With that in mind as the next objective, Skottsberg organized another trip to the Southern Hemisphere, this time with specific goals being the Juan Fernández Archipelago and Easter Island, the latter lying much farther westward into the Pacific Ocean. The crew consisted of Skottsberg, his wife Inga, who was a botanical assistant at the University of Uppsala, and an advanced zoology student, K. Bäckstrom, who also served as photographer. They departed Sweden on 5 October 1916 on the appropriately named ship *Suecia*, this time heading through the Panama Canal and down the western coast of South America, eventually arriving at Valparaíso. From there they sailed to Robinson Crusoe Island, entering Bahía Cumberland on 1 December 1916. They stayed in the archipelago five months, which covered the austral late spring, summer, and early fall, subsequently visiting coastal regions of Chile and Peru, sweeping westward to spend a month in Easter Island, and finally returning to Sweden on 8 December 1917.[37]

Skottsberg was always very productive from each expedition. He collected not only plants but also animals and rocks and minerals, the latter of which were sent to collaborators in different countries who were experts on the non-botanical samples. Many of the investigations were edited by Skottsberg in a series of chapters in three volumes published from 1920 through 1956, volumes 1 and 2 being dedicated to geology and botany, and volume 3 dealing with zoology.[38] These were printed at different times in small runs, depending on the speed of each investigator, and when all were finished, they were consecutively paginated and distributed, with libraries then binding them together. It can be a bibliographically confusing set of chapters, because, for example, there are articles in volume 1 that were published later than some of those in volume 3. In any event, the contributions by Skottsberg covered flowering plants and ferns (with Carl Christensen), pollination and reproductive biology, biogeography, and vegetation, in short a complete and intensive survey of the plant world of the archipelago.[39] He built upon the earlier studies of Hemsley and Johow, but he took understanding to a much higher level. It was upon this foundation that all other studies of an evolutionary nature were able to make substantial progress during the twentieth century, and even today they are indispensable publications for anyone contemplating work in the archipelago.

A very special event relating to Skottsberg happened to me in 1994. I was invited to present a talk in the Carl Skottsberg lecture series in Göteborg, Sweden, which was associated with the botanical garden where Skottsberg was once director. Unsurprisingly, the topic of my presentation was on the plants of the Robinson Crusoe Islands. A highlight of the event was meeting the daughter of Skottsberg, Ann-Mari, and visiting their family summer home in Axvall. Although she had not accompanied her parents on these trips to the islands, she was well informed about them. She presented me with his personal copy of the book on his expedition to the islands in 1916–1917,[40] a gift that I will always treasure. Following in Skottsberg's footsteps was very inspiring for me, as we built upon the research platform he had established and added new levels of understanding for the plants of the islands.

The Touristic Period (1918–1959) was less of a scientific phase than an interest in these islands due to curiosity and entertainment. Into the 1920s, the world experienced a time of tranquility and wealth, which supported traveling for knowledge and entertainment. Chile by this time had also become more prosperous, and many citizens wanted to experience these islands as part of knowing the heritage of their own country. Robinson Crusoe Island has never had opulent facilities to house and feed high-class tourists, but often luxury ships would anchor in Cumberland Bay and allow voyagers to go ashore and hike to accessible localities, particularly Selkirk's lookout (Portezuelo), which provided good exercise and a nice panoramic view of both sides of the island. The tourists also wanted souvenirs to take home, and these usually consisted of handicrafts made

Events related to my visit and lecture in Göteborg, Sweden, 1994. Top, cover of the published text of the talk; bottom left, socializing with Ann-Mari Skottsberg, daughter of Carl Skottsberg, and Gunner Weimark, Director of the University of Göteborg Botanical Garden; bottom right, the Skottsberg family summer home in Axvall.

from goat skins or horns, walking sticks from native woods, dried lobsters, and other items.[41] They also were interested in purchasing small native tree ferns, which rarely survived in gardens in the much hotter Mediterranean climate of Valparaíso or Santiago, let alone on the long trips back to Europe.

The last phase of botanical activity in the Juan Fernández Archipelago can be labeled the Modern Period (1960 to present) during which several major expeditions took place. The first was organized by Otto Solbrig from Harvard University (United States) with Carlos Muñoz Pizarro from the Museo de Historia Natural in Santiago. This was a large expedition with eight U.S. participants and four Chileans, the largest that has ever been mounted with a purely botanical focus. The group left Valparaíso on

Scientists who have worked on the flora of the Robinson Crusoe Islands during the Modern Period. Top left, Otto Solbrig; top right, Carlos Muñoz Pizarro; middle left, Tod Stuessy; middle right, Daniel Crawford; bottom left, Philippe Danton; bottom right, Christophe Perrier.

6 December 1965 and spent eleven days collecting on Robinson Crusoe Island, with most participants leaving on 19 December and two staying on until 27 December. The collections were very valuable, consisting of 166 samples of marine algae, 1,624 lichens, 800 mosses and liverworts, ca. 150 ferns, and more than 700 flowering plants.[42] Few publications emanated directly from this trip, except for a series on the *chonta*, the endemic palm, *Juania australis*.[43] The collections were distributed to many herbaria, especially in the United States, which provided materials for publications by other workers in subsequent decades.

The second set of major explorations during the Modern Period was conducted by our collaborations principally between The Ohio State University, the University of Vienna, Austria, and the University of Concepción. These began in 1980 and continued with eleven more expeditions to our last trip in 2011. Because of the strong inventory of the flora established by previous workers, especially Skottsberg, our objectives were not to redo the inventory but instead to ask and attempt to answer evolutionary questions. It was a good system in which to do this: there were only two major islands, they were in an east–west orientation to the major source area in southern South America, they were dated radiometrically,[8] and the flora was reasonably small (124 endemic species), suggesting that we might be able to investigate many of these groups and be able to synthesize hypotheses regarding the origin and evolution of the entire flora and vegetation. It was a bold set of objectives, but over more than forty years, we have been able to contribute a much deeper understanding of the archipelago.[44] See Appendices 1 and 2 for lists of our expeditions and publications over the years.

The third set of major botanical expeditions to the Juan Fernández Archipelago, more than a dozen, have been conducted by Philippe Danton and Christophe Perrier of Grenoble, France. These enthusiastic workers have published several conservation-oriented overviews of the endemic plants of the islands,[45] and just recently they have published a new flora for the islands.[46] These contributions have been particularly valuable because some of them have been in Spanish and French, which opens the information more to the people of the islands and the Chilean park service.

CHAPTER 5

Bizarre, Unusual, and Interesting Plants

The floras of oceanic islands are often disharmonic.[1] This has nothing to do with music. It means that the elements of the flora are assembled into vegetation that is very different from continental regions. The plants of oceanic islands are there because of one very important factor—their superior ability to disperse by air, water, or with birds. What gets to an island has made a very improbable journey across the ocean, and for those that do arrive, most fail to become permanently established. As the landscape on a new oceanic island fills with successful plant voyagers, the vegetation often has a very different appearance from that of continental areas. There is a new mixture of botanical elements, some bizarre and unusual species, and others of particular evolutionary interest. In this chapter I want to present some of these strange and noteworthy plants that did make the journey to the Juan Fernández Archipelago and became successfully established. Robinson Crusoe Island is 667 kilometers from continental Chile, which is a challenging distance for a propagule to travel. The other island, Alejandro Selkirk, is even more remote, 181 kilometers farther to the west. Most plant groups have arrived and evolved on the older and nearer island first, followed by dispersal to the younger island. A few plant immigrants have made the trip from the continent directly to the far island, and still others have arrived from the western Pacific, but these cases have been less frequent.

To me, bizarre means really strange, and this depends on the viewer, obviously. The more experience a person has with different aspects of life, the less bizarre things seem. But there are some plants in the Juan Fernández Archipelago that are a bit strange, no matter who is looking at them. Two good examples are from the sunflower family (Asteraceae), which is the largest family of flowering plants on the planet with more than 25,000 species.[2] Economically important members would include lettuce, endive, artichoke, and hundreds of valued ornamentals, such as *Aster*, *Chrysanthemum*, *Coreopsis*, and *Dahlia*. Species of this family are found commonly in temperate or cooler regions of the world, in contrast to the also large orchid family, which frequents tropical environments. In the sunflower family, most of the representatives are herbaceous, with some becoming shrubby and a very few achieving status as large trees, even to thirty meters tall, such as *Vernonia arborea* in Malaysia. Members of this family are good dispersers, and they have often arrived to oceanic islands, frequently evolving into shrubs or small trees. Two illustrations of this phenomenon in the Juan Fernández Archipelago are the endemic genera *Dendroseris* and *Robinsonia*.

Dendroseris just looks weird (see color insert, **2**). Although the continental relatives of the group belong somewhat distantly to the herbaceous sow thistles (*Sonchus*),[3] after the original colonizer to Robinson Crusoe Island arrived some millions of years ago, the lineage began to change and became woody, accompanied by speciation into different ecological zones on this island. *Dendroseris* is the largest genus within the archipelago. Eight endemic species evolved on Robinson Crusoe and Santa Clara Islands, and three on Alejandro Selkirk Island. All these species are bizarre in appearance, because they are rosette trees, having long, thin trunks and leaves clustered at the top, where the flowering heads also are produced. The flowers are cream-white to yellow-orange, depending upon the species. The most striking species is *D. litoralis*, confined to low, rocky outcrops near the sea, but now extremely rare in nature. Fortunately, it adapts to cultivation and has done well in gardens in the island, in continental Chile, and in selected botanical gardens in Europe. The leaves are large and leathery, and the flowering heads, bright orange, hanging down when mature, are pollinated by hummingbirds.[4] These lively birds flit from one flowering head to another, and they sometimes hang onto the outside of the heads when they are probing for nectar. The birds interact, moving quickly and squeaking loudly and aggressively when an intruder attempts to share the same floral resource.

Some species of *Dendroseris* resemble small palm trees. The most commonly seen is *D. pinnata*, which has long divided leaves attached to the top of a very thin stem. Plants usually grow solitary in the upper montane forest at elevations above 450 meters. They stick out from among the other native species, and the rosette of deeply divided leaves waves in the wind, perhaps being an adaptation to avoid structural damage in a windswept environment. This species, and its close relatives *D. berteroana* and *D. regia*, are monocarpic, which means they flower once with a large cluster of whitish heads, set fruit, and then the entire plant dies.

An interesting evolutionary aspect of *Dendroseris* is the occurrence of three species on the younger Alejandro Selkirk Island. The eight species on Robinson Crusoe Island show marked morphological differences and also different habitat requirements, which suggests that this group adaptively radiated when the island was younger, larger, and more environmentally diverse. One might suspect that the three species endemic to the younger island would have evolved by dispersal of one immigrant from the older island during the past one million years (the age of the younger island) followed by adaptive radiation to yield three new species. Phylogenetic studies, however, show this not to be the case.[5] Each of the species on Alejandro Selkirk Island is a close relative of one of those on Robinson Crusoe Island. During the past one million years, dispersal occurred three different times from the older to the younger island, and after arrival, transformational speciation (anagenesis) occurred independently in each of the three lineages. These three species are more closely related to their progenitors on the older island than they are to each other. This emphasizes the importance of having a phylogenetic understanding of relationships to better understand processes of evolution.

Another bizarre genus in the Juan Fernández Archipelago is *Robinsonia*, named after the islands in 1833 by Alphonse de Candolle who worked in the herbarium and botanical garden in Geneva, Switzerland. The genus is endemic in the archipelago,[6] with seven species being restricted to Robinson Crusoe Island and only one to Alejandro Selkirk Island. These species are also rosette trees, the species (see color insert, **3**) varying from about one fourth of a meter tall (the rare *R. saxatilis*) up to about three meters (*R. thurifera*). The stems are thin, even thinner than in *Dendroseris*, but they also tend to branch more toward the top. A special feature of the genus is that the plants are dioecious, that is, they produce only male or female flowering heads, rendering each individual unisexual. One can distinguish the two sexes easily based on floral parts as well as by the female heads being smaller than the male ones. From a conservation perspective, viable populations need to contain a mixture of both sexes, which is usually the case.

Species of *Robinsonia* show some ecological differentiation. After four million years of island landscape alterations, plus additional human disturbance to the vegetation during the past 400 years, the ecological zones that likely were very distinct when the island acquired its initial vegetation are much less so today. Even so, *R. gayana* (named after Claudio Gay) and *R. saxatilis* are restricted to open, rocky basaltic ridges, and *R. gracilis* is found above 700 meters on the highest ridges at the crest of montane forests. *Robinsonia evenia* and *R. thurifera* both inhabit the dense native upper montane forest, but due to differences in flowering time (phenological differences) they maintain distinctness and do not hybridize. *Robinsonia macrocephala* is extremely rare, and it may be extinct. Likewise, *R. berteroi* (named after Carlo Bertero) is also extremely scarce, with only one plant recently having been observed at the top of the highest peak, El Yunque.[7] Previous to this discovery, the species was considered extinct.

The evolution of *Robinsonia* within the Juan Fernández Archipelago is also of great interest.[8] Because of the size of the genus in terms of numbers of species (the second-largest genus in the archipelago), it also must have been an early colonizer to Robinson Crusoe Island, followed by adaptive radiation that eventually yielded seven species. These are less diverse morphologically among themselves than those within *Dendroseris*.[9] The single species on Alejandro Selkirk Island, *R. masafuerae*, is related phylogenetically to *R. evenia* on the older island. It is likely that a propagule of *R. evenia* arrived to the young island after it had formed, and in isolation over the past one million years it transformed anagenetically[10] into what is now recognized as *R. masafuerae*, with distinct morphological and genetic features.

Bizarre examples in the flora of the Juan Fernández Archipelago also are known among the tree ferns. On Alejandro Selkirk Island is the conspicuous *Dicksonia externa* (see color insert, **4**). This species forms groves of individuals in dense thickets of intertwined trunks, at the top of which are clusters of large leaves. Currently this species grows, appearing like rows of soldiers, in the southern part of the island in higher elevations (above ca. 800 meters). The appearance of the landscape in this sector is very bizarre, looking very otherworldly, largely because of the dominance of this one species. Another curious tree fern is *Lomariocycas cycadifolia* (*Blechnum cycadifolium*),[11] found on both major islands. This species has a cluster of cycad-like leaves (fronds) perched around the shoot apex and supported by a trunk one to two meters tall. A group of these ferns on an open ridge amid the upper montane forest is a very spectacular sight, and definitely unusual in comparison to landscapes on continental Chile.

The distinction between bizarre and unusual is not precise, but in the context of this book, I would like to share information about two most unusual species. The first is *Lactoris fernandeziana* (see color insert, **5**). This has no common name, unless one just wants to call it "Lactoris." It is a genus with only one species, so the genus equals the species and vice versa. *Lactoris* is sufficiently distinctive morphologically that it has been placed in its own family, Lactoridaceae, by researchers.[12] *Lactoris fernandeziana* is a small shrub, with greenish stems and small oval leaves, not a very imposing plant and certainly not bizarre. The flowers are simple and greenish, hardly noticeable, and again, not attracting immediate attention. As far as I am aware, there is no living plant in cultivation anywhere in the world, so the only place to see it live is on Robinson Crusoe Island. It is not found on Santa Clara Island nor on Alejandro Selkirk Island. If you want to see and photograph it, you have to obtain a permit from CONAF and be taken there by the park guides, to look but not to collect. It is estimated that between 1000 and 2000 individual plants exist on Robinson Crusoe Island,[13] and these are all in very secluded places, hence the need for guides, both for your own protection and that of the species.

Lactoris fernandeziana was first described as new to science by Rodulfo Amando Philippi,[14] who made major contributions to understanding the flora and vegetation of the island. Specimens were collected by Germain in 1854, and Philippi published the new genus and species in 1865. His observations at the time indicated little understanding of what the relationships of this new species might be. It is not easy to see affinities, in part because the flowers are so small. There are six stamens (the male parts of the flower) and three carpels (the female parts), and there is an outer whorl of six asexual appendages, called tepals, because there is no distinction between petals and sepals. Arrangement of flower parts in threes and sixes usually indicates belonging to the monocots, which include things like lilies, palms, aroids, hostas, tulips, irises, and the like.

These plants almost always have parallel orientation of the vascular traces in their leaves, but *Lactoris* has net venation, typical of the dicots (most of the rest of the flowering plants). The pollen of *Lactoris* is also interesting, in that when dispersed from the anthers by wind,[15] it is in a tetrad, that is, four pollen grains stuck together.[16] This does occur in other species of flowering plants, but it is not a common condition. Most plants start out with four young pollen grains resulting from meiosis, but they separate at maturity and are dispersed individually. Efforts to understand affinities of this unusual species have suggested ties to the Magnoliales and more recently Piperales, and some workers have even placed it in the dutchman's pipe family, Aristolochiaceae, based primarily on nucleotide sequence data.

Lactoridaceae are an ancient family of flowering plants. Fossil pollen (as the tetrad *Lactoripollenites*) have been found in sediments off the western coast of South Africa at seventy to ninety million years of age,[17] in Australia in sediments fifty to seventy million years old,[18] in southern South America about ninteen million years,[19] and possibly also in Canada, the United States, India, and the Antarctic Peninsula. These reports show convincingly that the family was broadly distributed in the Southern Hemisphere as early as ninety million years ago, and oddly enough, it now survives refugially only on Robinson Crusoe Island at four million years of age. The best explanation for this anomaly is that continental populations over millions of years of geological time have slowly become extirpated, with a successful dispersal event occurring during the past four million years (the age of Robinson Crusoe Island), followed by extinction elsewhere. It is always possible that some populations in continental areas remain undetected, in part because the plant is not particularly showy or attractive, and less likely to catch the eye of botanical collectors. Only time will help us understand better the biogeographic puzzle of *Lactoris* being restricted to Robinson Crusoe Island.

A second unusual genus is the fern *Thyrsopteris* with the single species *T. elegans* (see color insert, **5**), the sole representative of the family Thyrsopteridaceae[20] and another ancient lineage, now confined to both major islands of the archipelago. This is a tall fern with fronds (leaves) that arch high over one's head. The leaves are delicate, some containing brown, fertile regions with spore-filled sporangia. The relationships of this taxon to other tree ferns are remote, and molecular phylogenetic analyses suggest possible affinities to genera in several other fern families. What is particularly interesting is that fossils from Jurassic and Cretaceous deposits resemble *Thyrsopteris*, which suggests a very old evolutionary line. One fossil comes from continental Chile in Cerro Guido, Magallanes, at a stratigraphic age of seventy million years,[21] and other fossils from the Southern Hemisphere and elsewhere are also known.[22] A fossil leaf portion called *Thyrsopterorachis* has been described from the Upper Cretaceous of Japan (one hundred to sixty-six million years old).[23] These fossils suggest a much broader world distribution of the lineage, especially in the Southern Hemisphere. *Thyrsopteris*, as with *Lactoris* described above, apparently has gone extinct elsewhere in the world and now survives refugially in the Juan Fernández Archipelago. The population on Alejandro Selkirk Island must have developed from dispersal from plants on Robinson Crusoe Island, after the younger island formed one million years ago.

Another unusual member of the Juan Fernández flora is the genus *Gunnera* (see color insert, **6**). Three endemic species occur in the islands, two on Robinson Crusoe Island and one on Alejandro Selkirk Island. They are very robust plants, a leaf of which can serve as an umbrella. Of particular interest are the petioles, the stalks of the leaves. These have a warty exterior, but the inside tissue is soft and edible. In mainland Chile, petioles of a related species, *G. tinctoria*, are sold in the market as a vegetable. Of particular botanical interest is that inside and toward the base of the petiole are contained colonies of the cyanobacterium *Nostoc* (earlier regarded as a blue-green alga), which provides nitrogen for the plant in turn for receiving water and a protected place to live, very advantageous for both.

In addition to bizarre and unusual plants of the Robinson Crusoe Islands, there are also many interesting and attractive endemic species (see color insert, **7**). *Pernettya* (*Gaultheria*) *rigida* (Ericaceae) is an endemic shrub to one meter tall, with cup-shaped, nodding, whitish flowers. The species is relatively common in open areas on Robinson Crusoe Island. *Ochagavia elegans* belongs to the pineapple family (Bromeliaceae), and it forms clusters of plants in rocky crevices and exposures. The flowers are a lovely red-pink color, also attractive to hummingbirds. *Wahlenbergia fernandeziana* (Campanulaceae) occurs near the ocean on coastal rocks, which provides a nice contrast with the gray lava and basalt. There are five endemic species of this genus in the islands, three on Robinson Crusoe Island and two confined to Alejandro Selkirk Island.[24] Further study is needed to fully understand the evolutionary processes that have led to this interesting pattern of diversity. A most interesting tree is *Rhaphitham-*

nus venustus (Verbenaceae), with its pendant deep blue–purple flowers. This species is also pollinated by hummingbirds, which flit quickly from one flower to the other, sipping nectar from each.

Another fascinating endemic plant is the shrubby *Chenopodium sanctae-clarae* (see color insert, **8**) of the goosefoot family (Chenopodiaceae). Species of this genus are usually herbaceous, our example being another case of the evolution of insular woodiness. There is nothing particularly spectacular about the appearance of this plant, but it tells a conservation story, which affected me deeply. This species occurs on the Earth *only* on one small islet, Morro Spartan, which is just off the northeastern side of Santa Clara Island, and only about sixty plants exist. When we inventoried these individuals, it finally hit home to me how precarious survival was for this species. If anything were to happen to these plants, such as the inadvertent introduction of a harmful insect predator, the species would become extinct. It occurred to me that a person with clippers could bring this species to extinction within only a few hours. What became poignant in this moment, as I stood looking at these plants, surrounded by the beauty of the deep blue ocean and the profile of Robinson Crusoe Island in the distance, was the need for human stewardship of this critically endangered species. This is true not only for this species here on this rock, but also for species elsewhere in the islands and everywhere else on the planet. We, among all the 8.7 million or more other species that inhabit this globe,[25] have the ability to destroy other life forms as well as to save them. *Chenopodium sanctae-clarae* speaks loudly to this responsibility. It is perhaps fitting that Santa Clara Island refers to Saint Clare, a follower of Saint Francis of Assisi, the protector of animals and the natural environment.

CHAPTER 6

Arrivals and Departures

I am always amused when listening to colleagues who describe working in the Canary or Hawaiian Islands, first jetting into a modern airport, checking into a nice hotel, discussing in which restaurant to eat in the evening, and during the day pulling off the road for lunch at a quaint restaurant with a fantastic panorama of a valley or the ocean. I know, they work hard, very hard, but the logistics of getting to and from those islands are simply nothing by comparison to the Robinson Crusoe Islands. To illustrate some of these difficulties, I have chosen to tell you about two trips. The first is a boat trip from Valparaíso to Robinson Crusoe Island in the beginning of January 1980 (our first expedition), and the second is our return from the island by airplane at the end of January 1997. A major theme of these arrivals and departures is patience, patience, and still more patience. Chapters 8 and 11 explore even more daunting challenges of working in the isolated Alejandro Selkirk Island.

Whenever one travels to a new location, there is real excitement. This is part of the reason we love to travel so much. New lands stimulate the senses, with new sights, new smells, and new foods. Before our first trip to the Juan Fernández Islands, there was lots of anticipation. Roger Sanders, my postdoc on the project funded by the U.S. National Science Foundation, and I had flown from Columbus in the dead of winter down to Santiago in the middle of summer. As any North American snowbird knows, when flying from the frozen north in January, such as from Chicago to Miami, there is this wonderful sense of relief and freshness upon descending the steps from the plane. Nowadays with a jetway, this sensation has to wait until exiting the terminal with luggage, but the effect is the same—soothing in one sense, but very stimulating in another, because it suggests a new approach to life for the next weeks or months. If everyone had the time and means to leave the cold during January and February, few persons, other than winter sports aficionados, would be left in the cold northern regions. It would be a complete seasonal mass migration south similar to Canadian geese in the fall.

Roger and I arrived in Santiago on 30 December 1979 to begin the very first expedition. We flew into Santiago rather than directly to Concepción because there was no international air service to the latter. After the plane had taxied to a stop in front of the terminal, the doorway was opened, and we descended the aluminum stairway to the tarmac, smiling at seeing the palm trees and feeling the fresh, warm breezes across our face. At this time, the terminal was still very small, and hence there was very little wait to pass through immigration, find our bags, and work our way through customs. After exiting the airport terminal, we were approached by several independent fellows offering to take us into the city, but we opted for the regular airport taxi service for a ride to the central train station in the middle of the city.

There were two ways to get from Santiago to Concepción, by bus and train, and both could be taken overnight, hence saving an entire day. On this occasion we chose to take the train. The traffic in Santiago is busy, but it is quite disciplined, which made for orderly travel downtown to the train station, along the broad avenues, around traffic circles, past high-rise gleaming office buildings, and finally into the older section of the city with lower buildings and residences, until arriving at the cavernous train station. What I hoped to find were two tickets to Concepción for that evening in the sleeper car (*coche dormitorio*), costing forty-five dollars round trip, and fortunately these were available. This meant we could conveniently leave our luggage in the baggage check (*custodia*) at the train station. With this organization, we had the rest of the day to take a taxi to the main downtown shopping area, the *peatonal* (shoppers-only street) and do some window-shopping, have a coffee and lunch, and get the feel of the city.

When I came through Santiago in 1977, I had little time in the city proper, so this new visit was a special treat for me. Roger had never been to the Southern Hemisphere, and hence all was new and interesting to him. Academics generally relate well to travel, as they are naturally curious, and travel teaches about different peoples and customs, and of course, language. I was fluent in Spanish, but Roger only knew a few words and phrases, which meant he stuck close to me as we walked along in the center of Santiago. We strolled as the Chileans do, slowly, enjoying the sun, men with hands clasped behind their backs. People enjoy this sense of belonging, not being packed like sardines, but being close. Who in a restaurant likes to be eating alone? If one enters a restaurant and there is already one couple at a table, generally we like to sit somewhere near them, perhaps not exactly next to them, but close. It feels strange to be sitting on the other side of the restaurant, separated by empty tables.

After a pleasant afternoon in Santiago, we had an early dinner at about 7:00, just sandwiches (I had melted cheese and avocado [*palta*]) and Coca-Cola, and then hopped in a cab, another of the typical black-and-yellow small taxis, back to the train station. After checking on the schedule and finding it all in order for departure at 10:00, we retrieved our bags from the *custodia* and took a position under the old, arched steel roof to await the arrival of the train. This was a classic train station, similar to what one might see in a movie set in Europe during the 1950s. The diesel train arrived by 9:30, inching very slowly back and back until nearing the end of the track, followed by a loud wheeze of air and then stopping completely. After about ten minutes, we asked the conductor if it might be possible to board, and he, dressed very smartly in a dark blue suit with gold trim on lapels and matching cap, said yes, now was the time.

The train from Santiago to Concepción offered regular seats in the six coach cars, but there were also the two sleeper cars of stainless steel and mahogany made in Breslau, Germany, in 1929 before WWII, still in remarkably good shape and taking one back to the past when such cars were a high standard of travel. This particular train made occasional stops, but few of them, and persons just got in, climbed into their berths that were stacked up as two bunk beds on both sides of the aisle, and pulled the dark gray curtains on rings along a chrome bar to provide privacy. I doubt anyone changed clothes into pajamas (I certainly didn't), in part because it can get somewhat chilly in the evening, and also if the need arose to rush to the bathroom in the night, it could be accomplished quickly. There was a blanket to snuggle under as well. Each compartment had a small light inside, too, which allowed for reading or writing. The car had several cabins with wash basins affording privacy for couples, but we never used these for expedition personnel. There was one community bathroom at the end of the car.

The train left the station punctually at 10:00, lumbering along with a strong diesel engine and slowly picking up speed. Each sleeping berth had a window with a curtain, and at night one could see a few buildings, switching and maintenance areas, and sidelined railroad cars, all interspersed by dark landscape. There was the muffled clickety-clack of the wheels on the rails that was soothing, helping to lull us into deep relaxation and finally sleep. Roger had the bunk below me, so I wished him good evening, rolled over and pulled the blanket close, and soon drifted away to my own personal world of dreams. I woke up a few times during the night, not unusual for me when sleeping in a new setting, and once for a short scamper to the bathroom. Climbing down and back up into my berth required some dexterity so as not to disturb Roger sleeping below. At about 6:30 in the morning, the porter/conductor came through the car with a continental breakfast of rolls and coffee on a small tray, which helped wake us up; he announced that we would be arriving to the end stop in Concepción in short order, on schedule.

The train arrived at 7:00 and slowly braked to a halt. There were a few persons waiting to greet new arrivals, but otherwise it was quiet, still early in the morning. Stepping down from the sleeper car with our bags, we passed into the main station with its square tiles arranged in diagonal black-and-white rows and, high above, all around on the walls, was the famous mural by Gregorio de la Fuente completed from 1943 to 1946, which shows the history of Concepción from the indigenous Araucanos, the Spanish conquest, the earthquake of 1835, the Lota coal mine region, veneration of the family, and the positive value of industry. In the early light of the morning the mural was cloaked in subdued tones, but it was a powerful reminder of the centuries through which the city had passed. As we left the station and to the curb with waiting taxicabs, the sun was still slightly hidden behind the hills to the east of the city, radiating its thin yellow light along the silhouetted, pine-covered ridges. It was a bit brisk, since Concepción is close to the sea and along the Bío-Bío River, both of which provide more humidity than found in Santiago, and it is also 350 miles farther south. There were folks going to work, walking briskly with sweaters and some with woolen caps, bags with lunch or other essentials, some with woven ponchos, and all with purpose. A few cars sped along the broad avenue in front of the train station. It was a new day and there were things to be done, just as we also had to do. We were a bit tired from the evening train ride, but no matter—an expedition awaited us, and many organizational aspects had to be decided.

The cab took us on the fifteen-minute ride to the Arco de Medicina of the University of Concepción, and we asked to be dropped off behind the building that houses the Botany Department. There is no vehicular access to the broad walkway in front of the building, which is the principal route for students entering the university campus. The Botany Department occupies the first, second, third, and fourth floors on one-half of the building and the Zoology Department has the other half. The bottom floor (*planta baja*) contains laboratory classrooms, a lecture hall with tables and chairs, bathrooms, and a small room for the security man (*el portero*), who keeps an eye on who comes in and out of the building. This was particularly important now at the end of 1979, because Chile had passed through a very difficult political time, with the overthrow in 1973 of the elected communist government of Salvador Allende by the military regime of Augusto Pinochet, who was now very much in power and seeking to eliminate all opposition. Military leaders tend to see the world in black-and-white terms—they are not trained to seek compromise. If there is a military objective to defeat the enemy, the campaign is not successful until the mission is completely accomplished. You are either with them or against them. Chileans, used to more than fifty years of democratic government prior to Pinochet, were not all supportive of these repressive communist-hunting measures, and occasional resistance took place through sabotage of bridges, electric plants, and governmental offices. The university was therefore on alert for possible disturbances at all times, which made the job of the *portero* more important than ever before.

Because universities in general tend to be bastions of liberal thinking, the Pinochet government was carrying out purges of professors who were deemed unsuited for teaching young minds about politics, sociology, philosophy, and the like, and these departments country-wide were being eliminated. Many intellectuals left Chile during this period for other South American countries, especially Venezuela and Colombia. In fact, we were to learn that a few years before our visit, the most distinguished botanist in Concepción, Professor Mario Ricardi, was tortured, exiled, and now lived in Mérida, Venezuela, never to return (he passed away in

2005). What Roger and I learned very quickly was that it was better not to talk politics with our botanical colleagues—it was simply too dangerous. Informants were everywhere, and our colleagues did not know us well or trust us yet. There were cases within the University of Concepción where a professor made some negative accusation against the government in class, and he just disappeared without a trace. This was not common, but it happened enough to put everyone on guard, close guard. No point in getting our developing relationship off to the wrong start. There would be time enough for such confidence if the collaborations continued into the future. Furthermore, the main objective for us was the science, to learn about the evolution of the plants of the Robinson Crusoe Islands, and scientists the world over know to keep other distractions out of the picture for the greater good and interest of solving scientific questions. This is one reason why scientists make excellent international ambassadors between countries,[1] and why international scientific projects need always to be supported.

We entered the building that housed the Department of Botany and headed directly to the stairway leading to the other floors. The building had no elevator. There was a distinctive smell, as all buildings have, perhaps deriving from the wax on the wooden floors that extended to the right and left of the tiled entry. Roger and I hefted our bags up the stairs to the first floor and presented ourselves to the secretary of the Department of Botany. After introductions, she called Professor Mario Silva in his laboratory on the fourth floor, and we arranged our bags and waited for him to come and greet us, which he did in a few minutes. It was good to see Mario again, smiling, enthusiastic, energetic. It is always a relief to see a familiar face after having traveled a long distance (in this case 8,608 kilometers) to a distant and unfamiliar land. There was Mario, smiling, extending his hand and following with an embrace. For friends, Chilean men have a clear greeting protocol: first a handshake, then an embrace with a pat on the back, each moving their heads to the left, and then another handshake. Nice, but it took us a few encounters to get the hang of the ritual. For ladies, one kisses her right cheek. In Europe this varies, for example, in Austria it is one kiss on each cheek, and in Switzerland three kisses, right cheek, left cheek, and finally returning to the right cheek. Interesting how every culture has its protocols for greeting, and it is very important to learn these as quickly as possible. Of course, the citizens of a country realize that visitors may not understand these routines, but every person wants to be accepted, and getting into the swing of these cultural norms makes life easier for everyone.

After chatting with Mario and introducing him to Roger, Professor Clodomiro Marticorena appeared and we made the same round of greetings, introductions, and chatting. Compared to Mario, Clodomiro was more eccentric and quicker to judge, whereas Mario was smoother and more calculating before he spoke. Mario had more administrative experience, and he also ran a large laboratory, whereas Clodomiro was more of a one-man show, an excellent show, but he had a different type of academic approach. He did publish with others, as he did with us, but his personality was more guarded than that of Mario's, which is one of the reasons academic life is so fascinating. People are not hired for their personalities—they are hired for their professional competence. We academics are acutely aware of what people have published and in which journals, such that a very eccentric professor can command very high prestige despite difficulties in dealing with him or her. This just comes with the turf, so to speak. We all then adjourned to the small departmental seminar room and got out our notes and information to see where we were with preparations. Mario was not planning to go on the trip, simply because he was a phytochemist and had no field skills. Both of these colleagues were fundamental for success of the Robinson Crusoe Island project, with Mario being the planner and organizer and Clodomiro the expert on the Chilean flora. On this trip, Clodomiro would be the head of the Chilean contingent, and I would be the co-head with Roger on my team. The group was going to be rather large—me, Roger, Clodomiro, Roberto Rodríguez (fern expert), Oscar Parra (algal specialist), Eduardo Ugarte (plant ecologist), Jorge Arriagada (graduate student), and then three persons from the film and promotions department of the university, Violeta (wife of Jorge), Gabriel, and Manuel. In addition to funds obtained for the trip from the U.S. National Science Foundation and Comisión Nacional de Investigación Científica y Tecnológica (CONICYT) of Chile, the University of Concepción contributed funds in support of the film crew, the objective being to produce a documentary to use on Chilean national television (which, in fact, did happen).

The trip to Robinson Crusoe Island was to take place by boat from Valparaíso, which lies on the coast directly west from Santiago. From information that the Chileans had obtained, there was little food available on Robinson Crusoe Island, especially for our large group of ten people. Food, therefore, had been purchased in Concepción, and it had been laid out on the tables in the large laboratory room at the end of the ground floor, which at this time of the year was not being used for coursework. Because the semester finishes near the end of January, lab work had already stopped for the term. Summer vacation in Chile runs from the middle of January to the middle of March, a pattern similar to summer vacation in the Northern Hemisphere, but in different months of the year. To hold food and supplies for our trip, twenty large, sturdy, wooden chests (crates) had been constructed, each with a hinged top and lock on the front. These were painted yellow with the university name stenciled in blue on the sides and top. The only problem was that when fully loaded, they were very heavy, and it took two persons to move them. The room looked quite full with these boxes on so many tables, plus other supplies being organized—medical items, rain gear, plant presses, plastic bags, etc. Roger and I had brought personal items, sleeping bags, rain jackets, altimeters, cameras, notebooks, and literature, but the major items had been coordinated in Concepción.

The missing link with these preparations was that the boat schedule was not yet fixed. Worse than that, the plan originally was to go with the Chilean Naval boat *Aquiles*, but this particular voyage was now suspended. Mario's organizational skills were challenged now, and he plunged into the fray with all the energy he could muster. There was also a commercial boat, the *Carlos Darwin* (appropriately named), that also made the trip from Valparaíso to the islands, with the purpose at this time of transporting villagers returning to Robinson Crusoe Island, especially high school students who had been spending the semester in Valparaíso and wanted to return home during summer vacation. It was possible for us to get passage for this voyage, but the schedule had not yet been fixed. We agreed that this seemed the best alternative—we just had to wait.

As the days wore on, eventually all the preparations had been made, and there was nothing left to do but wait. The silver lining to this circumstance was that it provided Roger and me with lots of time to conceptualize the project even further. With the very good botanical library in Concepción, we were able to chase down needed articles and books and slowly get much deeper into the evolutionary aspects of the flora, especially the tree Compositae. In Columbus, Roger had done an outstanding job of putting together an identification key to all the indigenous flowering plant

Comando de Transportes Navales

Valparaíso, 27 de Diciembre de 1979.

Señor.
Mario SILVA O.
Instituto Central de Biología
Universidad de Concepción
Casilla 1367
CONCEPCION :

En respuesta a su atenta carta fechada el 10 de Diciembre de 1979, lamento informarle lo siguiente :

a) Por necesidades del servicio, el viaje del AP. "Aquiles" al Archipiélago Juan Fernández ha sido suspendido.

b) No ha sido fijada fecha ni Unidad para un futuro desplazamiento, pero de todas formas su requerimiento se mantendrá presente y en caso de presentarse una nueva oportunidad le será avisada.

Este mando está consciente de los contratiempos que se producen por estas situaciones, por lo que ruega a Ud., comunicar a su Instituto mi sincera preocupación por este problema.

Sin otro particular se despide atentamente de Ud.

ADOLFO CARRASCO LAGOS
CAPITAN DE FRAGATA
COMANDANTE TRANSPORTES NAVALES
SUBROGANTE

ARMADA DE CHILE - COMANDO TRANSPORTES NAVALES - COMANDANTE

Letter from the Chilean Navy (Armada de Chile) to Mario Silva informing him that (a) the planned voyage of the ship *Aquiles* to the Robinson Crusoe Island had been suspended, and that (b) no date for a new voyage had been set species in the flora, all from available literature, and Roberto in Concepción had done the same for the ferns. For the basic inventorying, therefore, we were well prepared. What we needed was to sketch evolutionary relationships within and between the two major islands and make hypotheses that could be tested through gathering of more data, especially phytochemistry, morphology, and cytology. This waiting period, however frustrating, did allow us time to get our preliminary research hypotheses firmly in place.

But there is only so much strategizing and planning that can be done before patience starts to wear thin. The weeks were passing, and there still was no definite date for the boat to leave for the islands. We wanted to get into the field and start experiencing the islands and their bizarre flora. Days passed, no news of the ship, more days passed, and still more, now nearly two weeks since we had arrived. Our Chilean colleagues had plenty of other obligations to keep them occupied, but Roger and I were going from the hotel each day to the university, especially to the herbarium and library, but we were anxious to get started with the expedition. We were also spending money that could more profitably be spent on the island.

Our hotel in Concepción was the Alonso de Ercilla, a three-star accommodation that had been recommended by Mario. He knew the owner, and he often put up visitors there. It was clean, reasonably priced, and it was within easy walking distance of the university. It was also not far from the center of town, the Plaza de la Independencia. The service in the hotel was good, and they could place calls to the United States, but one had to give the number and then wait for an international line, which might take fifteen minutes or more. There was a small wooden telephone booth with a stool, receiver, and door for privacy, but if closed very long, it got quite stuffy (a good reason to keep the call short and efficient!). The floors of the hotel were attractive parquet wood, heavily waxed and shiny. The walls were thick so that the hotel was quiet (another of Mario's requirements, with which I completely agree). The rooms were painted a light green, and they were small, with one or two single beds, a closet, a small desk, and then the small bathroom with shower. In 1980 there was no TV in each room; on the first floor was a single television with easy chairs around, serving as a communal lounge. One could order a Coke or beer and relax in the evening, but one watched what everyone else was watching. The maids in the hotel were short ladies, all wearing tan uniforms with white collars. Breakfast was served in the same salon, continental

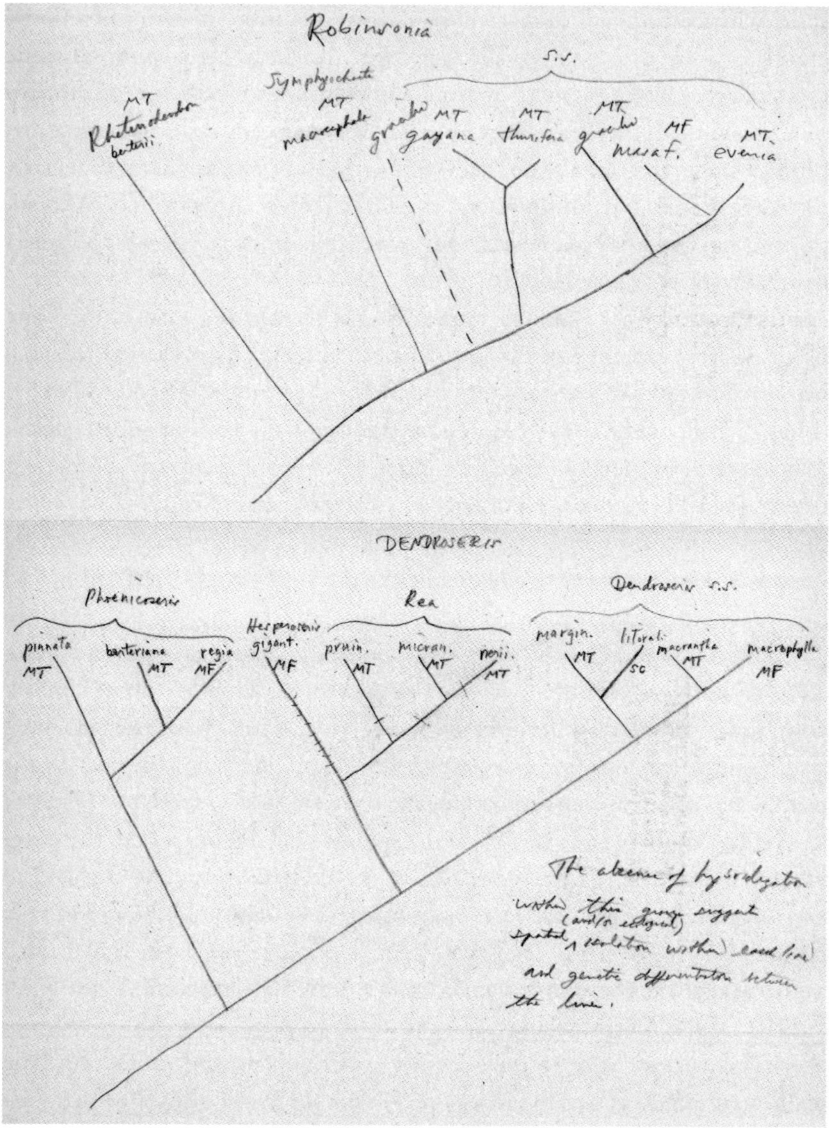

Sketches of preliminary hypotheses of relationships among species of *Robinsonia* (top) and *Dendroseris* (bottom), based on morphological observations and existing literature. MF, Masafuera (Alejandro Selkirk Island); MT, Masatierra (Robinson Crusoe Island).

style, with coffee, milk, toasted bread, butter, and marmalade. Because I always enjoyed a little more for breakfast, Roger and I purchased some Gouda-type cheese, which kept well, and brought that to the table with us. In fact, we established a small stash of provisions in our room, large, plastic soft drink bottles that could be opened and resealed, mineral water, and some wine, too. Roger did not drink alcohol, but I have always enjoyed a glass of wine in the evening, especially red wine, and particularly a quality Chilean red wine. A good traveler needs to sample the fruits of another country and take advantage of the positive aspects. In fact, one might regard that almost as an obligation, sort of a contribution to global understanding! The water throughout Chile is treated and safe to drink, but it didn't always set well with me, so my approach was always to drink mineral water, and I continue this even today in Santiago as well as elsewhere. It just requires a little planning ahead and stopping by a store on the way home from the university, stocking up on whatever might be needed, including also cookies, chips, candy bars, melons, whatever would keep well in our room without refrigeration.

The word finally came—the ship would leave Valparaíso on 22 January, slightly more than three weeks since we had arrived in Chile. Mario contracted with a local trucking company to ship our expedition gear from Concepción to Valparaíso, to have it ready for loading on the boat the day before it was scheduled to leave. I never knew exactly where the equipment and supplies were to be stored, but I assumed that Mario knew what he was doing. Life teaches us to worry about things that need to be worried about—and for me, this was not one of them. Our job was to get all personal items ready for an overnight bus trip from Concepción to Valparaíso to be there in the morning of the twenty-second, so as to supervise the loading of our equipment and supplies into the hold of the ship. The bus left at 10:00 in the evening, departing punctually from the Turbus office in the city. In Chile all things are reasonably punctual, not at all like in Peru north to Mexico, where regard for punctuality is more casual. The reason probably is due to the high number of European immigrants into Chile during the nineteenth and twentieth centuries, from Spain, Germany, France, Switzerland, Italy, and Yugoslavia, among others. Although not all these countries adhere to punctual societal events either (Italy, for example), they are much better than Latin American countries that are heavily populated by indigenous peoples, who in general maintain a more relaxed approach to watching the clock. If Chileans say the bus leaves at 10:00 at night, that means you had better be ready to go at that time. It was ready, we got our personal gear stowed away under the bus, and it lumbered out of Concepción in the night toward Valparaíso. I don't know how these drivers managed to stay awake on these nighttime runs, but we never had a problem. The only somewhat annoying distraction was that whenever the driver exceeded the allowed speed limit, an automatic buzzer sounded in the cabin (based on speed of wheel rotation), and he had to slow down. If you were not sleeping too soundly, this would keep you awake.

Before arrival in Valparaíso early the next morning at 6:00, the Turbus attendant passed out a small tray with coffee and rolls as a wake-up small breakfast, similar to what we had received on the train, and that signaled we were getting close to our destination. From the bus terminal, we got into three different taxis, each of which had a rack (*parilla*) on the top, so that the ten persons and baggage could be transported to the dock where our ship was waiting. The harbor of Valparaíso is very large and industrial, as it handles worldwide shipping. We had information on where our ship was supposed to be docked, and with a little searching, the taxis made it to this location. There before us was the *Carlos Darwin*, a Chilean ship named after the famous biologist, Charles Darwin, who had made a very important around-the-world trip from 1832 to 1836 on the *Beagle*[2] that was so stimulating for Darwin's development of the theory of evolution. This was a very good omen, I thought. Perhaps we, too, will develop original ideas on the evolution of the plants of the Robinson Crusoe Islands,

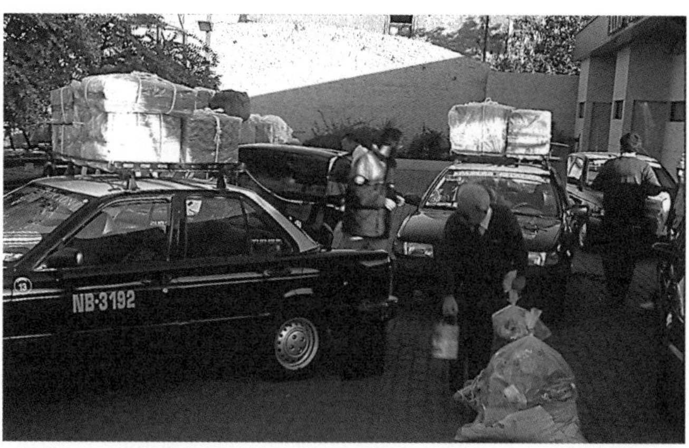

Loading up taxis with expedition gear for the trip to the dock in Valparaíso

Busy harbor at Valparaíso, Chile, the largest shipping port of the country

surely nothing as earthshaking as Darwin's ideas, but perhaps nonetheless significant. A scientist always dreams of making important discoveries, but rarely do we worry about achieving fame and fortune, especially biologists working in the field of biodiversity and evolution. Of course, one never knows exactly what might come from any original research, and one can always hope, but the first steps are to get the work moving forward, add as much insight as possible along the way, ask significant basic questions, and finally publish the results for the benefit of society as a whole. Still, shipping out on a *Carlos Darwin* was a far better beginning than on a *Benedict Arnold*, *Lusitania*, or *Titanic II*, so we enjoyed the moment.

The *Carlos Darwin*, which took us from Valparaíso to Robinson Crusoe Island on the first expedition

Having located the boat of our scientific destiny, our first task was to take taxis to the address that Mario had given us, find our materials, and arrange for them to be brought to the dock later that afternoon. Valparaíso is an old city, and some of the streets are very narrow and winding, now one-way, which made our taxi ride very interesting. Luckily, when we arrived at the warehouse, it had just opened, and we quickly made the required arrangements. With this important task accomplished, we had the rest of the day available for talking, eating, and relaxing, which we did in the center of the old town. At 4:00, we took taxis back to the dock and waited for the truck with our supplies, which arrived as scheduled at 4:00.

The *Carlos Darwin* was no *Love Boat*, far from it. It was a basic ship for transporting supplies and people from the Chilean mainland to the islands and back, so there were no movie theaters, dinner buffets, jogging tracks, or luxury cabins. In fact, there weren't any cabins at all, just bunks, and these packed together in sets of four, at least in the "first-class" area. In the regular area, there were only cots spread out in a large general space below deck. The hold of the ship was quite large, and here went all supplies destined for the island, and this included lumber for construction, foodstuffs, furniture, and now our numerous wooden crates. Our ship had a large crane mounted on it, which allowed for stacking and roping of our cargo on a wooden pallet and lifting it up and down into the hold. It took several trips to get all materials positioned below and tied down so that no movement would occur if weather turned nasty. With these supplies finally stowed securely on board, we were allowed to walk up the gangway and into the ship, and down to find our bunk quarters. Everything was on a small scale, but we couldn't care less—what mattered was getting to the island safely with our equipment and beginning our research work.

The *Carlos Darwin* set sail in the evening of 22 January and headed out into the darkness toward Robinson Crusoe Island. There wasn't much to see westward except the dark ocean and the last gray glimmer of afternoon light. Some small waves were striking the bow of the boat, and Valparaíso was still visible eastward, its harbor lights growing fainter as we continued our journey. The distance is 667 kilometers over a sea that can be very calm, such as on this evening, or sometimes very angry with high waves. Each of us had a simple bunk, which was to be our home until arrival in two days to our destination. Sleeping under any new circumstance can be trying, but we were quite tired after the all-night bus ride the evening before, so I did drift off for several hours, if not more.

In the morning, the weather was excellent, the sea was very calm, and we were anxious to explore the ship further. The boat was filled with supplies for the villagers in the island, including cattle that were simply tied on deck. The weather was not cold nor was it raining, but the animals did their normal digestive actions, which made the deck slippery and not completely appealing for strolling about. There was a small, very small, shower below deck, which I tried with some success, and of course a bathroom. The galley was quite small, with only three long tables and stools in the eating area, all attached to the floor for safety. Food was available essentially whenever one wanted, but only for passengers in the upper-class bunks (in which we were traveling). The other passengers had to bring their own food, which they did—bread, cheese, salami, chips, fruit, etc. There wasn't much to do on board, except chat about this and that. Because Roger and I had spent so much time already in Concepción making research plans, we were worn out on those topics, but it was good to talk to our Chilean colleagues, as this was the first time we were all together. None of us had ever been to the archipelago, so the anticipation of arriving to Robinson Crusoe Island was high. For me, having read considerably about the islands and their fascinating flora, it was getting really exciting, realizing that in one day we would be arriving to our destination. Travel to new destinations is so much more fulfilling when one studies about the location beforehand. This formulates ideas and questions that can be answered when actually there—it focuses the trip into a stronger learning experience. I confess to having taken this approach to family vacations as well, getting prepped up on what we were to see and do, the historical background, natural history, etc., and unrealistically trying to get the other family members (wife and two sons) as excited as I was about the trip.[3] What has been the history of this region (in this case, island)? How did Chile come to make this a national park? What kinds of plants (or animals) are unique to this area? All these aspects Roger and I had covered in detail, and we were ready to get into the work, especially after the long three-week delay.

The awaited moment came in the afternoon of the second day—there in the distance was the low silhouette of an island, looking very much like the submarine from Jules Verne's *Twenty Thousand Leagues Under the Sea*,[4] a dissected landscape riding low in the water. But that was really it,

Approaching Robinson Crusoe Island on the *Carlos Darwin*. Top, the island seen from a distance; bottom, lava cliffs on the northern side of the island.

the island about which I had read so many years earlier. We were finally actually going to see it, learn from it, conduct research on it, and enjoy it. The sea was deep and dark, the sun was shining, and the tips of the waves shimmered like pieces of silver mirror that had been sprinkled over the water. We were now all on deck to get a glance of this new land, wondering how it really might be, and how well our work would go. As we approached the island it grew up out of the water until we began to get a sense of its size. Most notable were the many sheer cliffs and precipices along the northern side of the island, interrupted by four distinct broad valleys. Some of the sheer lava cliffs had intrusive dikes of basalt that looked

San Juan Bautista. Top, location of village adjacent Bahía Cumberland (Cumberland Bay); bottom, pier (in 1980).

like the drippings from a child's sandcastle. The *Carlos Darwin* moved along the island until turning into a large bay, Cumberland Bay (Bahía Cumberland), with the village of San Juan Bautista spread out before us, with buildings placed on slopes of a broad valley below high peaks. The anchor went down loudly as the huge chain clanged through the outside portal as it went, and then the engines grew quieter.

I had seen many pictures of San Juan Bautista from books and articles, so seeing this small village held no great surprises. But the feeling of finally having arrived safely to this very spot after so many years of planning, seeking of funds, sketching of research ideas, selecting personnel, obtaining permits to work in the archipelago (it is a Chilean national park), and coordinating everything so that we could get to this point, was immensely satisfying, even on less than adequate sleep for the last two nights. The *Carlos Darwin* was too large a boat to moor on the relatively small pier, which at this time was only wooden, so after laying down anchor, smaller boats were needed to transport materials to the pier. The passengers were allowed to depart first, followed by unloading of equipment and supplies. Quite a few villagers had come down to greet arriving friends and relatives, and in our case, Señor Bernardo Ackermann, the chief of the national park, met us and made introductions to some of the CONAF guides. Ackermann was a short, pleasant man of German ancestry. We learned later that he was cohabiting with a very young and attractive girl, and though he had a wife and family on the continent, this was not viewed as out of place in the island. There were lots of interesting relationships in the village, which we became aware of over time, but that just added to the fun of being in a very different part of the world.

The first challenge was to get our materials off the pier and to our designated housing. Sr. Ackermann indicated that we would be staying in the cabin on CONAF property, which was up the hill at the base of Damajuana Peak on the left side of the central valley. At this time there were no personal motorized vehicles in the village. Only the *carabineros* (national police) had a vehicle, which meant that everything had to be moved around in carts pulled by horses or burros. Our group consisted of ten persons, with nine men, so this made it possible to get carts loaded and

Moving supplies and equipment in 1980 from dockside to our cabin at CONAF headquarters, the mule being assisted from behind by Roger Sanders and Eduardo Ugarte

moved up the slope to our cabin. This took a while, because we could only commandeer one cart and burro, as other people in the village had similar needs in dealing with their arriving items. After several hours, we were standing in our cabin with boxes scattered all about. This would have to be organized better, but at least we had arrived. We were now on Robinson Crusoe Island.

One can also arrive to the island in small five- to twelve-seater propeller planes that leave from Santiago, and this we did more often than by boat, but a description of arriving by that mode will be presented in more detail in the beginning of chapter 11. In fact, our return trip from this first expedition was done by airplane, because by the time we had finished our research work, the *Carlos Darwin* had long returned to the continent. The only way of getting back to Santiago (and then to Concepción), keeping to our schedule, was by airplane. This emphasizes the importance of getting the proper connections to work with the timing of the research, and sometimes things just don't go quite as planned, despite all good intentions.

Which brings us to the topic of departures. Leaving the islands has its own set of experiences, whether returning to the continent by boat or by small plane. The most dreaded aspect of leaving from Robinson Crusoe Island is the possibility of delay due to inclement weather. One such occasion took place on Monday, 27 January 1997, which was our scheduled date of departure at the end of the eighth expedition. The trip had been very successful, but it was now time to return to the continent. Because it had rained all night, we woke up on that day with the premonition that no flights would be coming in early, if at all. As the morning progressed, the sun came out briefly, lifting hopes, but then it rained again, bringing hopes back down. At 10:00 in the morning, Dan Crawford and I went down to the airline office to get the official word on flights that day, but the office was not even open. And—now more rain. Pedro Aqueveque and Eduardo Ruiz visited the office at 11:00 and this time it was open, but there were no flights that day—maybe tomorrow. The problem with heavy rain is that the airport runway, at that time only of packed and oiled dirt, would become soft and dangerous. Furthermore, the steep, bulldozed dirt road up from the ocean (Bahía Padre) to the landing strip was also not passable when too wet. We would simply have to stay put all day, waiting patiently. After a long trip is finished and the decision is taken to head back to the continent, thoughts start to turn toward obligations at home, and any delay becomes extremely irritating. But here in the islands, one just has to hunker down and be patient. Having lost not a single collecting day on this particular expedition due to rain, we now paid the price at the very end.

For dinner on that day we were invited to the house of Bernardo López (one of the CONAF guides). This was especially appreciated because his wife, Juanita, is an exceptionally good cook. Her dishes are always tasty and with ample portions, which are essential for replacing calories lost on the trail during the day. Dan Crawford and Tom Lammers on one trip each lost twenty pounds without any diet whatsoever. This is the famous "Juan Fernández Diet." If you walk and climb for eight hours a day, you can eat and drink virtually anything and the pounds just roll off. The walk up La Pólvora ("The Gunpowder") Street to Bernardo's home was good exercise, but it again started drizzling. Horrible weather. We improved our spirits with a punch *apertivo* (from a recent wedding party), followed by a good egg soup, then an excellent dish of pasta with rabbit and tomato sauce. Wine and juice complemented the food. The discussion at dinner covered many topics, especially our departure. After a very satisfying evening, at 11:15 we walked down La Pólvora to the main street, then strolled parallel to the bay, and in the center of town finally walked back up to our house at the Servicio Aeronáutico. Once inside we chatted a bit more, but then turned in for the night.

I was the first to arise at 8:00 on Tuesday, 28 January. Being anxious about our departure, after a quick breakfast, I went down to the airline office at 9:00 and on the way ran into the agent, José María. In his opinion there could be a flight today, because it had rained less at the airport (La Punta), and it now was sunny there, making an afternoon flight possible. This was welcome news, but I knew from experience that this might or might not happen, but at least it was something positive to take back to the house. I spread the word, with the plan being to get everything ready by 1:00 in the afternoon in case we might be flying that day. The news was well received, spirits were uplifted, and we began to pack up our sleeping bags, tents, kitchen items, plus giving a final cleaning to the bathroom, kitchen, and areas in the front and back of the house. It was an upbeat flurry of activity. The plan was that at noon José María would call Don Osvaldo, who lived next door to us, with the final flight information. Due to our industry, we were ready by noon, keeping our fingers crossed for a positive report. At 1:00, there was no news, nothing. But having a slight delay would not necessarily be unusual in the island. We fidgeted and waited, making nervous conversation about nothing very important, just waiting it out the best we could. As 2:00 rolled around, we were still waiting, now somewhat impatiently. Because the flight of the larger plane (the Nomad) takes three-and-a-half hours from Los Cerrillos Airport to Robinson Crusoe Island, the latest it can leave the continent is 3:00 in the afternoon. This puts it at the airstrip by 6:30, and by 7:00 it can load up and return. This is the latest it can depart because the runway on Robinson Crusoe Island has no lights.

After an interminable additional thirty minutes, at 2:30 came the final word from Osvaldo—no flights today. We were a little skeptical that this was the final word, so for verification we walked down to the airline office, learning that due to the rain and soft conditions of the runway, Santiago Figueroa, the pilot, decided not to leave the continent (probably a proper cautious decision). After talking to the lady at the airline office (Valeria, the wife of José María), we decided to unwind from the tension of the day by visiting Remo, the bar-restaurant near the water's edge, for a beer. This was always pleasant, sitting near the water, listening to the soft waves, and letting the alcohol achieve its desired effect. You might ask how we could be sad about having to stay another day in this wonderful island, but once a traveler decides to head home, that becomes the mental focus. Our entertainment consisted of watching three people cut the grass in front of Remo—it took them more

Getting ready to leave San Juan Bautista, Robinson Crusoe Island. Top, the airline office on La Pólvora Street (with Delbert Wiens, left, and Daniel Crawford, right); bottom, supplies and collections organized on the village pier awaiting trip to the airstrip, with (left to right) Patricio Peñailillo, Ana María Humaña, Daniel Crawford, and Marcelo Baeza.

than an hour to cut a patch of grass about five square meters—a true overkill of effort. One person used the weed eater (gasoline powered), one a power mower, and the third raked up the old grass. We were pleased when they finished the job so that we could again hear the sounds of the waves rolling softly to shore. The workers had done a good job, but it took so many people to complete it. On the other hand, what was the rush? Perhaps we worry too much about efficiency instead of simply enjoying every moment of the gift of life, a useful lesson but hard to hang onto when returning to busy lives on the continent.

Because of this additional delay, we had to prepare our own dinner, which meant buying wine, pisco (there are priorities!), bread, onions, and parmesan cheese, plus butter and marmalade for breakfast the next morning (and perhaps also lunch depending upon when the flight might be leaving—*if* it might be leaving!). After shopping on the main street and returning to our house, we had to unpack kitchen items to prepare dinner. The fare was creative and simple, but somehow it tasted just fine. Marcelo and Gabi made a good salad of beans, eggs, and onions mixed with *cholgas* (a type of mussel[5]), a bit unusual, but under the circumstances, excellent! Dan prepared a very nice pasta mostaccioli. To wash all this food down, we polished off five liters of wine before and during dinner, and then started on the bottle of pisco. We chatted about nearly any old topic. When you are just waiting without any other real focus, the conversation is not very profound or scientific. Alex, our meteorological neighbor, came to chat and to give us his view of the weather tomorrow. It should get better, he said. Well, due to the changeable nature of the island region, we knew that meant next to nothing. Dan had stomach trouble (could it have been the *cholgas*?) and filled the bathroom with such an odor that it drifted out into the house with many chuckles. It was no laughing matter, however, and he went to bed early. He had chills later in the night.

On Wednesday, January 29, we woke up at 8:00 after a more or less good sleep, except for Dan who was still not feeling well. Don Osvaldo came at 9:00 with news that we would be going today. Hallelujah! We went to the edge of the fort that looked down to the village, and we could see that the airline office on La Pólvora Street was not open yet. How then could they have called the mainland? It sounded suspicious, too good to be true, especially because it had rained more during the evening. After a quick breakfast, Pedro and I walked down to the now open airline office, and they did confirm the high probability of leaving today. The problem for them was that the number of passengers wanting to leave the island was increasing as each day passed, and scheduling was getting backed up. Our challenge now was to be packed and down on the dock by 11:45 that morning, which meant shifting into high gear. Back at the house, we informed the others to wrap up all kitchen utensils, sleeping bags, and remaining personal items. Carlos, one of the few who now in 1997 owned a private vehicle, came at 10:30 in his four-wheel drive, orange Chevrolet pickup. While we loaded our bags and boxes, Eduardo, Dan, and Pedro packed up our small refrigerators in the lab, and by 11:00 all was ready for transfer to the dock, where the pickup could drive out to near the end. There we unloaded our items on the pier and paid Carlos 3,000 pesos (about fifteen dollars)—a bargain for us. On a final trip back up the hill to the house to retrieve our backpacks and give the house a final check, Don Osvaldo and Alex were waiting to ask for a rain suit for the new worker of the aeronautic service. This I carried in my backpack, and I had planned to take it on the trip around the island to the landing strip, in case the weather turned bad, but we could hardly refuse their request after having used their house for three weeks at no cost. After profuse thanks and final farewells, we made the final walk down to the dock at 11:30, the scheduled time. Iván Leiva, the park chief, and Oscar Chamorro, one of the CONAF guides, were there to see us off. This was an official duty, most probably, but I always felt deep ties to both of them.

It was time to return to the continent. Leaving the island is nothing like getting into a cab and whisking off to the airport. Here it is a community event. Many people came down to the pier to say goodbye to visitors, friends, or relatives. We knew it would be at least a year until we could return, so these sorts of departures were always a bit sad. After the municipal launch was loaded with bags and lobsters (lobster can be sold for 20,000 pesos in the continent—forty dollars), we loaded ourselves on and got ready to depart. There were about twenty passengers, leaving us to wonder how all would be flying today. The municipal launch pushed off from the pier, chugged away, and left San Juan Bautista. We waved again, looking at the pier with our good friends, as they slowly got smaller, and we felt a hollowness. They always helped us, kept us safe, and we taught them about the rare plants of the islands. They are unselfish and cooperative people, and it has been a privilege to know them. But it was now our time for transition. As we turned the bend from the bay out to sea and along the northern side of the island, our thoughts started to turn to home and to the laboratory where data could be obtained, analyzed, and interpreted in context of our evolutionary questions.

It was cloudy now, but the sky seemed clearer farther out to sea. All around was dark blue water, backdropped by the hum of the Yamaha diesel motor, with the pilot and his grandson chatting happily, and the stark lava slopes displaying their shoestring lava strips and corrugated cardboard-like layers. The sea was calm and caused no problem for anyone. By the time we arrived at Bahía Padre the sky had completely cleared with a bright sun. We unloaded everyone from the launch onto the pier and then moved our things to a convenient spot where the jeep would come down the hill. We waited, shared some salty crackers, and waited some more.

Now, in 1991 Dan and I were in exactly this same departure situation, but we took advantage of the waiting period to walk quickly around the bay about one hundred meters to obtain a last collection of the endemic small shrub, *Wahlenbergia berteroi*. As we got to the rocks by the sea, to our horror a cat stood in front of us, scarcely able to see or walk. It was a ghastly sight with its mouth open, eyes all eaten by flies, these still packed densely into each socket, and the cat's legs moving imperceptibly forward. I have never seen a creature so near death but still alive, and still with the will to keep living. My first emotional reaction was to think how we might save this poor creature, but it was so dehydrated, so eaten, and so near death, and we were on our return trip to the continent to continue the rest of our own lives. We had no realistic choice—with a large rock we put it out of its misery. This was a disturbing event, and I still find it so when I think about it. There is no fresh water in this part of the island, which no doubt led to the cat's desperate situation. Quite frankly, I have never loved cats, being much more of a dog person. I need a buddy, a close friend, not a vase with legs. Furthermore, feral cats in the islands prey on the endemic birds, and they, as the goats, should be exterminated. But this cat really penetrated my soul, profoundly so. If circumstances had been different, might there have been a way to save it? Doubtful, I think, very doubtful, but there is always this nagging thought that, well, perhaps, somehow, some way, we should have tried harder.

Returning to our departure of 1997, as we waited on the shore at Bahía Padre, quite suddenly a small rain shower came up that engulfed all of us. Dan, Eduardo, Pedro, and I took shelter in a large concrete drainpipe on the beach. All our bags got wet, but they later dried without dif-

Bahía Padre at the western end of Robinson Crusoe Island. Top, panorama of the bay; bottom, equipment unloaded on shore from the launch, with airstrip jeep in background. Left to right: Ulf Swenson, Daniel Crawford, Erik Tepe, Pedro Aqueveque, Gabriel Bernardello, Gregory Anderson, and Marcelo Baeza.

ficulty. The plant collections and equipment were wrapped in plastic, so this was no worry. After the rain let up, down the hill came the dilapidated, tan Land Rover of our airline to pick us up. Some of the other passengers with gear went in this first trip, but due to a weak battery, several of us had to push the vehicle to get it started. Rain came again, and Eduardo and I took shelter again in the cement culvert. After five minutes, this drizzle stopped, and we heard the small Land Rover coming down on its second trip. This time Dan and I went with a full load of our supplies. Because the driver of the jeep (doubling also as the pilot!), Santiago

Figueroa, had left the jeep running, we thankfully didn't have to push it this time. Up we went very slowly, bumping along the zigzag road with spectacular vistas of the dry western end of the island.

Upon arriving at the airport, we saw an amazing and untypical collection of four planes, two of Transportes Aéreos Robinson Crusoe, one of SAE (Servicio Aéreo Ejecutivo), and one of Lassa (which was the old Taxpa). This quantity of aircraft explained why so many persons had left the village with us. While we waited, the SAE plane took off successfully, which gave us confidence that the runway had dried off sufficiently from the rain to allow safe takeoffs. We helped load the small plane from our airline, which had three passengers (other tourists) plus some of our gear, but we climbed into the larger Nomad with the rest of our baggage. Our bags and boxes were numbered as usual, so that back on the continent we could more precisely check that we had recovered everything. Soon the jeep came again up the hill from Bahía Padre with Pedro and Eduardo, and they climbed into the plane with us. It was time for departure, and we were sitting happily, crowded as usual, seat belts fastened, boxes everywhere, and expectantly awaiting take off. The plane taxied one hundred meters to the landing strip, making a ninety degree turn into position, followed by some checks of the pilot, including revving the engines. Finally the brake was released, and we accelerated quickly, picking up speed toward the end of the runway, and suddenly the island dropped out from beneath us. But as nearly always, the wings felt the lift and sent us soaring upward and along the southern side of Robinson Cru-

The landing strip area ("airport") at the western end of Robinson Crusoe Island. Top, the terminal building; bottom, our small plane getting loaded for the return flight to the continent.

soe Island.[6] There was some turbulence at takeoff, but after greater elevation was achieved, everything calmed down, and we settled in for the return trip to the continent.

Once airborne, we opened our snack bags with the usual ham and cheese sandwich, Coca-Cola, and *alfajor* cookies,[7] plus three hard candies. We also were given a juice drink, sort of a peach flavor, but more appreciatively, a fifth of whiskey (Grant's) appeared. First we ate the snacks, and then got out the whiskey for a toast of good fortune. The trip had been a complete success—all objectives had been achieved. After finishing food and drink, Pedro, Dan, and Eduardo began to snooze, then transitioning to snoring, not loudly, but with heads back, interrupted by occasional snorts. I was also tired, but I worked on finishing diary notes. Fidelina organized her handbag for the return to civilization, and Marcelo enjoyed his remaining whiskey, no doubt with thoughts of obligations awaiting him back home. There was nothing left but to land on the continent in Santiago at Los Cerrillos Airport and catch an evening bus (Turbus) to Concepción, thus ending still another expedition to the Robinson Crusoe Islands.

CHAPTER 7

Island Living

I am not going to kid you into believing that our stay on Robinson Crusoe Island was anything like that presented in Defoe's novel. The island is not deserted, nor has it been for most of the past 400 plus years,[1] but it is not like downtown Manhattan, either, or even like Santiago or Concepción. We are talking about a small village of about 900 persons in a very remote island in the Pacific Ocean that still has not been commercially developed to receive hordes of tourists. This island is less remote than Alejandro Selkirk Island (see chapters 8 and 11), but life on Robinson Crusoe Island does have its challenges. To give you a sense of how we lived while doing field research there, this chapter shares some aspects of our housing, health problems, and finally our fiestas. In effect, this lets you know how we dealt with the physical and mental dimensions of human survival that allowed us to have success with our scientific research.

The only place to stay on Robinson Crusoe Island is in the village of San Juan Bautista. There are now some small camping areas for backpackers, but these have no water or bathroom facilities, so these spots are for those who enjoy really roughing it. In our ten expeditions to Robinson Crusoe Island, we only saw young persons with tents twice. In other words, camping was not common. The village has one upscale hotel, the Hostería (Crusoe Island Lodge; regarded as a luxury boutique lodge). It is not actually in the village proper, being nestled at the mouth of the Quebrada Pangal on the eastern side of Cumberland Bay, about a one-kilometer walk from the village. It is very nice, with a pool, and deck with an overlook of the ocean. We never stayed or ate in this spot for the simple reason that

La Hostería, the only international lodge on Robinson Crusoe Island, located at El Pangal, on the eastern side of Bahía Cumberland

we could not afford it. It cost about seventy dollars per night per person, and with our often large contingent of scientists and students, this would have been more than our budget could handle. Furthermore, it is located far from the village and CONAF headquarters, which would have made our work nearly impossible. We did occasionally walk to the Hostería from the village to have a beer or Coke and enjoy the scenery and good conversation. The owner, a European, did a decent business with more upscale tourists, but we rarely had contact with them. The hotel provided trekking activities, snorkeling, diving, and fishing for the guests, but again, we never entered this world of greater luxury, nor did we ever meet anyone from the Hostería on the trail. Other than the Hostería, there are some very modest pensions, a number of which were destroyed by the tsunami in February of 2010, but one can still find a room with some family at a reasonable price. Due to our large group and work objectives, we were fortunate on our expeditions to find living quarters in five different places. Getting installed went smoothly, because CONAF helped locate possibilities prior to our arrival.

The first place we stayed in San Juan Bautista was in a cabin located on the CONAF administrative property at the base of Damajuana Peak. This was originally designed for the CONAF guides, but since they had homes and families in the village, it was usually vacant and used for visiting scientists. Initially it cost nothing, but as the years continued, there

CONAF cabin where we stayed during the expeditions of 1980 (both January and November), 1990, and 1991

were nominal fees, which was fine. We stayed in the CONAF cabin on our trips of 1980 (in January and November), 1990, and 1991. The great advantage was that it was very close to the working area where we processed materials, so that one could get to work quickly and go to the cabin for meals or other necessities. It had two sleeping rooms with two bunk beds in each, giving a capacity of eight persons. One entered the front door through a small porch, which had a roof that provided protection from rain. The main room was living and dining room combined, where all meals were taken and where we relaxed, read, or wrote notes. At the other end of the cabin was the small bathroom with a toilet, sink, and bathtub (with shower). No hot water existed, and the water coming from the hills was very, I mean *really*, cold. There was a small kitchen, with a door to the outside (the back door), and it had a refrigerator, sink, gas stove (from a propane tank), and small counter area. Nothing luxurious about the cabin, very rustic, but it was clean (when we paid someone to keep it that way for us!) and very convenient.

The second place we stayed was in a fisherman's home in 1984 way up the street La Pólvora, nearly to the top, which meant a relatively long hike up from the main street of the village and to and from CONAF headquarters. There was a cement walkway, so there was no physical problem of access, only that sometimes at the end of the day it seemed very challenging to make the climb. The house was loaned to us for our trip in 1984, and it was very simple. In fact, this was about as far from luxury

The long street, La Pólvora, extending from the main street of San Juan Bautista up the hill to the fisherman's house where we stayed in 1984

as one could get in the village and still be in a house. The outside was a collection of different types of boards, which did not look very appealing, but inside the living room, it was painted and adorned with some pictures and family mementos. The fisherman, whom we never met, was working on Alejandro Selkirk Island at this time, and therefore it was vacant. CONAF had made the arrangements.

This house suited us well enough, but there were inconveniences. There were two sleeping rooms, without beds, and the floor in these rooms was just packed dirt. To deal with this, we pitched tents in the house, which provided a cleaner sleeping floor. Dan and I took one room, and Hugo Valdebenito, Eduardo Ruiz, and Alejandro Landero took the other. The last member of the expedition, Patricia Pacheco, had been placed in a room with a family across and down the street a bit. As the nights rolled on, we came to realize that our room was infested with fleas, which enjoyed visiting each night. Somewhat surprisingly, we got used to the sensation of these little critters crawling over our legs, and they did not bite often, at least not me—perhaps they preferred Dan, who shared the tent, or perhaps others of our group in one of the other tents. In any event, this was not the main difficulty with the house—it was the bathroom.

The bathroom had a sink, toilet, and bathtub, all the basics. But having the basics did not necessarily mean that all worked properly. The sink had running water (stone cold), but there was no water in the toilet or from the shower over the tub. We had to develop a successful protocol.

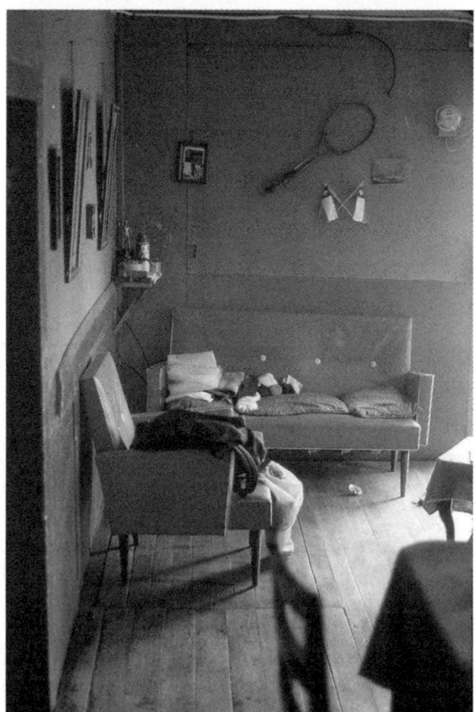

The fisherman's house on Robinson Crusoe Island in which we stayed during the expedition of 1991. Top, front of house; middle, back of house, with Dan Crawford; bottom, living room.

The sink allowed hands and face to be washed up, or a sponge bath, so that was a good start. One could sit on the toilet, but there was no water to flush the waste away. We located a rubber hose that fit on one of the spigots of the sink, and this worked to fill up the toilet and flush it. For a shower, however, we had to get more creative. We found a tin can, punched holes in the bottom, and hung it with wires from the shower pipe (with no head) over the tub. We could use the rubber hose to fit inside the shower can, and when the sink water was turned on, it sprinkled over the tub—a useful shower. However, there was no shower curtain, but this was a minor inconvenience, except that the floor of the bathroom was just packed dirt, so that water splashing out of the shower wet the floor and made it muddy, not ideal for stepping out of the shower when clean. We fixed that problem by putting layers of cardboard (from our plant presses) on the floor, to serve as bathmats. The more serious problem was that the tub did not drain out at all, the water just rising as people showered and shampooed. However, this proved in one sense to be beneficial, because we obtained a small pail that was used to dip water from the tub and use that to flush the toilet. With that innovation, the rubber hose could be left in place over the shower, using the other spigot only in the sink to wash hands or face or brush teeth.

Once we got the hang of the bathroom system, it functioned quite well. Taking a shower in the early morning, however, was exhilarating, to say the least. The water was icy cold by itself coming down from the hills, and the wind whipped around the house entering the bathroom through many spaces between the boards. The wind entered fiercely, which when combined with the cold water, made for a very brisk experience. One did not dilly-dally when showering at any hour of the day, and certainly not in the early hours. The strategy was simple: first do a quick wet-down all over, and then apply soap everywhere, followed by rinsing of the arms, legs, front, and then back, in that order. This was followed by shampoo and then rinsing that off. Very cold water actually hurts the head a bit, so this was done as quickly as possible. As the day progressed, the water lines coming from the hills, being black, absorbed some of the heat of the sun (if there was any that day), and the water arriving in the village was then a bit more tolerable. But I can tell you, after taking a super cold shower in the morning, heading to the kitchen for a hot cup of Nescafé never tasted better. Many of our group decided it was easier to brush teeth or wash hands and face during the day from another spigot behind the house,

House of the Servicio Aeronáutico de Chile that we used during our expeditions of 1990, 1996, and 1997. With Patricia Stuessy and Marcelo Baeza (in the background).

which also relieved any congestion at the sink in the bathroom. As days wore on, this house, which initially had seemed almost unlivable, became our home. It is impressive how humans, really all creatures, are able to adapt to their surroundings.

Our third home in the village was in an odd location, but it proved extremely useful. The Servicio Aeronáutico de Chile maintains a weather station on the island, actually on top of the old Fort Santa Bárbara, constructed by the Spanish in the middle of the eighteenth century to fortify the island and keep out competing European powers and pirates.[2] There were three houses, two housing personnel for the weather station, and a third used for occasional visitors. The head of the station, Don Osvaldo Jara, generously allowed us to use this house for two of our expeditions (1996 and 1997), which really solved lots of problems. The challenge on these trips was to set up an electrophoresis laboratory and run isozyme gels immediately after the plants were collected. In this manner we could see the patterns of genetic variation in the island, and this eliminated the need to bring back fresh or frozen plant materials, which had proven so difficult on the two previous trips (1990, 1991; for more details, see chapter 13). This rather large house, at the moment unoccupied, would be ideal for our needs. Furthermore, it was free (unbelievable), which loosened funds for other unanticipated costs, such as sending two of our people on the boat to Alejandro Selkirk Island for several days. There was also a bathroom that worked, and an electric water heater for the shower, which meant warm water—such luxury! The kitchen was small but adequate, again with a propane tank gas stove. The refrigerator didn't work, but it served to keep food away from the flies. Two rooms were set up as sleeping quarters, and I pitched my tent inside as usual (to keep out the mosquitos—I just cannot sleep with these buzzing around my head. If I am under a net, and I know they cannot get me, it is fine, but if I am exposed, and they start jumping on my ear, nose, and cheek, I can't relax). The final plus to the house was that two rooms were left, one for the electrophoretic laboratory and the other for pressing plants for herbarium vouchers.[3]

Our fourth home was entirely different. On the two expeditions to Robinson Crusoe Island of 1999 and 2000, with support from the National Geographic Society, and dedicated to vegetation analyses, we were a much smaller group. In 1999 it was myself, Pato López, and Josef Greimler (Sepp), the latter a faculty member in our department specializing in vegetation studies. In 2000 we had the same group plus my son, Alan Stuessy (age 21), whose expenses I had paid to come along and share the experiences. It was time for him to learn what his father did on these "vacations" to the islands. We also had no laboratory to set up, and we were not planning to collect specimens, all of which greatly simplified the logistics, including finding a place to stay. CONAF had arranged a private cabin for us with two bedrooms, a bathroom, and a kitchen and living room combined (very small). The kitchen again had a refrigerator that didn't work, and the stove also didn't work, but we had brought along an electric tea kettle that heated up fast, and this was sufficient for making coffee or tea in the morning. For breakfast we had bread, marmalade, margarine, and salami, and an occasional can of sardines (I was the only person who could stand this in the morning). Because we were three persons (four in 2000), we asked to use the adjacent vacant cabin to give everyone his own room, and this was granted (at no added charge). These accommodations cost 20,000 pesos per day (forty-three dollars for all persons), which was more than at CONAF or the Aeronáutica, but still quite reasonable. We contracted to have dinner with Juanita López at about 8:00 each evening, for a price of 9,000 pesos per day (twenty dollars) for the three of us. With more ecotourism in the islands, prices were starting to rise. We took care of purchasing our own breakfast and lunch items, now easily available in the village. On our first trip in 1980, we brought all our own food and drink with us, because at that time, there simply wasn't always food avail-

Staying in the home of Bernardo López in 2010. With Dan Crawford processing plant samples in silica gel.

able between arrivals of the ship from the continent. Now there were many smaller stores and more items available.

The fifth and last home on Robinson Crusoe Island was in Bernardo Lopez's house, and this we occupied during our trip of 2010. He had been constructing two additional small houses in front of the main one, and these were now about ready when we arrived. This allowed us to use the sleeping rooms of his main house, there being three rooms with bunk beds in each, which allowed space for all of our six participants. We were kicking out Bernardo and his wife, obviously, but it was a good money-making situation for him. The total bill for lodging, food, and laundry for six persons for the sixteen days was 1,330,000 pesos, or about $2,100 (on a per-person basis about $350). Although this was more than we had paid previously, it was still extremely reasonable. It was also more luxurious, as we now had a TV connection for evening relaxation, an international telephone line, and a bathroom that had hot water from a *califont*. This is a gas-fired small heater that, when ignited, heats all water that passes through it. The modern versions turn on automatically when water starts to flow, but here we had first to light the pilot. These are efficient machines, obviating the need for always having a hot tank, even when no one is using the water, such as during sleeping hours.

By 2010, the island had started to change based on influences from the continent. Motorized vehicles began to be popular, with pickups and motorbikes now becoming part of village life. Perhaps more impactful was the arrival of telephone, television, and the Internet. Previously, people walked constantly around the village to chat with friends or to conduct business, but once the telephone arrived, this personal connectivity substantially reduced. When television became available, world news was more accessible, and the wives stayed more at home in the afternoon to watch soap operas instead of meeting with friends for coffee. The Internet further encouraged people to stay at home, with them now glued to the computer screen. Men were even showing less motivation for competitive sports. Even a remote island such as Robinson Crusoe had now been impacted by the digital age.

Occasionally, when it was necessary to stay in the field for more than one day, we pitched tents. We brought several two-man tents with us on the expeditions, enough to house all participants. The CONAF guides had their own equipment, nothing fancy, but adequate for the terrain. Staying in a tent on any part of Robinson Crusoe Island was no big problem, as temperatures never dipped much during the night, especially in the lower valleys. The thick vegetation at higher elevations in this island also cut the impact from the wind. On Alejandro Selkirk Island, however, where elevation in the high mountains is more than 1,100 meters, with frequent fog and strong winds, a compact, well-tethered tent and warm sleeping bag are essential for overnight stays.

One of my greatest worries on all expeditions to the islands was that one of our group might become seriously ill or suffer an accident on the trail. On Robinson Crusoe Island there is a clinic, and it is fairly well equipped with general supplies, including antibiotics, and a medical per-

Use of tents during fieldwork in the Robinson Crusoe Archipelago. Left, in Valle Inglés on Robinson Crusoe Island, with Dan Crawford and Hugo Valdebenito; right, at 1,100 meters on Alejandro Selkirk Island, with Eduardo Ruiz and Alejandro Gatica.

son is stationed there. On Alejandro Selkirk Island, however, there is simply nothing. Zero. And, there is also no landing strip for airplanes. In a medical emergency, a boat has to come to pick up a victim and return to the other island, each way taking about thirteen hours. By that time a plane would be waiting to fly to Santiago and a hospital, but only if weather permitted. There are no helicopters with range to the far island and back to the continent, the only airplane being some sort of amphibious aircraft, but most of these are also of limited range. Furthermore, such a plane can land only if the sea is calm, which is not often the case around Alejandro Selkirk Island. The bottom line is that if one works on the far island, it is imperative that medical expertise and equipment be brought along, and even then, in a real emergency, it would be a terrible situation. Consider an appendicitis attack, something now relatively routine to treat, but not on Alejandro Selkirk Island, or some sort of heart attack, strangulated hernia, or serious infection.

Danger was often around us on our excursions on both islands. Frequently we were moving along a basaltic ridge, because these offer the most efficient way to climb up into the high forest, but these ridges fall off sharply to the right and left several hundred meters. A misstep and fall could easily result in broken bones at minimum and more likely death. On the positive side, most ridges are fairly solid, which aids placing feet carefully and watching each step. When climbing up a rocky slope in forested areas, it is more difficult. In these moist glens there is no cliff to fall down, but the lava rocks often come loose, resulting in slips of footing and worse, sending sizable rocks down the slope to whomever might be climbing below. We had a number of these boulders come loose, and some came close to team members, but again, due to concentration and luck, they always managed to jump clear of these tumbling and bouncing threats. A call would go out: "*Piedra*!" ("Rock") and everyone was on alert to move quickly away.

Although thankfully we never had any major health problems on our twelve expeditions, a number of minor issues did arise. The most serious were two broken fingers, Dan's in 2010 and Marcelo's in 2011. That these came on our last trips might reveal that we were growing older and perhaps becoming less resilient. In Dan's case, the accident was almost comical, although the results were certainly not. He and I had just purchased some food at one of the small markets in San Juan Bautista, and we had just turned the corner up the sidewalk of La Pólvora to Bernardo's house where we were staying. As we walked, along in the middle of the street came a very old friend, one of the CONAF guides, Ramón Schiller. It was great to see him, and we wanted to greet him warmly, but between us and him was a small drainage ditch, where water rushes down during the heavy spring rains. I hopped over the ditch and embraced Ramón enthusiastically, but at the same time I heard a whump behind me. Turning around, it was Dan, who had missed his step and fallen flat down, his extended hand softening his fall, but in the process cutting one finger deeply, blood now flowing out. Dan also embraced Ramón, but then we looked at Dan's finger with concern. Ramón said that we needed to go to the clinic (La Posta) to have it bandaged up, and I agreed fully. Dan, the old, tough farm boy, said it was nothing, but the dripping blood was telling another story. The clinic is up the hill toward CONAF headquarters, and we walked up there as fast as possible, but it was locked tight. No one there. We then went to the private house of the park chief, Don Iván, who came out and listened as we explained the situation. Iván knew where the physician lived, which was nearby, so we waited about fifteen minutes until he returned with the doctor, and we followed her to open up the clinic. She disinfected and then bandaged Dan's hand, but this naturally bothered him a bit for the rest of the trip, requiring daily changing of bandages. We did not know it at the time, but with an X-ray later at home, he learned that there was a small fracture that had now started to heal. Marcelo's broken finger was similar, but this took place on Alejandro Selkirk Island in 2011 on our first field day. Fortunately, our medical student (and nurse) Daniela Bacho did a great job of fixing him to allow continued work during the entire trip. He, also, had a small fracture as revealed later by X-ray in Concepción.

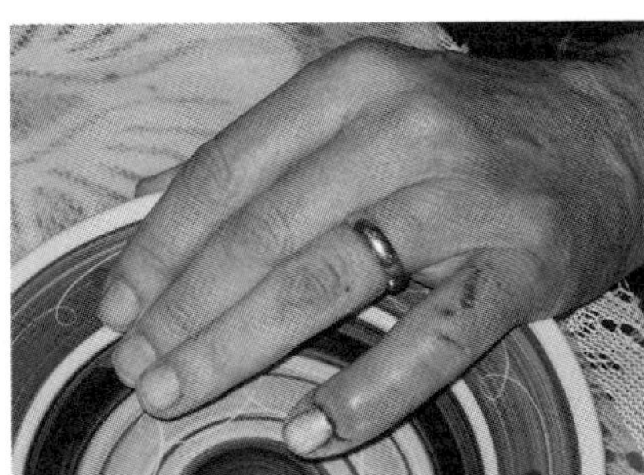

Dan Crawford's accident on Robinson Crusoe Island. Left, medical clinic (La Posta) with Dan receiving treatment; right, his fractured little finger after two weeks.

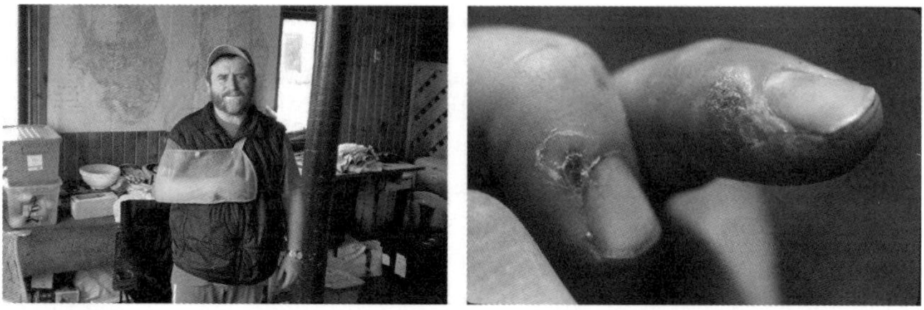

Other health issues during expeditions in the Robinson Crusoe Archipelago. Left, Marcelo Baeza on Alejandro Selkirk Island in 2011 with his right arm in a sling to support a fractured finger; right, hand of Tod Stuessy with a bacterial infection during expedition to Robinson Crusoe Island in 1980.

Other health issues were of a lesser nature. Pato López had an impacted wisdom tooth on Robinson Crusoe Island, and it required extraction. There is a small dental clinic, and it did have the tools and anesthetic for a tooth extraction, but as is well known, recovering from this made the following several days less than ideal for him. On our second trip of 1980, my hands developed some bacterial infection that just didn't want to heal. Eventually it did go away, and I never had this condition again. In addition to these maladies, there were muscle aches and pains, bruises, twisted ankles, tendonitis, shoulder pains, heat rash, colds and coughs, and occasional diarrhea. I would hasten to add that the water in both islands is completely drinkable because it comes through plastic tubes from springs in the hills. It is also not treated, which gives it a great taste, but there is a chance of contamination from goats, especially on Alejandro Selkirk Island. After a heavy rain, the water turns a bit brownish from sediment. In the early trips, Roger and I always treated water with iodine crystals,[4] but later this was dropped. Iodine was then taken on trips as an emergency measure, should we be stranded. A fallback measure was to drink rainwater that had accumulated in the leaves of *Gunnera peltata* or *G. masafuerae*, but this was rarely necessary. Overall, when one considers how many trips we had, how many persons were involved, and how dangerous some of the excursions really were, we were fortunate to have had no serious injuries. We were well prepared, but we were also lucky. Perhaps we were spared tragedy because when one is in dangerous situations, there is little interest to fool around or joke, and dedication to survival becomes the focus, which fortunately was always the case.

Techniques for getting emergency water on the trail. Left, collecting stream water for purification with iodine solution, with Roger Sanders; right, drinking from accumulated rainwater in the large leaves of *Gunnera peltata*, with Oscar Matthei.

And now a few comments about parties (fiestas). I am not certain if dedicating part of a chapter to fiestas on the island during our expeditions is a good idea. These are professional scientific efforts supported by granting agencies; they give money for research work—not parties! But I will tell you from deep conviction that having a safety valve every two weeks is an essential part of a field expedition. There are several very good reasons for this. First, the fieldwork is physically demanding, such that each person gets pushed to the limit. The hours are long, from 9:00 in the morning to 8:00 at night, and there are no weekends. The only break in the field effort is when it rains, but then other work has to be done—pressing plants, running electrophoretic gels, preserving leaves in liquid nitrogen or silica gel, comparing data from one day to the next, completing field notes or logs, and planning for the next outings. Second, a good fiesta with lots of wine and pisco lets expedition members unwind and get to know each other better. If someone doesn't drink, can he or she still have a good time? Well, of course, but that eventuality rarely happened! There are Chileans who don't drink at all, not even wine, but this is uncommon. And finally, we always invited the CONAF guides who had been helping us, in some cases saving us, plus the park chief. When on the far Alejandro Selkirk Island, we invited all the fisherman families, some twenty or so, to join in (see chapters 8 and 11). There it is impossible to have a party without inviting everyone on the island because they all live close together and can hear the music, smell the grilling fish, and see the wine bottles, to which the families pitch in and bring food and drink, too, such as Coca-Cola, potato salad, canned fruit, and cakes. All gets spread out on tables in the open and the fun begins. To give you a flavor of these fiestas, and not without some small tinge of embarrassment, I relate an example from our expeditions of 1984 and 1996.

A very memorable party took place in 1984 in the rustic fisherman's house. Here we organized an excellent dinner and fiesta, including pisco, wine, and lobster prepared by our cook. After dinner we danced, each person moving with the rhythm. A woolen cap was produced, and when it was passed to you, it was time to dance in the center until you passed it to someone else. The pisco and whiskey flowed as the evening progressed, and Patricia wisely exited quietly to her own room in another house across the street. Dan was downing shots of pisco with the park chief's brother (Marcelo González), and bit by bit, the brother was showing signs of fatigue, eventually leading to his flopping on the couch and sleeping until morning. After this celebration, our simple home seemed even more like we really belonged there.

Another memorable fiesta, a bit larger, took place on Saturday, 27 January 1996, on Robinson Crusoe Island. We had been staying in the house of the Servicio Aeronáutico, and that is where we organized the party. During the day, we worked hard with more plant processing, trying to finish everything before the fiesta began. Ulf went to the electric plant (five times, because it was often closed) and sorted specimens; Eduardo and Pato were charged with getting items for the fiesta; Gabriel and Greg finished up their floral biological investigations; and Dan did the final electrophoretic runs for the genetic analyses.

The fiesta started early at 5:15 in the afternoon, when several of the guides arrived (Bernardo, Guillermo, Cote). Cote had brought us the lobsters the previous evening, and Sra. Peta, our cook, was preparing them. Parties in the islands start very slowly and informally, but they can last a long time, so no rush! The CONAF folks brought a grill, a gasoline drum cut in half longitudinally, and we put it on the front lawn of our Aeronáutico house. Cote, Guillermo, and Bernardo chopped wood with Greg trying his hand, too. I went down to the village to help Eduardo pick up the wine we had already bought (one large five-gallon jug of vino blanco, 4,000 pesos [ten dollars]), but the store had a padlock on the door—and

Enjoying a fiesta in our fisherman's house in 1984 on La Pólvora Street, Robinson Crusoe Island. Left to right: Tod Stuessy, Daniel Crawford, Eduardo Ruiz, Marcelo González, Gastón González (park chief), Patricia Pacheco, and Alejandro Landero.

we had prepaid! We next visited a store that had pork, where we had agreed to buy five kilos, but it wasn't yet available—come back at 8:30 tonight, they informed us. Well, miraculously enough, Eduardo did return at this time, and it was ready to be put on the grill! He also was able to purchase more wine.

The fire for the fiesta was started at about 8:00, and the wine began to flow with pisco and juice and/or Coca-Cola (piscola). I really prefer the juice to the Coke. The lobster was prepared first, and we had purchased ten from Cote at 20,000 pesos (fifty dollars). Other people began to come too: Ramón, two friends of CONAF (young women), two Chilean university students doing projects in the islands, Osvaldo (our landlord), and a friend of his, Gonzalo, from Santiago. The meat was put on—goat pieces, two rabbits, and the five kilos of pork. People stood and ate while holding plastic plates, or sat on the front porch of the house. There was much more food than anyone could eat. We had two potato salads, one with onions and another with mayonnaise, and a great ensalada chilena. There was juice, wine, pisco, and Coke for all. After about 10:15, we got out our cassette player, this time not with Frankie (not the right mood) but the very alive and happy Colombian cumbia,[5] and Bernardo began to dance some—the mood becoming very pleasant, but there were not enough women to keep the dancing going. Also, some of the men were getting pretty drunk—especially Pedro, Ramón, and Guillermo.

The fiesta continued to midnight, at which time we decided to visit the disco down by the waterfront. We all walked down, on the way discussing the recent soccer match where Chile beat Denmark 3–1, to everyone's great satisfaction. The disco was on the main street, a wooden, unpainted building, but quite attractive inside with a bar at the far end. There were alcoves for sitting and drinking while waiting to dance. By 2:00 in the morning the place was filled and jumping—really alive. It was strongly driven by the teenagers, but older folks were there too—like us! I had one piscola and then switched to pure Coke; the music was a combination of Latin rhythms, the popular Luis Miguel, and Argentinean rock and roll. There is no hang-up here—everyone dances with everyone else. By 3:00 Dan and I were ready to leave, heading along the main street. When we reached the fairly steep cement walk up to our house at the Servicio Aeronáutico, I was wondering how we could get up there together safely, as we were both a bit unsteady, but suddenly Gabriel Bernardello appeared from behind, having left the disco shortly after us. He gave us a hand in getting up the walk and into our house. I didn't sleep too soundly, waking up now and then, but as an overall assessment, it was a successful party for our people and for the guides who have always made it possible to complete research in the islands. After two weeks of hard work and serious physical effort, a fiesta is perfectly in order, a well-appreciated contribution to a successful expedition.

CHAPTER 8

Real Isolation

Our expedition to Alejandro Selkirk Island in 1986 was quite the experience, a real adventure in part because it was so new to us. I had visited the island for a few days each in 1980 and 1984, going on boats that brought supplies to the fishermen and their families and carried back lobster and fish. These short trips for me were helpful, providing an introduction to the island and its flora and vegetation, but this was just a modest beginning. Organizing for the trip in 1986 to Alejandro Selkirk Island began in 1985, and this was a challenge, because there is no permanent village and hence no supplies or facilities. Fishermen stay on the island for nearly eight months of the year (October through May), about twenty families, but they have their own supplies and small houses, not enough to share for an expedition lasting one month with eight participants. The plan was to purchase supplies in Concepción, pack them and ship them to Valparaíso for the boat trip to Robinson Crusoe Island and then on to Alejandro Selkirk Island.

We had a good group of expedition participants. I was the leader, the only professor involved with this trip. From Ohio State in my laboratory came Thomas Lammers, a Ph.D. student doing a thesis on *Clermontia* (Lobeliaceae) in the Hawaiian Islands. Although his work had nothing directly to do with the Robinson Crusoe Islands, Tom's thesis dealt with evolution of plants in another archipelago, so having the chance for him to visit a different island and flora was a good opportunity. Hugo Valdebenito, originally a student with Mario Silva in Concepción, was now a Ph.D. student in my laboratory (and that of Dan Crawford) working, in

Participants on the 1986 expedition to Alejandro Selkirk Island, assembled at the bus terminal in Concepción. Standing in foreground in front of other passengers (left to right): Eduardo Ruiz (1), Alejandro Landero (2), Jaime Sepúlveda (3), Thomas Lammers (4), Hugo Valdebenito (5), Leonardo (Leo) Gaete (6), and Michael Doyle (7). Patricia Pacheco (between 3 and 4) from our laboratory was also there to see us off.

part, on the evolution of the genus *Erigeron* (Asteraceae) on Alejandro Selkirk Island, which meant that he, for certain, needed to be on the expedition. This was central to his thesis investigations. Eduardo Ruiz was an undergraduate (*licenciado*) student studying phytochemistry (flavonoids) with Mario Silva, and he later came to Dan's laboratory at Ohio State for two years of work, investigating genetic variation in endemic plants of the Juan Fernández Archipelago, and finally returning to obtain his Ph.D. in the Department of Botany in Concepción, where he now is professor. Alejandro Landero was also a *licenciado* student in the same Department of Botany, as was Leonardo (Leo) Gaete. Alejandro had been on the trip in 1984 to Robinson Crusoe Island, and I knew him to be an excellent field person—very strong, a good climber, and highly motivated. Leo is now a professor at the University of Chile in the Instituto de Ciencias Biomédicas. Jaime Sepúlveda was in his final year as a medical student in Concepción. Jaime was not yet a certified physician, but he had experience in clinical work and could do many tasks, including giving injections, which we deemed extremely important. In retrospect, it might have been a bit naïve to think that a student of this level could provide the medical support needed in serious situations in such an isolated place, but fortunately we never had tragic circumstances. He is now a physician working in Santiago at the Colegio Médico. Finally came Michael Doyle, a graduate student of Thomas Elias, Director of the Rancho Santa Ana Botanic Garden in Claremont, California. Michael was interested in collecting fungi, and this added a dimension to our collecting activity that we had never emphasized previously. He would also collect mosses and lichens, adding to our general understanding of the non-vascular flora. After the expedition, he pursued Ph.D. studies on *Gunnera* of the Hawaiian Islands, continuing his interest in plant evolution in isolated archipelagos.

It was an excellent team of young, energetic biologists, who were going along on this expedition at no pay, although all expenses would be covered. The trip coincided with summer vacation, which meant no interference with academic programs, and in fact, this expedition would be valuable experience for continued biological studies. My advantage was the simplicity of leading the group without having a co-leader (another professor) with whom I would always have to confer and compromise, which had been the case on the first two trips.

On 5 January, Tom, Hugo, Patricia, Michael, and I assembled at the Columbus airport to begin the trip. The first three were studying at Ohio State, and Michael had flown in from California to join up with the rest of the group. We flew first to Miami, arriving at 7:30 in the evening. Miami International Airport is very cosmopolitan, where many Latin American airlines bring passengers for a visit to the United States. Most of these visitors sightsee and shop in the Miami area, which causes them no linguistic challenges because so many persons in the region are bilingual. Some travelers continue on to Orlando, New York, or other destinations across the country, or even to European cities. After getting checked in for the second leg of our trip to Santiago, we had time to have a small dinner, do some window shopping, and chat about the expedition. At midnight, the plane loaded up, took off, and we settled in for the long overnight flight to the Southern Hemisphere. In the morning we would have a short stopover in the airport in Buenos Aires, and then over the Andes to Chile.

Upon arrival at the airport in Santiago, we passed through immigration, retrieved our bags, and went through customs, but the main order of business was to get tickets on the train for the overnight trip to Concepción. Patricia would visit relatives briefly in Santiago, joining us later in Concepción. Because we had considerable baggage, we decided to take the train with ample storage capacity. Unfortunately, all sleeper bunks were taken, which meant we had to accept seats in Super Salon, not as comfortable as a bunk bed, but the seats at least did recline. After an uneventful and somewhat restless night, we pulled into the Concepción station in early morning.

Tom, Michael, and I, with all the luggage, caught taxis to the university to the building housing the Botany Department to begin expedition preparations in earnest. Hugo would visit his family and then join us in the afternoon at the university. Getting materials together in Concepción for the trip was the usual exercise, this now being the third time we had so done. Much advance planning had taken place by letter (email did not exist), with Mario and his assistants taking the lead in getting food and beverage items purchased before we arrived in town. Permits for collecting from CONAF also had to be obtained, and Mario had completed these efforts. Mario had a good working contact with the owner of a local food market on Plaza Peru, just off the university campus, and we purchased many items from him. He packed products with anticipation of these being in the hold of the ship for four to five days, especially providing protection from invading seawater. The Chilean participants had all been called to Concepción a few days prior to our arrival, which provided sufficient people to get tasks accomplished.

We did our organization in the usual laboratory room, large enough for laying out items and going over them carefully with the checklist. Did we have enough for all participants? How long would the chocolate bars last? How was the medical kit coming along? Jaime was doing a good job with getting the medical items together, including having prescriptions written by a physician. The most important items would probably be for treating infections, that is, different antibiotic pills, creams, and some more potent injectables if need be. We had medicines against diarrhea and constipation, for sore muscles, and bands for twisted ankles. What we had was not a complete clinic, but our medical chest was well stocked with pharmaceuticals and basic equipment, which did turn out to be important for us, as well as for some of the fishermen in the island.

After two days of packing and organizing twenty-one wooden crates and many boxes, nearly two tons of materials, all this equipment was picked up at the university by a small van that Mario had contracted for transporting everything to a warehouse near the dock in Valparaíso. I wor-

The port city of Valparaíso, Chile

ried about letting our materials depart without any personal control, but each item was numbered, and I had the inventory for checking after we arrived. It was now time to get to the bus station in Concepción for the overnight ride to Valparaíso. The trip was uneventful, and we arrived early in the morning to this coastal city, the busiest shipping port of Chile, having served this role for nearly 500 years. After stepping down sleepily from the bus and gathering up all backpacks and personal luggage, we headed out of the terminal and went immediately into a nearby simple restaurant for a good breakfast. Having a sandwich of hot ham and cheese, or even scrambled eggs with ham, hits the spot in the cool air of a seacoast morning. This, washed down with some coffee, gets the motors going. There was no rush; we could enjoy the breakfast together, each person thinking about the adventure awaiting us, and each coming to the challenge from different backgrounds. I felt very positive about this group of participants—three of them had worked in the islands before with me, so this meant that they, and I, made up one-half of the crew, a good ratio for getting things done that required some experience.

As the light of the morning became brighter, and it started to warm up a bit, we decided to hail taxis, head down to the dock, and see if our ship was already there, and in general, to check on the situation to be certain all was in order. Valparaíso is a very old city, and the older parts have winding, very narrow one-way streets, that remind one of prior centuries when ships from different countries stopped for trading, diplomatic pur-

The *Río Baker*, which carried us from Valparaíso to Alejandro Selkirk Island and back on our expedition of 1986

poses, and other commerce, in a time when this was the only convenient way to travel the globe. Our taxi pulled up by the number of the dock where our ship was supposed to be at berth. There was a ship there now, a whitish one, quite a bit rusted in spots, and called the *Río Baker*. We learned later that this ship had been used for many years among the southern channels and islands of Chile, and apparently was still giving proper service (we hoped!), although its appearance did not engender huge amounts of confidence. As we took turns commenting on our "cruise ship," a well-dressed fellow came to the dock and shipside, armed with a briefcase. I approached and asked if he knew anything about the *Río Baker*, and when it might be available for loading up our gear for the trip, and also, if he knew when departure might take place. He smiled, was very cordial, and said that yes, he did know these details because he was the captain! Well, this was a pleasant surprise, because then I could get answers to these and other questions, too. He said that the boat would leave at about 9:00 that evening, and the hold of the boat would be opened at 5:00 for loading. We should have all our equipment ready on the dock next to the boat at that time. Good. This gave us a clear idea of how to structure our day. I mentioned how enthusiastic we were to be going to Alejandro Selkirk Island, and how much we had prepared for this expedition. He looked at me directly and said: "I am sailing to Robinson Crusoe Island, not Alejandro Selkirk Island." At first I thought I had not heard properly, so inquiring about this, he responded, again politely but firmly, "No, I am not authorized to sail to Alejandro Selkirk Island on this trip." Well, I explained that all our preparations were designed for the far island, and this had been our understanding from the beginning. He was resolute and stuck by his orders from the shipping company, nothing to discuss further with him. But, as I was obviously not going to let this matter go, he offered to take me to the office of the shipping company (Empremar) to see what the problem seemed to be. We had already prepaid for our transport of personnel and equipment, so I could not imagine what might have gone wrong. Further, Mario had handled the reservations and payment, and he does not make these sorts of simple mistakes.

Fortunately, the shipping office of Empremar was only two blocks from the dock, and I, Eduardo Ruiz, and Hugo Valdebenito walked there with the captain. The other persons stayed dockside with our backpacks and other personal items. Traffic was light in this part of the city at this time of day, so the dock area was quiet, but I was not quiet inside, having great concerns about why there appeared to be confusion. This was not a trivial matter—the whole objective of this trip was to focus on the far island, which was very much needed at this point in our research.

The office of the shipping company was on the ground floor of a three-story building, air conditioned inside, with a medium-sized room containing five desks and persons occupying them, the employees shuffling papers as they worked. This was a typical business office, nothing special, nothing fancy, and its appearance seemed compatible with that of the boat itself, functional, without luxury of any type. The captain kindly introduced us to one of the clerks, and the discussions began. Papers were produced that indicated that we had, indeed, contracted to go on board the ship from Valparaíso to the islands, but only to Robinson Crusoe Island. I pointed out that one always goes to the near island first, but then in our case on to the far island, so just because it said "Isla Robinson Crusoe" on the request form did not mean that we did not intend to travel farther.

Well, discussions went around and around without any resolution. I was obviously not going to let this go, because the whole point of this planned expedition was to go to the far island for one month. We had to achieve this objective, period. They countered: "Well, sorry, but you had reserved and only paid for transport to the near island." The only way they would go to the far island is if we paid more money, and at that point

I began to smell a rat. Was this simply a convenient extortion scheme to squeeze more money from this American gringo? I frankly wasn't sure, but I was getting suspicious that something was just not right. The shipping clerk finally offered: "We would be willing to take you to the far island and pick you up at the end of one month for the additional sum of $2,000." Essentially we would be hiring the ship to make these additional visits. The *Río Baker* was going to Robinson Crusoe Island, in any event, and so we would have to pay for two additional round trips between islands, once to drop us off and once again to pick us up. Furthermore, if the sea were rough upon our arrival, the captain could not unload our materials, and we would have to return to Robinson Crusoe Island. We still had to pay for the trip—it was our risk and a very big one.

I do not deal particularly well in situations that to me seem unjust or unfair. I have a pretty deep tolerance of difficulties of life that can roll off me, but in circumstances that seem to me wrong and totally unjust, I lose patience. Predictably, I was starting to slide into this mindset the more I chatted with the shipping company. I turned to Eduardo and Hugo, and we went outside the office to confer, trying to get our bearings and decide how to handle the situation. I was determined to get to the far island and complete our research objectives, and if we had to pay more money, we would have to do it, somehow. I did have enough cash to pay such a bill, but then I wouldn't have enough for the trip itself, which was obviously not possible. This was long before credit card charges were accepted in most businesses, and certainly not here on the waterfront in Valparaíso. Level-headed Eduardo suggested that we call Mario Silva in Concepción to try to find out what had happened and what to do, and upon returning to the shipping office, they did agree to allow us to make a call. It was Saturday, however, so we were not certain how hard it might be to locate Mario, but we did have his home telephone number. The maid who answered informed us that Mario was no longer at home, having left already with his family on holiday to La Serena, north of Santiago, and he was presumably right now on the road. He would be of no direct help, therefore (this was before cell phones made life easier!). We also had the telephone number of another botany professor, Oscar Parra, who had been on the first expedition in January of 1980. Oscar was a strong person, and I trusted his judgment, and fortunately Eduardo had his home telephone number. Mercifully, I got Oscar on the phone and sketched out the complete problem with him, and I asked if he might talk to the shipping clerk directly, which was done. They chatted for a few minutes, and then the receiver was returned to me. "Well . . . ," said Oscar: "As far as I can determine, what has been set up and paid for is transport only to the near island. For you to go farther, you will have to pay more for it." There it was—more cash or no deal. I laid it out clearly for Oscar that I just didn't have money for this, as it was not part of the planning. Could the University of Concepción possibly make this additional payment through a bank transfer, and I would repay this later upon returning to Ohio? We had enough money in the grant to cover this additional expense, but I just didn't have any in hand. Oscar said that he would have to call the Vicerrector de Investigación, whom he knew personally, to see what could be done. We should wait there in the shipping office for his return call.

Latin American universities do not have the same deep pockets as American universities, but they are not poor. As in all situations in life, there is always money in society, the challenge being how to get access to it. I had no realistic idea of whether Oscar might be successful, or what sort of response we might be getting, or conditions attached, and I was trying to think how I might get some money wired down from Ohio State or even from my personal account in Columbus to make this additional payment. Eduardo and Hugo saw no alternative other than the university, especially because holiday time had started and the more senior professors, who might have been able to advance personal funds, had already taken off for family holiday, a very typical countrywide event as soon as the schools let out for the summer months.

The three of us sat on a blond wooden bench alongside one wall of the shipping office. Hanging on the wall above were all sorts of certificates, maps of regions along the coast, and photos of ships, presumably from their own line. As the day was heating up, we were well protected inside the office, which had several window air conditioners blasting away, making a racket, but a welcome noise as we knew these machines were saving us from the humidity and heat outside. On another wall was a large, wooden octagonal clock with large black hands, marking the time of our existence. It didn't make any sound, which we probably couldn't have heard anyway, due to the air conditioners. In this situation, I could do without a tick, tick, tick as we waited patiently for a response from Oscar. Several other calls came to the office, but none was for us. With each pick up of the receiver, we would look for a sign that it pertained to our situation, but these were other communications, the clerk searching for other documents and then having further consultations on the phone with the client.

At 12:30, only half an hour before most Chileans head for lunch, and when presumably the office would close until 3:00, another call came. It was Oscar. He first asked to talk to me. He had been able to contact the Vicerrector de Investigación, who in fact, was in his office at the university tidying up items before heading off for vacation also with his family. The Vicerrector had agreed to pay the additional cost to get us to the far island, as long as I would agree to pay him back after returning to Ohio State. What excellent news! And what a relief to have a solution to what was a threat to our entire trip. Of course I agreed to this arrangement and told Oscar that he could count on me for this repayment, and if for some reason Ohio State balked (although I had no reason to imagine why), I would cover the expense personally. While I am a person of my word, neither Oscar nor the Vicerrector knew me well enough to assess this quality, but I think the importance of the international collaboration between our universities, that there would be publications resulting from the expedition, and that students from Concepción were on the trip, all contributed to a positive answer. This having been assured, I passed the receiver back to the shipping clerk for him to discuss with Oscar exactly how the money was to be paid. For my part, I wanted to have a paper that stated we would be going to Alejandro Selkirk Island, and I presumed that they would be providing a document also to the captain, who by this time had returned to the ship. A mechanism for payment was agreed upon, and after hanging up the receiver, the clerk got out another form and typed up the order for us to rent the entire boat from the near to far island, twice. If any other passengers wanted to make the trip, they could do so, and their payments would be reimbursed to Concepción. He finished typing up the document, and he stamped it with the seal of the company, making it official. This to me was now one very valuable document.

I felt good about getting this problem cleared up, but in the back of my mind there was still the doubt as to how this misunderstanding came about. It would have been impossible that Mario misunderstood our objectives, but how, then, was this handled so poorly? I never did learn what really happened, but at this early moment I was put on caution in dealing with the shipping line. Our job right now was to head back to the dock and have another chat with the captain, to be certain he was agreed to take us to the far island. Eduardo, Hugo, and I left the shipping office,

Lunch in Valparaíso before heading back to the dock. Left to right: Eduardo Ruiz, Hugo Valdebenito, Jaime Sepúlveda, and Thomas Lammers.

now feeling the oppressive weight of the heat and humidity, and this time keeping an eye out for traffic. We walked back to the dock where our colleagues were patiently waiting in the hot sun, not overjoyed that it had been three hours since we had left them watching our luggage. The captain was by the gangplank, and I went over to talk to him, which he was willing to do at that point. I showed him our stamped document and indicated that he now was authorized to take us to and from the far island. He said, "Fine, I will be happy to do that, and I am glad that we resolved the problem." Again, I just didn't know if I was being shaken down for another $2,000 or not, but whatever, the expedition had to take place and be successful, and that was that. This crisis passing, we picked up our hand luggage, walked through the gate to the street, and hailed two cabs to find a restaurant for lunch.

At the lunch table, I spent time explaining to the entire group what had happened in the shipping office. They were amazed how this could have occurred, and each person had a theory about whether it was legitimate or not, but the bottom line was that we had little choice. The problem had been resolved, and our job now was to ensure that the trip was successful. Slowly, our spirits lifted again, and we also talked of objectives, sharing experiences from previous trips. After lunch there was work to do, going to the warehouse and having all our equipment sent down to the dock. We agreed that three persons would take charge of this, Leonardo (Leo), Alejandro, and Hugo, and the rest of us would head to the dock. We took any extra personal items with us in our taxi, so that the other three had more hands free to fetch the equipment.

There was not much activity back at the dock, but at least now there were more crew members doing various tasks aboard the *Río Baker*. The captain came off the ship down the walkway, and I asked him again if all was in order to sail to Alejandro Selkirk Island, and he answered, "Yes, all is now in order." There are times in life when you just have to trust other people. One can be suspicious but not all the time, or life becomes a paranoiac maelstrom, in which one worries about everyone all the time. Basically, most persons want to do good work, they want you to trust them, and they want to be viewed as trustworthy. So, I let it go. It was time now to worry about getting all materials loaded safely on the boat so that we could sail out of the harbor that evening toward the islands.

At shortly after 5:00 in the afternoon, a flatbed truck appeared just outside the gate to our dock number, and it was our boys bringing all the equipment. We had lots of items, many yellow wooden boxes with the blue stenciled lettering that we had used on the three previous expeditions, but still holding up in good shape, and then lots of boxes with contents inside wrapped in double sheets of plastic and tied with strong rope on the outside for security and ease of handling. We all pitched in to get items off the truck, with two persons taking either end of the wooden crates, muscling them to dockside. These crates were heavy, perhaps 100 pounds loaded, which for two persons was manageable, but not for only one person. As the items were brought to dockside, I kept track of their assigned numbers to be certain everything had arrived from when we had packed them in Concepción. In they came, and each number got a checkmark on my sheet. Yes, it looked good so far, until everything had been stacked ready for loading in the boat at 6:00, or whenever the captain so ordered. But wait—there were three extremely valuable items missing—drums of gasoline (two of them) and one drum of kerosene (for the drying stoves). Without kerosene we could not dry plants on the far island, or let's just say it would be very tedious and time-consuming, having to change newspapers constantly. Worse, without the gasoline, CONAF was not prepared to take us in boats around the island, because this would deplete their own supply, which they needed for routine reconnaissance work mandated by the park service.

Expedition cargo arriving to the dock in Valparaíso

We now had to solve this problem, and I asked Hugo to take charge, giving him substantial amounts of cash to ensure that the job was successful. When I wanted Hugo to help me do something, I would stand close, look him in the eye, put my hand on his shoulder, and say: "Hugo, I know you can do this—I have every confidence in you." He hated to hear this, but I delegated the responsibility onto his shoulders, because he always got things done. He also knew the Chilean culture and had a good manner with people, all requisites for this particular challenge. He took Eduardo and Alejandro with him, and they ran out the gate and hailed a cab to plunge into the city. Somehow they would have to find a store with empty drums, take these to a gas station (*gasolinera*), and bring them back to the ship by 6:00 for loading into the hold. As elsewhere in Latin America, there were many, very many small shops along the crowded streets of Valparaíso. The trick was to find out where to go, and for this, you have to ask. I presume our boys began with the taxi driver, and that led somewhere, perhaps to a gas station, that they might know where to obtain drums, or perhaps they had some used ones we could buy from them. Kerosene also would probably be available at a gas station, but not all of them would have it.

In the meantime, back at dockside, at 6:00 the captain signaled the order to open the hold so that items could be loaded into the ship. There was a crane mounted that could reach comfortably to the dock, where two wooden pallets were supported by thick ropes for loading. Lots of

Preparations on the dock in Valparaíso. Left, Jaime Sepúlveda, our medical student, checking his supplies; right, loading part of our equipment into the hold of the *Río Baker*.

other supplies destined for Robinson Crusoe Island were ready for loading: lumber, carafes of wine, foodstuffs, some well packaged as if from a supplier, and others more rustically bundled, obviously personal items. I looked down into the hold, and it seemed completely dry. All our equipment was in double plastic bags, but having no water at all would be much preferable to having flooding that would test our packing quality. We didn't need further surprises at this early point. In went a substantial quantity of lumber, destined for someone's house on the island. I doubted that any supplies were being shipped to the far island, because no one had known that the boat was going to that destination, or did they? Again, that slight feeling of being squeezed. Jaime took advantage of this time to recheck our medical supplies and to pull out needed items to do baseline checkups of all participants during the voyage of two days to Robinson Crusoe Island.

At 7:00 our turn came. We helped load first the yellow wooden crates, and after one pallet was ready, it was hoisted up, over, and down into the hold. There, crew members moved items from the pallet and arranged them securely (we hoped) for the trip. While this transfer was taking place, the other pallet had been loaded up, so that it could next be hoisted over and down as well. After our crates came the boxes, and these needed to be roped around a bit, so that nothing slipped off during transit. I began to wonder how we would manage if our equipment suddenly slipped off a pallet, or worse, if a rope simply broke and all crashed down into the hold, breaking open and scattering material everywhere. Well, at least if that happened, we could probably regroup things, but I realized how much risk there was going to be before we had all our equipment sitting safely on Alejandro Selkirk Island.

It was now 7:30, and all our material was safely loaded into the hold of the *Río Baker*—all except three drums of gasoline and kerosene. This was not at a time when one could get out the cell phone and check on progress of our crew members attending to this need. We just had to wait. I told the captain that we were expecting delivery of these three drums any minute, and if he could please give us some more time. He said okay, but obviously there would be a limit to how much he would be willing to delay his schedule.

The sun was lowering now on the horizon, which cast a softer light over the dock area, and also the temperature began to drop somewhat. From occasional smaller boats that would pass nearby, there was the occasional lap of water against the *Río Baker*. This is a part of the day that I have always liked, a pensive time of day, an opportunity to assess what has transpired and plan for the tomorrow, in effect the start of the rest of your life. But the normal reverie of this time of day was marred considerably now because of the anxiety of waiting for our drums. We all stood around, joking, looking at our watches, and fidgeting. If necessary, we would go without the drums, but it would make the expedition more complicated. The captain waited courteously till 8:00 and then came out to say that he really now had to get the hold battened down and make other preparations. And just then, exactly then, a pickup truck appeared outside the dock gate with three persons and three drums of liquid. They had managed to get the drums, fill them up, and bring them back to the dock at the last moment! The captain produced a trolley to heft the drums, and our boys and a crew member helped muscle them from the truck to the ground. In short order the drums reached the side of the ship, where each was roped safely and hoisted up and down into the hold. Good old Hugo! He had made it, he had done it, and we had now survived two serious crises even before we had left land! I hoped that this would not be typical of the entire voyage.

After all supplies were secured in the hold of the *Río Baker*, the Captain closed it tight, and we were allowed to come aboard with our per-

Arrival to the dock in Valparaíso of our three drums of gasoline and kerosene. On the right side (left to right): Alejandro Landero and Hugo Valdebenito.

sonal items. This was a very utilitarian ship, similar to the *Carlos Darwin* on which we had sailed on the first expedition of 1980, nothing fancy, but with good luck, it would be adequate to get us to the islands and back. The trip from Valparaíso to the islands can be very bad if a storm picks up, but I, thankfully, never experienced this. The ship swayed and rocked from side to side, which didn't bother me, but I also always took Dramamine, which dulls the senses a bit and helps keep seasickness away, usually causing sleepiness, which was fine for the evening passage.

After we had slipped anchor and sailed out of the Valparaíso harbor, Jaime Sepúlveda of our group appeared and asked if he could take basic metabolic data from me, as he was going to do from all expedition members. Jaime was a sharp fellow who was taking his responsibility seriously. I watched him taking blood pressure, and he seemed to know at least this simple technique. I asked if he knew how to give injections, and he said, yes, that was no problem. He had worked for a semester in the university clinic as part of his training, and he had considerable experience with patients and routine methods.

After thirty-six hours, in the morning of the second day, we came upon Robinson Crusoe Island. This was a familiar scene now, a low-lying island

Robinson Crusoe Island as seen from the deck of the *Río Baker*

reduced by tectonic subsidence and the persistent action of waves, wind, and rain. But it was a comforting sight to us—one imagines the elation of mariners coming to this island in the seventeenth century, there to find fresh water, vegetables, goat meat, abundant fish and lobster, and cool, ferny glens in which to rest and recuperate. Our plan was to spend the rest of the day in the village of San Juan Bautista on Robinson Crusoe Island and then leave in the evening for Alejandro Selkirk Island. This was the captain's new schedule, and it seemed fine with me. Our objective was to sail to Alejandro Selkirk Island, and the sooner the better. There is nothing particular to report in this chapter regarding our activities in the village, but it was not completely trivial (see chapter 9!). Suffice it to say that the *Río Baker* departed that same day in the evening on time at 8:00, and we had an uneventful additional night sleeping on the ship. In the morning, the sky was beautifully clear, the air crisp, a very good omen for being able to go ashore, and after a few hours on deck, we sighted Alejandro Selkirk Island in the distance (see color insert, **9**).

Arriving to Alejandro Selkirk Island is not without challenges. Because the island is geologically young, approximately one million years,[1] it still reflects its volcanic ancestry, which means it slopes quickly downward into the ocean. There is no protected bay, only the ellipsoidal shape of the island surrounded by the Pacific Ocean, and there is no pier for unloading or for receiving small boats. But at this moment foremost on my mind was the condition of the ocean—was it sufficiently calm so that our equipment could be loaded off the ship and to the shore? If not, we would have a huge problem. This would mean having to return to Robinson Crusoe Island to see if there might be another way to get back to Alejandro Selkirk Island. Sometimes the municipal boat made the trip, although this was a very small craft and not the most secure way to sail. And who knows what other compromises this boat might already have, and even then, would I have enough money to pay to make the trip? As we approached the island, the sea seemed calm enough to me, but it all depended on the captain. We waited, waited, and finally the word came—yes, it would be all right to unload—what a relief!

The only way to get to the settlement area (Las Casas) is with difficulty. The main ship can only drop anchor off the coast, but the water is very deep, which makes anchorage less stable. Any materials arriving to the island must first be offloaded from the main ship into a lobster boat. This can only be done if the sea is calm, obviously, just as arrival to shore can only be accomplished successfully when the sea permits. On our two quick trips to the island in 1980 and 1984, I had seen this system and knew how important it was that all our equipment be protected against water damage. If there were slippage in the ropes holding the pallet, all contents would drop either into the sea or down into the boat, probably getting seriously damaged. Neither alternative would be acceptable. Arriving to the island and having some of our equipment or provisions destroyed would have been a disaster. Luckily, this didn't happen.

Small boats, generally serviceable wooden lobster boats, of the same type as used on Robinson Crusoe Island, come from the shore to the main ship, get loaded up, and then return to the shore. Without a pier, the boat can only come directly to the shore and be pulled up on land, but there are many rocks on both sides of this dredged entryway. To help in this procedure, a thick rope had been fastened firmly on shore, and then strung out and attached to an offshore buoy. The lobster boat has a notch in the prow, into which the rope can be run, such that the boat can now be pulled along the rope until arriving to shore exactly where the rocks had been cleaned away. The timing of arrival must be calculated with a high wave, so that the boat is moved as far on shore as possible. As the wave recedes, people must jump out, get a foothold on logs placed horizontally to the shore, and keep the boat from tipping over if another wave comes again.

Unloading our expedition gear in front of the settlement of Las Casas, Alejandro Selkirk Island. Top left, lowering of our equipment from the *Río Baker*; top right, supplies safely in the lobster boat; bottom left, transit of the boat toward shore; bottom right, safe arrival on land.

The boat is then hooked to a winch higher up on the shore, protected under a small metal roof, with villagers cranking the boat up farther. After contents of the boat are taken out, and if the boat is not going to sea anymore at that time, the winch is used to move the boat even higher on shore, and finally shoved to the side so that other boats can be used. In 1986 the winch was hand cranked, but in 2011 a motorized version was installed, made possible by electrical power from a new generator in the village.

Once our equipment was stacked safely in piles to the side of the arrival slot for the boats, we had to decide where to take the crates and boxes. The main CONAF authority assigned to the settlement at this time was Eduardo Paredes (Don Uto), a legendary guide who, it was rumored, had sired more than sixteen children in the continent from different women, and who also was the father-in-law of our principal guide, Bernardo López, who had come with us from Robinson Crusoe Island. Don Uto suggested that we put all materials in a rather large shed no longer in use, this being very close to the shore. We organized our crew members for the work, two per crate, with smaller boxes being handled individually. This activity took about an hour, including arrangement of items into some sense of order for ease of access. This shed was nothing more than quite a few pieces of metal fastened together, but there was a door with lock and key, which would provide some security. Obviously, if a theft occurred, the culprit would have to be here in the settlement, a situation that would lead to discovery in short order. Stealing was virtually nonexistent in the islands.

We had to discuss a plan for sleeping and working with Don Uto. He said we could have two small sleeping rooms in the green-painted building where the CONAF kitchen and radio were kept, and then we would have to put up tents for the other six of us. We had brought a very large tent with capacity for eight persons, so this would be erected nearby and serve as the basic dormitory for me and the crew. Close to this on level ground we set up another tent, this one a workspace with canvas roof and sides of mosquito netting. I knew from my previous two short visits how intense the flies were in the island, and trying to work on plant specimens when fighting hordes of flies would be difficult. This work tent, therefore, was the solution to that problem. For the next two hours we put up tents and got our personal items properly organized. Fortunately, it was a nice sunny day, and we were able to work efficiently. Some of the children in the settlement came close to watch, wondering what in heavens these visitors were all about. At this point I began to realize that our stoves, kerosene, and the diesel drums had been stored in the shed along with the food and equipment. It was convenient because there was plenty of room for us to dry plants in this space, but danger lurked behind this idea. The stoves must be in their own restricted area, such that if a fire broke out, we might lose presses but not our food and equipment. I noticed a very old stone small house nearby, in very bad condition, but the roof was intact and it was large enough for us to set up six drying racks for our plant presses, kept well away from our sleeping quarters or equipment depot. Don Uto gave us permission to use it. I tend to worry about lots of things, especially on an expedition, where a slip up can be disastrous. My wife says I'm prone to carry that behavior into our own home, always closing drawers, tightening faucet handles, and checking stove tops. I suppose she is right, but here on Alejandro Selkirk Island we needed no unnecessary risks.

Food activities on Alejandro Selkirk Island. Left, Marcelo Baeza covered by "friendly" flies as he selects a grilled fish snack during a fiesta in the settlement of Las Casas (from the 2011 expedition); right, butchering a wild goat for our supper.

Eating on Alejandro Selkirk Island is a two-handed exercise. This is due to the flies, which are large, friendly, European, and flit around everywhere. They have little, if any, fear of humans; in fact, I think they like people, preferring to crawl on the forehead, into the ears, along lip lines, on the hands, and across the head. No part of the body is immune from chummy communion with these insect buddies. When eating soup, for example, the right hand holds the spoon and does the real work, and the left hand keeps moving back and forth to shoo the flies away, occasionally one falling down into the soup for additional flavor. The huge quantities of flies in this island may be due to the large number of feral goats, the flies breeding on the dung, which occurs all over the island. It is comical, in a way, that when in the hills for lunchtime, just open up a can of sardines and you will immediately attract hordes of flies. It is a blessing that they do not bite, but they irritate and frustrate. We always tried to work inside one of the large tents with netting, to be free of the constant crawling and pitter-patter of these six-legged visitors. They will get into eyes, too, and don't leave your mouth open for very long. The real fly infestation occurs when butchering a wild goat. The only respite comes when the sun goes down and the flies disappear until the morning. I have no idea where they go, but all that matters is that this irritation goes away, and one can

Mule loaded and ready for the trip up to La Cuchara at 1,100 meters, Alejandro Selkirk Island. Left to right: Bernardo López, Thomas Lammers, Eduardo Ruiz, and Alejandro Landero.

enjoy a pisco and Fanta in peace, while ruminating over the progress of the day.

On the following day, we began our expedition with a trip into the deep quebrada (ravine) of Las Casas, just behind the settlement. These ravines are cool and damp, with interesting ferns and other plants. We explored some of the drier quebradas, too, all the while getting better coordinated and perfecting our collecting techniques. To maximize our collecting efforts on the island, we were going to break into teams of two or three people, so it was essential that everyone knew what was expected.

After ten days and considerable experience, we organized an excursion to the higher elevations, ascending the slope just north of Las Casas and then up the ridge to about 1,100 meters. I had done this before in 1980 and 1984, because it is the quickest way to get up into the hills, but on those occasions we had only a day or two while the boat loaded up lobsters for return to the mainland. This time we planned to remain a full week at this elevation, somewhat of an ambitious effort. The idea was to establish a camp at that level, so we would not have to go up and down each day, in the process losing valuable time (and getting more exhausted). Mules were used to bring food and equipment to the campsite, and after two days more food would be brought up the same way, and garbage and processed plant materials would be loaded for the trip back to the settlement, where two crew members would be pressing the plants and putting them up to dry. This way, all would be preserved properly, and we would not have a huge backlog of materials to process when we returned to base camp.

One of our guides, Aldo Recabarrán, a fisherman who was hired to accompany the expedition primarily as field cook, had also been a good camp organizer, having selected a site at La Cuchara ("The Spoon") at about 1,100-meter elevation near the upper reaches of the southern branch of Quebrada Óvalo. He picked a location at the bottom of a dry small quebrada, so that with a tarp overhead and tied down tight on all corners, it formed a cozy enclosure underneath, keeping us dry even when raining hard. Aldo assured me that no matter how hard it rained, our camp would not be flooded, and I accepted this, although I still had some worries (as always!). Here we had our meals. The food he prepared was quite good, nothing special, but it was warm, and he always had coffee ready, too. For one dinner, he had shot a wild goat and prepared that along with some noodles, which was pretty good fare at 1,100 meters, sloshed down with some red wine. For sleeping, we set up small tents, on the higher regions adjacent the quebrada (just in case it did rain too much), and this arrangement worked quite well. There are no bathroom facilities in such a situation, so each person took care of his needs privately. Off the other side and also into a ravine we set up the larger work tent, so that we could press plants and take notes without being bothered by the flies during the day.

Campsite at 1,100 meters, La Cuchara, on Alejandro Selkirk Island

At the end of a day, it was pretty nice to come to our "dining room" and find Aldo with some hot meal, often noodles with beans, or rice with beans, that we washed down with red wine. In the field, even the most simple fare tastes wonderful. After dinner, we settled down to chat about the day and what we had collected, followed by more red wine, and finally

Frankie. Yes, even here I had brought enough batteries and a cassette player to listen to "Strangers in the Night," "Angel Eyes," or "Fly Me to the Moon," good tunes, good feelings. So there we were, in this really isolated island at 1,100 meters, under a tarp with Aldo, Tom, Eduardo, and Alejandro, sipping wine, chatting, and listening to Frankie. Can it ever get any better than that?

In contrast to the reverie of drinking wine and listening to Frank Sinatra in the evening, after dinner on 28 January a shocking experience occurred. We were fiddling with the shortwave radio in the cassette player, and we picked up a powerful station from Lima, Peru. Presumably due to our elevation and favorable atmospheric conditions on that evening, the radio signal successfully transmitted southward over the ocean some 2,500 kilometers, reaching us in the mountains of the very isolated Alejandro Selkirk Island. The signal faded in and out, but we were able to learn that the U.S. space shuttle, the *Challenger*, had disintegrated just 73 seconds after liftoff from the Kennedy Space Center at Cape Canaveral, Florida, killing all seven astronauts aboard. What? Was this true? This was hard for us to grasp, and we had no way of confirming the news, but as we listened more to the report, we slowly began to accept that such a tragedy had indeed occurred. The tone of the rest of the evening became subdued, all of us having difficulty digesting that an accident of this magnitude could have taken place. We headed to our tents earlier than usual that evening, each person reflecting on the implications of the disaster.

Having a shower at this campsite was obviously impossible, but one could bathe, but wow—this was an unforgettable experience. After three days, I suggested to Tom Lammers that we attempt a bath in a creek nearby that carried sufficient water to get soaped down a bit, enough to at least feel slightly more acceptable (I am not certain who smelled worse at this point, certainly not me!). This water was cold, I mean really, terribly cold. At more than 1,000-meter elevation and often fog-shrouded, no water at this level gets warm, or even slightly lukewarm. I slipped off my now very soiled shirt, pants, and underwear, and poked my foot into the water. Whew—this was going to be a quick in and out, I thought, and so inch by inch I was able to immerse to my waist, and then a quick soap-down and rinse and out. The only real satisfaction of this bath was getting out of it! In those few minutes it seemed that I was almost cold-preserving all my future generations, not in −196°C cold storage, but it felt darn close to that! The wonderful feeling of getting out of that freezing water, and then drying out a bit and putting clothes back on (the same ones, obviously), compensated slightly for the agony of the partial immersion, but we did feel much cleaner, which was the goal.

One challenging moment for me came when we had broken camp at La Cuchara after a week and were descending the trail to the settlement. Aldo had packed up the mule that had been sent up for us, and with all major equipment loaded and secured, he then headed down before us. We followed fifteen minutes later with another guide who had brought the mule up, and after about two hours, we eventually made our way down to the houses by the sea. Upon getting close, I could tell that something had happened, because about a dozen persons were standing in an open area in the center of the settlement. After dropping off my pack and additional bags, I walked over to determine what was going on. Don Uto had just broken up a fight between Bernardo and our cook, Aldo. Both of these men are huge and strong, and apparently it was a battle of the giants, according to bystanders. Don Uto commands huge respect in the islands, and he is also extremely strong and aggressive, when he wants to be, so he was able to break up the fight and send Aldo toward his own home in the settlement.

Later that afternoon as we were working on pressing the recent collections, Bernardo came to talk to me. He pulled me aside and said very directly that I would have to get rid of Aldo from our expedition, or he would have to leave. He could not continue with Aldo under the circumstances. I had no idea what really had happened to cause the fight, nor do I know even today, but my course of action was obvious—Bernardo was key to our expedition, and we could not do without him. I had to fire Aldo, simple as that. Now, in the meantime, I had learned that Aldo, some months earlier, had taken his son and a rifle (many folks in the islands have simple .22 rifles for goat or rabbit hunting, each registered with the carabineros on Robinson Crusoe Island) and gone up into Quebrada Casas in protest to his wife, who had done something he didn't like. He had remained there for several days, until finally being coaxed back to the settlement. I have no idea what this problem revolved around, but I was getting the message that Aldo had a difficult temperament and was not averse to using violence to solve problems. Yet, I had to fire him if the expedition were to continue successfully. This I resolved to do the following morning, figuring that if alcohol had anything to do with temper flareups, it would unlikely happen in the morning.

I slept a bit uneasily that evening, running through my mind the tactics I was going to use on Aldo as I told him he was removed from the expedition. I had always gotten along with him fine, so that was a good beginning. He would realize, of course, that my firing him would be a result of the fight with Bernardo, but I resolved not to mention that at all. After the usual breakfast of bread, butter, marmalade, and coffee, and not mentioning my agenda to anyone, I headed to Aldo's house at about 9:00; it was a beautiful, sunny day, but my situation in that moment felt a bit insecure. The worst-case scenario would be that Aldo might get mad and try to hit me in anger, but I figured that I could survive that circumstance—I doubted that he would try anything more serious. Still, his record was not particularly admirable in dealing with conflict situations. I arrived to his small, boarded house and knocked on the door. After about 30 seconds the door opened and there was Aldo, shirtless, pleasant enough I thought, and he invited me to come in. He motioned me to a small chair in the very small living room, and he took the other one, and I began to explain the reason for my visit. Basically I said that after the recent excursion to the high areas of the island, we no longer needed his services as a cook, as much of the rest of the expedition would be staged by daily trips from the settlement, and Juanita would continue cooking for us in the CONAF quarters. Now, this was partially true, but of course, he could have been a useful guide for any other part of these trips. I had calculated what I owed him to this point, and I added another week of wages, as a bonus, with the explanation that I knew it would take him time to get reoriented to fishing again (which he had done before), and this would help tide him over until he could get started once more. In actuality, that could be done in one day—I knew that and he knew that, but this provided an escape valve for both of us. I passed on the money to him, and he thanked me. I then stood up, shook his hand, and politely departed, thanking him for the chat. As I left his home, a wonderful sense of accomplishment washed over me, and also a sense of having jumped out of the mouth of the lion. I no doubt grossly exaggerated the dangers of dealing with Aldo, but being in conflicts with these very physical men in the islands was not something that I was used to doing. As a professor, I commanded a lot of respect in Latin America, often being called *profe* (short for *profesor*), but if tempers were to flare, that would rate as very little protection. Upon reaching the house of CONAF close to shore, I found Bernardo and informed him that Aldo was no longer with the expedition. He was satisfied, and we shook hands. Bernardo has been the most important CONAF guide during all our expeditions to the archipelago. He knows the islands, the trails, the logistics of how long it takes to get things done in different parts of the islands, the techniques needed for handling the mules, the boats, and even more importantly, the people. Because of his size, strength, and stable nature, he commands lots of respect among villagers, supported by his father-in-law, who also commanded similar respect. We owe very much to all the CONAF guides for helping us with our research, but we owe Bernardo even more than that.

Another excursion, and perhaps offering the most spectacular vistas, was the ascent to the highest point in the archipelago, Los Inocentes, at 1,315 meters. We all were excited about this new adventure, but it was physically challenging. We took the ascent up Quebrada Vacas, through the left fork, and up the southern slope. This is not very dangerous, really, because the slope is covered with grasses, mostly introduced, but it can be slippery. To me, with my vertigo, going up was not particularly fun, but I could do it, taking care not to slip. The guides, Bernardo, Cote (José González de Rodt), and Min (Guillermo López), did a super job of helping us with our heavy packs, plus carrying considerable weight themselves, including two tents that we would need to sleep in the hills. Tom Lammers, who was a bit overweight, decided to turn back after slipping too many times, and we agreed that he would continue to explore and collect into the quebradas that have access from the ocean. No problem.

Quebrada Vacas, Alejandro Selkirk Island

It took the entire day to get all persons and equipment to about 1,100 meters, where the tents were set up for the night. It was very foggy, and thank goodness the guides were able to orient successfully. When the fog rolls in, which can happen very quickly, the day can go from completely sunny and clear to totally closed over within only a few minutes. Once that happens, you can't see more than ten feet in any direction. This means that you cannot see any landmarks, and the only information available is the compass (this was before GPS instruments). This emphasized the importance of always having a good idea of where one was with reference to the campsite, such that if suddenly cut off from view, one could slowly work back to the tents. If close to a deep and dangerous ravine (quebrada), sometimes you could hear the mournful bleating of a goat, which meant—stop walking toward that sound. Go another direction. Working under the supervision of the guides was essential.

After we made the strenuous trip to near Los Inocentes, the fog rolled in and the temperature started to drop. For dinner, the guides had heated up some beans over a Sterno[2] can we brought that heated to 165°F, very light and convenient on the trail. We all crawled into one of the tents to get warm and have some dinner, and I remember how delicious it tasted—beans, some bread, red wine, simply excellent. With the Sterno going, it was also cozy and warm, and after the long, strenuous day, it felt great to be eating, drinking, and relaxing. After dinner, we headed to bed, persons split into two tents, and after some chatting, we fell asleep easily.

In the morning, at 6:00, it was beautiful outside, completely clear and not a cloud in sight. Our guides commented that this could change at any time, but for now we needed to get going and head up to the top of Los Inocentes before fog moved in later in the day. We split into two groups, I with one of the guides (Min) going to the top, and another group heading south into the tree fern forest of *Dicksonia externa* for collections. Although the night had been tolerable, I had not slept too well, in part because I was excited about the ascent to the top in the morning, and hence I was a bit physically tired. Min showed me the general way we intended to make the ascent, and we started up the route. It was not technically difficult at all, and with lots of shrubs around, my vertigo was not a problem. I would stop and make collections as we went, but this also took some time as I was making notes, but Min tolerated these delays. After a bit, I began to feel quite tired, the poor night catching up with me. Min encouraged me to keep going, as we were almost at the top, and so we went

Los Inocentes on Alejandro Selkirk Island. Left, early morning at campsite below the highest point (1,319 meters) on the island, with (left to right) Bernardo López, Leo Gaete, Cote, Jaime Sepúlveda, and Min; right, view from top of Los Inocentes.

even higher, hanging onto the shrubs and pulling our way upward. I would stop to rest, and Min again said, "Keep going—we have only a little way left to go." This pattern repeated itself many times, and finally, at last, we broke through the shrubs onto an open, rounded area, with a fog layer now accumulating below, but for us it was sunny and spectacular. At this very spot, and only here in the entire archipelago, grows a member of the sunflower family, *Abrotanella crassipes*,[3] a cushion plant quite common in this one locality. I made a good collection of it and other species in the area, made notes, and then we started back. Min emphasized the importance of getting back before the fog started to lift higher around us, and, having been in the pea-soup fog the day before, I believed what he was saying and agreed completely.

When we got back to the tent site, the other group came shortly thereafter, and we had a good discussion on how our two groups fared, followed by the usual lunch of sardines, bread, and some chocolate. We then broke camp and headed back down the ridges toward the ocean. It had been a quick journey to Los Inocentes, but for the short period of time, we collected well and now had a decent idea of the flora at this higher elevation. This vegetation has been called the "alpine" zone (see color insert, **10**) by Skottsberg,[4] because there are no trees, only small shrubs and tree ferns, and an herbaceous layer, with species that grow mainly at this higher elevation. As the day continued, we moved slowly downward, using the same route we used for the ascent, finally arriving to the ocean by about 5:00 in the afternoon. The walk back to the settlement at Las Casas is easy enough, over lots of lava rocks, but there is one small inconvenience, the Paso Malo ("Bad Passage"). This is a narrow passageway cut into the lava, such that the waves come crashing into this very small cove periodically. The trick was to estimate the rhythm of these waves, basically wait for a large one, and then after it receded, scamper across to the other side. This was not a life-threatening situation, but one can get pretty soaked, and we wanted to avoid this now as we were so close to base camp. With coaching by the guides we all passed without difficulty.

Fiestas on Alejandro Selkirk Island are a community affair. It is virtually impossible to have a party and not invite everyone, simply because everyone can hear and see what is going on. The settlement is nestled at the mouth of Quebrada Casas, and voices, especially music, can be heard everywhere. We always held our fiestas on Saturday night, so that our CONAF guides would not have to work on Sunday, providing time to get back to normal before Monday morning duties. Also, our own expedition members needed a break in the routine, and if they overdid it a little (in a few cases a lot!), they could take an additional day to get rested for the next excursion on Monday. Our first party on Alejandro Selkirk Island was a relaxed affair, held in front of the green CONAF building. Some goat meat had been procured, not too difficult as there are approximately 5,000 goats still on the island,[5] but in inaccessible parts of the higher elevations. The islanders have small rifles, a few bows and arrows, and dogs to help corral the goats. We got out our cassette player and put on popular tunes of the day from the group Pachuco y la Cubanacán. It was basically Colombian cumbia music, fast moving, very happy sounding, and it attracted attention. Bernardo brought parts of a grill, one-half of an oil drum, with a grate that set inside, affording a grilling surface for the fish and goat meat. Wood was gathered by the shore, mostly driftwood, we donated some newspapers and matches, and the fire started.

There are lots of different types of parties in life, some rigidly formal and others so informal as to not even be considered a party, just hanging out. Many Chilean fiestas start this way, slowly, getting ready together, and everyone participating. There is no rush, there is no starting time, and there will be no end, either. People will just drift away, bit by bit, as the evening runs its course. Right now, the fire was just starting, and it was time to get the wine open. We had brought lots of bottles of wine on this trip, many in three-quarter liter bottles for daily consumption, but for the party, we had brought six large ten-gallon *garrafas*—serious liquid. The corks were sealed with wax, and once the bottles were open, the wine was poured into plastic cups and given to whomever wanted a drink. The fiesta here was very successful, but our "physician," Jaime, drank too much, and in the days to follow, he was reminded that the "doctor" needs to be always in control, a useful career lesson as he transitioned from university student to professional physician. The fishermen and families would come to Jaime for medical help, and he was for that one month the doctor (*el médico*) in the settlement. They would come to our tent in the night asking for help from Jaime for whatever ailment they had. We did have a good medicine chest, and he was authorized to dispense to the fishermen. Since graduation, he has had a very successful medical career, including administrative positions in public health in Chile.

After we had been on Alejandro Selkirk Island for four weeks, and having accomplished most of our objectives, thoughts began to turn toward going back to the continent. On this island in the Pacific Ocean, there is more of a sense of real isolation, which heightens concern about getting off it and back to the continent. This sense does not exist on Robinson Crusoe Island because of the village of 900 people, the presence of the carabineros, the Navy post, and all the other facilities. I began to ask about the boat schedule one week in advance, in part because I wanted to plan our collecting work for the next days, but also because I was still a little concerned about the squeeze play that I encountered at the start of the trip. At this time it was only possible to make radio contact from Alejandro Selkirk Island to CONAF headquarters on Robinson Crusoe Island, and this connection was clearest in the morning, especially if the weather was mild. Often we would hear someone in the radio room in the green CONAF house, speaking very loudly, and then "*cambio*" ("over"), passing the conversation back to the other person. The information at this time was that the captain would, indeed, be coming to pick us up in about one week, but more information would be provided as soon as the schedule was fixed. To me, this meant we had three more excursions into the hills, and then to pack up and get ready for departure.

As the week progressed, I continued to press our radio operator for more detailed information from the shipping company about the pickup schedule. CONAF radio on Robinson Crusoe Island was able to talk to headquarters at Viña del Mar (adjacent Valparaíso), and they could easily contact the shipping office for information. Well, this in fact was finally done, and the word came back: "The captain has decided to postpone his visit to Alejandro Selkirk Island but would come for us in about four more weeks." You can imagine the sensation this news caused! I wasn't sure it was correct information, either, but due to our hassles at the beginning of the trip, I had little confidence in this report. Anything might be possible. And, we could not be certain that the ship wouldn't come as scheduled, either, which meant that we had to be ready no matter what, because after the ship arrives to the island, it might unload supplies and then want our materials to be loaded up quickly for return to Robinson Crusoe Island. So, I kept to our original schedule, and we now knocked down all equipment except for the sleeping tent, which we needed until the last minute. All plant collections were packaged, again with two layers of plastic and strong rope, and all this was stored in the large shed that had housed our food supply. Any leftover food or wine we donated to CONAF to be used as they saw fit.

The scheduled departure day came and we waited. It was a nice day, really, with calm sea, but we had nothing to do now, just to wait and hope that the earlier news was faulty. We could survive on the island for another month, because fish were abundant, but we all had schedules to keep back home, so getting off was a high priority. We waited, had a nice lunch prepared by Juanita, and then took a siesta afterward. Well, evening came and no boat. Obviously, the schedule had been altered somehow, but I had no idea what the changes might be. We began to get a little anxious, and so, a plan was hatched to do what we could to get the ship to come for us. We had learned that under certain favorable weather conditions, it was possible to make contact with the continent directly, and the radio operator was willing to help with this. Our strategy was to get a call to Mario asking for help, to the shipping office, and even to the U.S. Embassy in Santiago. Now, this last approach caused me some concern, because it could result in a significant expenditure of time and money, should the embassy decide they needed to mount forces to come and rescue three Americans from the island (Tom, Michael, and me). Later I learned they did receive the notice, but decided not to act on it, at least not until further confirmations. Silva also received notice at his vacation house, and he did call the shipping office, which may have been key. After two more days, we received notice that the ship was planning to come and retrieve us tomorrow, now three days later than originally scheduled.

Early the next morning, we were up at dawn, treated to a very beautiful ocean sunrise and then a good breakfast prepared by Juanita. Alejandro and Hugo, serving as old-fashioned lookouts, hiked up the trail on the ridge to the north of the settlement to be able to get a better view

Sunrise on Alejandro Selkirk Island, as we anxiously awaited arrival of the *Río Baker*

of the ocean. At about 8:15, our lookouts gave a shout, and shortly thereafter a ship appeared on the horizon, black at first, but when coming closer, whitish and tinged with rust—this was our *Río Baker*, our salvation! In hindsight, it would have been extremely unlikely that we would have been left another full month on the island, simply because the shipping company could not just leave us without planned provisions after agreeing to the schedule, but when in a vulnerable situation, it is not easy to see things in a balanced way.

We were certainly ready to get back on the ship, and, in fact, the loading worked well due to the very calm sea: there were no waves to interfere with the transit of the lobster boats. The problems didn't begin until we had said our goodbyes and everything was safely on board ship. As we sailed away from the island, the captain came to me and said that we could use the same bunks now for our travel to Robinson Crusoe Island, but due to having other passengers on the return trip from that island to the continent, we would have to move to steerage. Well, Mario would never have reserved spaces in steerage for us, and in fact, we had "first class" bunks for the initial trip to the islands. I again smelled a rat with this captain, and so I protested and insisted that we be treated fairly. I even intimated that my crew had just had a long, hard month on the island, and they needed some decent accommodations, or I could not tell what they might do. Well, saying this to the captain was definitely an error of judgment, because he immediately said, "Are you threatening me?" Whoops, not good, terribly ungood, a mistake. Making matters worse, Alejandro had decided to take a shower and had removed his trousers and was walking on the deck in his underpants, inappropriate at best. The captain stood up in front of me, very formally, and pronounced, "Sir, you must maintain discipline of your group and you must keep them in order, or upon arriving in the continent I will ask the authorities to deal with the situation." Well, in the time of Pinochet, the last thing anyone wanted to do was get mixed up with the military authorities, and I doubt the captain had any taste for this either, but his pronouncement had its desired effect. I gathered our group together below where the captain could see and hear, and I told them that we had to behave or there would be serious consequences. The captain then relaxed a bit and the situation was defused. The rest of the voyage to the continent was without incident, and the captain said nothing more about the bunks, which we just kept occupying until reaching Valparaíso. It was rumored later that several of the other passengers lodged complaints against the captain, and that he had a drinking problem, but this may or not have been true.

CHAPTER 9

An Improbable Encounter

I had made a solemn pledge to never again get involved with a woman who lived more than ten miles from my house. Period. There was a logical reason for this resolve, and once hearing it, I think that you will understand and agree. Here is the story.

I had divorced in 1981 after twelve years of marriage, and I was now on my own in a two-bedroom apartment in Columbus. It had not been easy, having lost the home and custody of two children, and finances were difficult due to the dissolution agreement. As the years continued, I developed a number of relationships with mostly divorced ladies, and it was a good time overall, exploring my own personality as well as theirs. I had a respectable job as professor of botany at Ohio State University, which meant tenure and financial stability, which proved to be a magnet for many women who had suffered through financial problems with irresponsible first mates. As evidence, at one point I placed an advertisement in the personal column of the *Columbus Monthly*, and I included the point about "tenured professor." More than forty responses rolled in. I could have been ugly as sin; that phrase alone probably accounted for half the letters. Anyway, it strained my energy (and finances) to meet some of these very nice women, but it was an interesting experience, and we had good times together, but nothing led to a lasting relationship.

During this period of personal exploration, I remembered a woman who was a graduate student in phytochemistry at the University of Texas after I had graduated with my Ph.D. in 1968. Barbara Timmermann was

Argentinian, also graduating with a Ph.D. from Texas, and taking a research position at the University of Arizona in Tucson in 1981. She became a U.S. citizen and now is a Distinguished Professor in the Department of Medicinal Chemistry at the University of Kansas. She was divorced, had two young boys, and was enjoying academic life in the American Southwest. We struck up a conversation through letters and then telephone calls, and things were moving forward quite well. We exchanged photos and other items, and to my way of thinking, at least, she was very attractive and extremely intelligent.

But letters and phone calls can only take a relationship so far, so I made a decision to visit her in Tucson. There were two impediments to this plan. First, it was in the middle of the fall semester at Ohio State, and I was teaching an upper-level course in plant systematics. I had a graduate teaching assistant, Bob Jansen, who was very good and particularly knowledgeable regarding cytological approaches in systematics, which formed part of his doctoral thesis and this was the topic I needed covering. He could take over for the two lectures during one week, which would allow me to skip town to Tucson. Officially, a trip of this nature during the semester should be requested from the department chair with approval obtained from the dean, but considering the situation, I decided to stay under the radar and hope for the best. The second problem was financial. I simply didn't have enough money to fly round-trip from Columbus to Tucson, but—hope springs eternal in matters of the heart—Greyhound Lines had just inaugurated a new bus ticket for travel anywhere in the United States during one week for only ninety-nine dollars. I could afford ninety-nine dollars, and so, after checking with Barbara about her schedule, I purchased one of these around-the-country-in-seven-days tickets and set the plan in motion.

I left on Sunday morning from Columbus. There were no stops except in bus terminals, enough time to buy a quick hamburger and jump back in the bus again. I had to change buses occasionally, but this was not a difficulty, just reading schedules and asking people in the terminals. Mostly these buses were not very full, being the long-haul express routes. I had brought along several cucumbers, thinking that one can eat the whole thing, rind and all, which leaves no mess. Well, logic again makes sense here, because there was an unexpected consequence: eating cucumbers produces intestinal gas, lots of gas (at least in me)! Being in a closed bus under this circumstance is not conducive to relaxation, especially when trying to sleep. Fortunately, there were bathrooms on these buses, but having to run back frequently was not convenient.

After riding on buses for forty-six hours, I finally made it to Tucson early Tuesday morning, and Barbara was waiting for me at the bus station. I was very taken by her lovely appearance, and of course I already knew quite a lot about her personality from letters and phone conversations. We hit it off immediately, and the days just flew by, doing lots of interesting things in the Tucson area, visiting historical sights, museums, exploring natural areas, and just getting to know each other better and her two young boys. I had to leave on Friday, so that I could get back to Columbus by Saturday, the end of the full week of the ticket. So we said goodbye in the Tucson bus terminal, having had a wonderful three days together, promising to continue with correspondence, until the next time we could meet, perhaps before Christmas.

When December rolled around, and after final exams at the university, I had managed to save enough money to make a flight to visit Barbara again in Tucson. This time I felt more like a normal professor who comes into town with proper transport, coming out of the jetway rather than descending from a bus after two unwashed days. This trip to Tucson before Christmas would also be a short one, but I was anxious to see her again and get further into the relationship. Well, as luck would have it (bad luck), after a day there in her home I came down with a terrible flu with coughing and high fever, such that a doctor came to the house and pronounced me very contagious, to be quarantined from Barbara and the boys as much as possible. This was not according to plan at all, and somehow this turn of fate caused the flowering relationship to wilt, slowly, but inexorably, until by the time I was able to travel back to Columbus, it was dead. I knew it, and she knew it. Strange, really, but sometimes these things just happen, and no amount of science can predict the outcomes of human emotional interactions (at least not yet).

Based on these hopes and events, I resolved never to develop a relationship with a woman who lived more than ten miles from my house in Columbus. There are quite a few women in the city, after all, which has a population of over 800,000, sufficient enough to find just one with whom I might enjoy a mutually profitable relationship. So, there were good reasons for this logical, sensible, and rational decision. Any man would have done the same. There have to be limits in life, and this was an exemplary case.

With these personal convictions, there I was aboard the ship *Río Baker* sailing in January of 1986 from Valparaíso to Robinson Crusoe Island, and then farther on to the more isolated Alejandro Selkirk Island. Considerable details of this trip already have been sketched out in chapter 8. We had just departed the port in Valparaíso, our boat now chugging westward into the dark, black Pacific Ocean. It was evening, and I was occupied with getting settled into the routine of the ship, and also preparing for bed in my little bunk, small, but cozy. I reflected once again on how weird it was to be going again to these fascinating islands. What new experiences awaited us? With these thoughts, and with a calm sea providing a small rocking motion, I fell asleep.

The next morning was a nice day, clear blue sky and calm blue ocean, with the ship cutting through the water in front and leaving white trails behind. There was nothing in particular that needed to be done that day, so the order of business first was breakfast. I descended into the small galley, which had tables and stools attached to the floor. Some other passengers were already there, and I sat with some of our fellow expedition members. The breakfast was simple—scrambled eggs, coffee, and toast with marmalade. Pretty good, actually, especially the coffee.

After finishing the food, but lingering over coffee, I had been noticing that there was a certain young woman with a female friend sitting at the adjacent table. So, to be neighborly (exactly and only that, as you remember my pledge), I turned to them and asked why they were sailing to the island. The lady of interest was in her mid-20s, dressed in a blue denim coat, rolled up a bit at the end of the sleeves, very casual attire, but entirely appropriate for a passenger on a supply boat heading for a remote island national park. A thin band kept her brown hair in place, which was cut at shoulder length. I chatted about what we were doing, and she seemed quite interested, asking questions of a more than casual nature, I thought. She was going to Robinson Crusoe Island for one month, the time between boat visits, to learn about the island and to write poetry. Her name was Patricia Antonieta Ghivarello Pesce, of obvious Italian extraction, but a Chilean living in Santiago. It was pleasant chatting with her, and because there was no work that needed to be done that day, it offered a nice pastime during breakfast.

Thankfully, the sea remained calm during the entire day, which provided some rest, a chance for a nap, and the opportunity to do some further thinking about how we intended to complete the work on Alejandro

Patricia Antonieta Ghivarello Pesce

Selkirk Island. In the afternoon, I ran into Patricia again (easy to do on such a small ship) on a small platform to the side of the captain's bridge. We chatted again, and it was a pleasant diversion from the now somewhat monotonous trip from continent to island. This was my fourth expedition to the islands, and the second one by boat, so the novelty had diminished somewhat, but I always got excited about making the trip, and all of them have been so different, each with new experiences. What might happen on this trip? No way to tell, but certainly there would be surprises, and in fact, I had no idea of what was to come—life-altering events!

On the second day we arrived to Bahía Cumberland on Robinson Crusoe Island, and all passengers were on deck for the event. The boat was too large to moor fast to the pier, so we would have to get off by small boat. There were few passengers, so within an hour all were ashore. Our equipment was staying on the ship, because this boat was then sailing on to the more distant island, which was our destination for this expedition. We were told that the boat would be sailing in the evening, about 8:00, which gave us the rest of the day to walk around the village of San Juan Bautista and make contact with CONAF, later having an early dinner, and then getting back to the ship prior to departure. After dealing with the park service and making final arrangements, I suggested to our group that we make a hike to the Hostería and have some refreshments. From their terrace there is a lovely view of the bay, and this would also give us some exercise after having been on the boat for two days. As we were heading down the main street, Alcalde Larraín, toward the path that went to the Hostería, we encountered Patricia and her acquaintence, Marina, from the boat. In the spirit of being friendly, we invited them to accompany us and they accepted, joining our group immediately.

Arrival of the *Río Baker* to Robinson Crusoe Island. With Patricia Ghivarello and Tod Stuessy.

The trail to the Hostería was very easy and mostly on level ground, but at places it dipped into a ravine filled with introduced *Eucalyptus* trees, now quite large, and infusing a terpentine-like odor to the air. The trail was dirt with occasional stones, all quite easy walking, nothing like we had experienced in the high mountains of this same island. There was one narrow passage on a slope above the ocean, which if someone slipped could lead to injury, but everyone scampered over this spot without difficulty. On the trail, Patricia and I somehow ended up chatting most of the time, and she occasionally corrected my Spanish. My Spanish is pretty good, I had taken three years of the language at DePauw University, where I completed my bachelor's degree back in 1965, and I had also by this time traveled and worked in Latin America for nearly twenty years. In Concepción, no one had ever corrected my Spanish, in part because as a guest, this might seem inappropriate, and certainly no Chilean student would dare to do so with a professor, but Patricia, well, she made the corrections in a helpful, not critical, fashion, but it was an intimacy that seemed both very nice on one hand but somewhat odd on the other.

The stroll to the Hostería was very pleasant; I chatted with Patricia, and Marina conversed with some of the boys in the group. Those who had not been on any Juan Fernández expedition previously, that is, half of the crew—Michael Doyle, Thomas Lammers, Leo Gaete, and Jaime Sepúlveda—were very excited to be here on this island, remembering per-

Relaxing on the terrace of La Hostería, Robinson Crusoe Island. Left to right: Eduardo Ruiz, Hugo Valdebenito, Patricia Ghivarello, Marina (acquaintance of Patricia), Leo Gaete, Thomas Lammers, and Jaime Sepúlveda.

haps when they had read *Robinson Crusoe* at some point in their youth. Upon arriving to the Hostería, the path came in above the lodge and then wound down into the adjacent quebrada and up again to the outside terrace, which was our destination. It was an attractive wooden deck, with a superb view of Bahía Cumberland. We ordered Cokes and Fanta but not snacks. From previous experience, I knew it was quite expensive in the Hostería, not by international tourist standards, but in comparison with regular island prices, it was beastly overpriced. I must admit that there have been times on these trips, when trying to sleep at night with fleas crawling over my legs, or when taking a brutally cold morning shower, that I wished I were staying at the Hostería with all modern conveniences. But, this was just out of the question for logistical and financial reasons. For our crews, it was only a place to go for a break, a change, or even to celebrate something very special.

After a pleasant respite at the Hostería, we strolled back to the village to see what was happening with our boat schedule. One learns to keep checking on these things as they can often change unexpectedly. Everything still seemed in order, so we strolled to the other end of town past quaint houses, Sr. Green's villa, a small store, a restaurant, the Jehovah's Witnesses church, the cabin made entirely of wine bottles and mortar (at

Cemetery on Robinson Crusoe Island, showing monument to the fallen seamen of the German cruiser SMS *Dresden*. With Roger Sanders. This monument, and others in the cemetery, were badly damaged in the tsunami of 27 February 2010.

least on the outside), and finally the cemetery. People vary on whether they like cemeteries, with persons with historical interests tending to like them more than others. But here at least was a monument to sailors of the SMS *Dresden*, the German ship that was destroyed here in Cumberland Bay by the English during WWI.[1] It was a colorful grave site with cement pillars forming three sides strung together with anchor chains, and in the center a plaque dedicated to the fallen crew members.

Our leisurely stroll through the village was followed by sitting down in a small restaurant near the pier for a quick dinner, mostly something to keep us going, and Coke or beer. We wanted something before returning to the boat, but since no one knew how rough or smooth the night transit might be, we didn't want to fill up too much before shipping out. There is no such thing as a printed menu in San Juan Bautista, because what you can order depends upon what they have that day. Generally, there will be some sort of fish, sometimes lobster, occasionally chicken, and rarely beef (if a steer had been slaughtered recently). Rice and pasta were usually always available, as these keep fine without refrigeration. At this time, there was electricity for only eight hours per day. The plate of the day was rice and fried fish, which seemed fine to all of us. What we wanted most was

Sr. Green (and his son) and his pensión (guesthouse), San Juan Bautista, Robinson Crusoe Island. This house was totally destroyed by the tsunami of 2010.

the beer, frankly, so the food was the appetizer for the beer rather than the other way around.

Dinner was fun and carefree since we still had nothing to do but wait for the boat. Patricia and her friend had left us because they had reservations in Sr. Green's villa, a very simple *pensión* on the main street, but which also included food, eliminating having to run to restaurants for meals. Our group finished up at about 7:00 and decided it would make good sense to drop down to the dock and check again on the departure schedule. There seems always to be someone around who has new information, so by asking this person and that person, we got the same answer that 8:00 was still the planned departure time. This being the case, when the small boat came again to the pier, we asked if we could all be taken back to the ship for the night, and this request was accepted. Once on board, we hung around on deck, chatting about research objectives, especially for the newcomers. Our ship didn't get off punctually at 8:00, but finally at 8:30 it slipped anchor, and we began backing out of Cumberland Bay. It was nearly dark now, and the stars had started to come out, giving us an indication of likely good sailing on this thirteen-hour voyage to the far island.

If you have been diligent and have read the previous chapter, you know what happened the next six weeks, so there is nothing to repeat because Patricia was not in my life. Oh, occasionally I wondered what that brief encounter meant, but as I had no contact information for her, she would be swallowed up by Santiago again and that would be that—fine, I mused, remembering my vow. We had lots of work to do in Alejandro Selkirk Island, plenty to keep busy and lots of more important things to be worrying about. If you read the previous chapter, then you also know all the difficulties of waiting for the *Río Baker* to come again and pick us up. But it did come, and we did get aboard again, and guess who was on that same ship—yes, the same lady with the blue denim coat, that same Spanish-correcting Patricia. She had decided to come to Alejandro Selkirk Island for a quick look, just for the experience; at least this has always been her story.

The *Río Baker* needed time to get supplies off the boat for the fishermen who resided on Alejandro Selkirk Island and to load up lobster, so we waited patiently on shore until given the signal for us to come aboard, along with our gear and valuable collections assembled during the six-week period. Getting into the lobster boat and to the main ship is always difficult because of the lack of a pier, but with lots of persons helping by pushing and steadying the boat, and with help from a wave coming to shore, it lifts up and drifts backward until the outboard motor powers it toward the main ship. Once the lobster boat is held fast to the side of the ship, it is a simple climb up the partially rusted steel ladder and onto the deck.

Later that afternoon the *Río Baker* finally lifted anchor and began the journey back to Robinson Crusoe Island. After the disagreement with the captain over bunk rights was resolved (see chapter 8), we settled in for a smooth trip back to the other island. The night closed in over the ocean with lots of brilliant stars. There is zero pollution here in the middle of the ocean, so the view is truly spectacular. As I was admiring these natural wonders standing along the ship's railing, a person appeared at my side with a greeting—it was Patricia. We exchanged pleasantries and then started explaining what we had been doing for the past six weeks—our crew in Alejandro Selkirk Island and she on Robinson Crusoe Island. I suggested that we might enjoy going topside to where the lifeboats were kept, and where there were two benches where we could relax and talk. This we did, taking care up the ladder as the ship rolled slowly from one side to the other. As we chatted, the moon began to appear from under the sea, slowly and steadily, a sliver, a half-plate, and finally a beautiful full moon illuminating the entire surface of the water, a shimmering, flickering, silvery surface. This was truly a magical setting, but then it got even better—the flying fish started jumping alongside the ship. One after the other took to the air and zoomed in front of us. As I had never before seen flying fish, this was an amazement to me, as they were also to Patricia. We chatted about many things, including the meaning of life and other philosophical issues, not idle chit-chat but extremely interesting topics of obvious mutual interest. So there I was, overwhelmed with the natural beauty of the moment, with the full moon, shimmering sea, flying fish, and an engaging discussion with an attractive woman. Well, you are probably nodding your head at this point, shaking your finger at me, and whispering under your breath that I was about to break my resolution. No, at this point that was not the case, but I have to admit that this Spanish-correcting Chilean did have some charms that caught my attention, especially in that particular romantic setting.

Next morning, after a very tranquil and easy passage overnight from Alejandro Selkirk Island, the *Río Baker* arrived and anchored once again in Bahía Cumberland at Robinson Crusoe Island. The task now was for unloading some materials sent from the other island, but more importantly now to load supplies, people, and lobsters for transport back to the continent. Because this was estimated to take a good bit of the day, we disembarked for part of the time and again strolled around the village. One of our tasks was to visit CONAF headquarters for a debriefing on our expedition, and once again to offer thanks for the excellent guide services. This was also a time to purchase souvenirs, and on the main street there was one small kiosk that sold items of wood, coral, shells, and the like. None of this was fine jewelry, but on the other hand, they were inexpensive, actually about right for pocketbooks of students and struggling professors. Toward the end of the day, the call came for all passengers to return to the ship in preparation

Tourist souvenir kiosk, San Juan Bautista, Robinson Crusoe Island

for sailing to the continent. All had been loaded up, and the ship was ready to leave the island. This information was relayed by persons in the village, and those involved with the ship line came specifically to inform us that it was time to leave. Once on board we checked again that our collections were stowed safely in the hold, and all seemed in order.

I do not remember interactions with Patricia during the next two days onboard ship; but surely there must have been some. Nothing has stuck in my mind in this regard until we came into the Valparaíso harbor in early afternoon after the two-day return trip. All passengers were on deck to witness our arrival at the pier in the large, busy port. Patricia was there next to me, and we were chatting about getting back to the continent. She mentioned that she was being met by her mother and brother and asked if I would like to meet them. Now, that was certainly a personal invitation, I thought, beyond the usual hellos and goodbyes. To meet members of one's family is a more intimate affair, and it did get me thinking about what was happening exactly. However, it had been a very nice time getting to know her, and it would be of interest to meet her mother and brother, so I happily accepted. Why not? Yes, I know what you are thinking—why in heavens did I accept to do this in view of my strong resolve not to get involved with any woman more than ten miles from my home? But, then again, there are different kinds of involvements, now, aren't there? There are casual involvements, which in no way would challenge my personal guideline, and this was my logic of the moment—a clear, direct, worry-free decision.

As the boat was getting tied up properly, Patricia began waving to two persons waiting on the dock, obviously her mother and brother. The passengers were allowed to disembark first with hand luggage, and later the hold was opened for off-loading of materials. I had, therefore, a few minutes to meet her family. It was a pleasant greeting, her mother about fifty-five years old, well dressed, and full of smiles. Her brother, Antonio, was in his early 20s, blond, and very pleasant. They seemed very friendly and receptive to meeting this strange gringo from the States. The greeting lasted only a few minutes, but then I asked Patricia the question that no man should ask if he has no interest in pursing a relationship: "Might I have your telephone number so that I could give you a call?" Now you are clucking negatively again, poking your finger at me, and shaking your head disapprovingly. Well, what can I say? The collecting trip had been long and hard, the return trip had allowed me to get to know Patricia, and the setting had been conducive to opening up our personalities, and they seemed quite compatible. So, having the chance to spend some time with her in Santiago before returning to Columbus seemed like a good opportunity, sort of a reward for the hard times in the remote island. I deserved this, really, and so I made the request. She smiled, got out a piece of paper from her handbag, and wrote the number for me, and then they turned around and left, she smiling in a fashion that suggested she hoped I would call.

We had lots of work now getting our materials from thc hold of the boat and onto a truck for a trip to Concepción. Some of the boys went into Valparaíso by taxi and looked for a company that would make the trip for us. In Chile, this is not such a difficulty, as there are always truck drivers of medium-size trucks, who are available for runs to any part of the country, for proper payment, of course. Within an hour, the boys had located a truck, and it was now waiting by the gate to the harbor dock. The yellow wooden crates were stacked by the side of the ship on the gray cement dock, along with other plastic-covered boxes and bags. With this completed, we began moving all items from the dock to the truck, keeping a list of the numbers of each item to be certain that nothing had been left in the hold of the ship. For this physical effort, we were in good shape after such hard fieldwork in the island, climbing all the time, so the work went quickly. I asked for a volunteer to go with me to Concepción with the truck for security, and Leo said yes. The ride would be slower, as we would be leaving in early evening and getting into Concepción the next afternoon. The challenge was then to get the materials unloaded and into the Botany Department at the university. All of this was agreed upon, and the truck departed, I having given the boys the necessary money to pay for their bus tickets to Concepción plus food expenses.

The reason for returning to Concepción was to inform colleagues there, especially Mario and Clodomiro, of our progress during the expedition. Also, there we always separated the collections, dividing them into full sets, with half of the sets being left in Concepción and the other half being shipped to Columbus. Leo and I arrived the next afternoon and asked to be dropped off at the university. Fortunately, other members of our crew were awaiting our arrival and helped unload all materials into the large laboratory room where we earlier had staged our preparations. The separation work always took several days, because everything had to be opened and collections ordered numerically, so that sets could be reconstructed and divided appropriately. The material for Ohio then had to be repacked for eventual shipment by air freight from Santiago to Columbus.

After several days of work and discussions, I was able to head to Santiago with the collections, now extremely valuable to us. This was again another night trip on Turbus, now simpler because I only had my baggage and the boxes of plant collections, not any of the equipment or other crew members. Everyone had already headed back toward their homes in the United States or Concepción. Upon arrival the next morning, my job was first to get into a modest hotel in the Las Condes region, a suburb of Santiago, and then after getting settled in, to take all boxed collections by taxi to the airport for dispatching to Columbus by air freight. A separate office handled these shipments, located away from the passenger terminal, which made access much easier. I asked the taxi driver to just wait for me. There were papers to be filled out, but eventually all documents were completed, and I was free to get back into the cab and return to the city. After that important duty was completed, I was free to spend two days in Santiago before I had to fly back to Columbus. This would give me some time for sightseeing and also calling Patricia. Why not? I had time to spend, it was Saturday, and she might be interested in going out for dinner.

So I telephoned her, and another lady answered (the maid, I learned later) who handed the receiver to Patricia. She agreed to have dinner together, at 8:00 that evening, and I suggested a Mexican restaurant that I had located from the telephone Yellow Pages (*Páginas Amarillas*) in my hotel. I chose a Mexican restaurant because I had previously completed more than ten collecting trips to Mexico, and I knew the food, well enough to preside over the dinner in a knowledgeable way, and for a first date, this would leave a good worldly impression. She had never been to a Mexican restaurant, which made it even better for me to showcase my experience.

I arrived punctually at the gate of her house in Las Condes, which turned out to be only one block from the home of General Augusto Pinochet, who had taken over the country in the political coup in 1973 from the elected communist president, Salvador Allende. Her house was a two-story affair, with a nice garden in front with bougainvillea, hydrangea, roses, and lots of other colorful flowers. I pushed the button on the outside of the gate, and after about thirty seconds, the front door of the house opened and out came Patricia, now wearing a charming summer dress. She walked delicately to the gate and, taking out a bronze key, used it to open the large bronze padlock holding the chain in place to keep the black iron gate firmly secure. As it creaked open, it was so nice to see her, here in her own environment, where she lived with her family. I was invited into the house and up the winding marble stairs to the second floor, there an entryway, a parquet wooden floor, a large clock with broad octagonal wood frame, and a large tank with tropical fish. We entered the living room, a typical sort of more formal room, with tall windows at one side and wooden bookcases at the other. There was the usual sofa and four chairs, all arranged around a low, glass-topped coffee table. More interesting, perhaps, was that through the living room was the dining room, and a gray-haired gentleman was sitting at the table reading a magazine of some type. Upon seeing me, he got up to introduce himself, Juan Ghivarello, her father. We chatted briefly, but he then went back to his magazine, and we got ready to head to the street and into a taxi. As all of this was done in Spanish, I don't think that Don Juan paid much interest to me one way or the other at this point. Here was some friend of Patricia's, and they were going out for dinner, fine, have a nice time. For me, I could not help but notice that her father was concentrating on what seemed to be a scientific journal at 8:15 on a Saturday night, indicating that this must be an academic family. Point registered and filed for future reference, but of course I had no particular goal in mind.

The evenings in Santiago in the austral summer are very pleasant. The heat of the day drops, such that a sweater or light jacket is in order. As we sped along the broad avenues that make for major throughways in Las Condes, which lies on the eastern side of Santiago toward the Andes Mountains, the air was soft and gentle, with a slight crispness as the day was closing and evening was just beginning. It was fun to be looking at and chatting with this attractive and intelligent lady in this faraway land, and speaking another language, which by itself has always been a satisfaction to me. In my studies at DePauw University in the 1960s, when first I began to think in Spanish, a whole new world opened up where ideas and thoughts get filtered through a different set of words, which reflect concepts that are culturally based. Learning a language for simple conversation, especially for tourism, is very important and completely valid. But, if you get further into a language, and it begins to live within you, taking over how you formulate thoughts—it gets really exciting. One learns that you cannot, and should not, just translate from one language to another.

Each provides a different way of how people view their world,[2] and it requires relaxing into that view in order to be successful with it. So there I was, in a black-and-yellow taxi cab, speeding along a well-manicured avenue in Las Condes, and by my side was an attractive Chilean girl and we were communicating in Spanish. Yes, very nice, very nice indeed.

The Mexican restaurant, El Mexicano, was not the most authentic, but it was appropriate for the occasion. It was designed for special nights out, not a simple taco stand on the curbside, which is so common in many parts of Latin America (but not in Chile). The dining room was ample and quiet. At 8:45 now it had two other couples, but this would increase as the hours progressed, this being Saturday night and the Chileans starting their evenings a bit later. Chile is not like Spain, where restaurants open at 10:00 at night and people start coming about 11:00. Dinner in Chile is usually between 8:00 and 10:00. At about 5:00, a tradition is to have *once*,[3] a coffee and some cookies or small sandwich, which is enough to tide one over until the later dinner hour. Each family, obviously, has its own routine.

The menus came, and I was able to make suitable suggestions to Patricia, although she had some idea of what some of the dishes were like. My perennial favorite is *enchiladas suizas*, rolled tortillas filled with melted cheese, and often with chicken, a good red salsa, and sometimes in an upscale restaurant such as this one, some guacamole and sour cream on top; on the side, always fried rice and frijoles (pinto beans). Chilean food is not at all like this, because the cuisine derives from Europe, especially from Spain, Italy, and Germany, and even Great Britain. We ordered margaritas to start, always appropriate to move the conversation forward, and we began to chat. There was still the need to exchange information about the trips to the islands, and we both had much to share. We each had been there for more than a month on different islands, and we had very different experiences. She, relaxing and writing poetry, and later I learned making friends with the locals and a Romanian expatriate who was also on Robinson Crusoe Island, but that is another story. I, on the other hand, was toiling away up and down the dangerous hills and cliffs of Alejandro Selkirk Island—a marked contrast—but at that moment in the Mexican restaurant, it really didn't matter. Nothing mattered except that as the margarita started taking effect, I was in Santiago, having a very agreeable dinner with someone who was becoming more fascinating by the moment. The mariachis arrived, and again, due to experience, I asked for a small serenade of the usual songs, “Bésame Mucho,” “Acércate Más,” and “Malagueña.” I enjoyed the music, which always has a strong emotional impact on me, and I think that she also enjoyed the attention. During the dinner, I mentioned that I was going to remain in the city for another day, which would mean we could visit a museum tomorrow before I had to leave for the airport and the night flight back to the States. Yes, that would be possible and welcome, she responded.

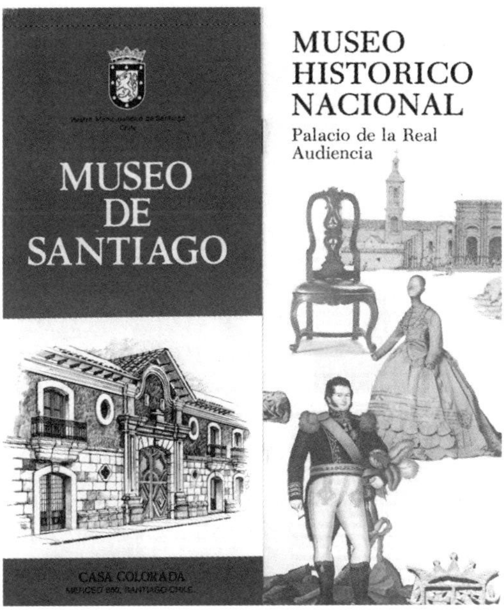

Museums I visited with Patricia on our first outing in Santiago, Chile, in 1986

The next day was again delightful, but I will not enter into details for it was just a couple having fun seeing things, visiting museums, and talking about them together, other than to mention that my interest in this lady increased. At this point I was still just thinking, well, this is a nice diversion, and soon I will be back in Columbus, and this will be just a sweet memory in passing. A gentle experience in life that gets compartmentalized in the brain with all the other similar memories, and not to be used very much more in the future. Remember, I had a resolve to maintain—only ten miles from my home.

I did arrive safely at home in Columbus, the collections arrived by air freight, and I settled into my usual personal and academic routine. After a week or so back in Ohio, a chatty letter came from Patricia indicating how much she enjoyed our time together, and that when I might return to Chile, perhaps we could see each other once more. Hmm, it felt good to get this letter, and I thought, in characteristic selfish male fashion, well, why not keep the fire going a little bit so that when in Santiago I would have someone with whom to have a good time? As Ricky Nelson had phrased it so well, a “Travelin’ Man,”[4] a lady in every port. I didn't have any ladies in other ports, nor did I have one in Santiago, but it seemed like a wise decision to keep the communication going, so I sent back an equally positive letter. Correspondence kept going, and I eventually sent her some pressed tree leaves with bright fall colors, which seemed appropriate from a botanist. In Chile, because of the absence of species of maple that turn bright red in the fall, the autumn colors there are much more subdued, lots of yellow hues due primarily to aspens and poplars. Patricia seemed to enjoy these letters, and little by little we were getting to know each other better.

In the meantime, funds had been allocated within the University of Concepción to invite international professors in different disciplines to give short courses of several weeks in support for the graduate (*postgrado*) program. I, along with my colleague, Dan Crawford, had been invited to the Department of Botany for one month of lectures, discussion, and research planning, including further work in the Robinson Crusoe Islands. Dan had joined the project in 1984 and by this time was a very important part of our research efforts. The academic schedule is always difficult to organize between Northern and Southern Hemispheres. They wanted us to come in November, but that was not possible due to our class schedule. We requested August, which they accepted. This was before classes opened at Ohio State on 1 October, but it meant leaving our enjoyable summer and spending it there in the winter of Concepción, where it doesn't snow, but it can be cold, rainy, and windy, not all that pleasant. I informed Patricia about this visit, and I asked if there might be any way that she could visit Concepción during this month so that we might have some additional time together. Of course, I could always visit her briefly in Santiago on a weekend, or in passing through the city coming or going, but these would be very short stints at best. She responded that after talking to her family (mainly her mother), she could come to Concepción for the entire month. She had a cousin who lived in the city who helped her find suitable living quarters in a house for female students near the university campus.

Years later I learned that her father, Don Juan, was not delighted by this idea, but it was her mother, Tessa, who supported the plan. Her mother never had a job outside the home, and I think she wished that she had been given more independent opportunities, and as a result, she supported her daughters (Patricia has a sister, Verónica) doing things that expanded their horizons. Years earlier Patricia, Verónica, and a female cousin had made a trip to Peru and Ecuador together, traveling to Machu Picchu and on to the Galápagos Islands. At that time, it was unusual for unaccompanied girls to make such a trip, but Tessa backed that plan. She was a wise, caring, and thoughtful lady who always made a very good impression on me. What I didn't appreciate at the time was how independently minded Patricia really was, a trait that was not fully revealed to me until years later!

Dan and I finally made it to Concepción at the beginning of August of 1986, and we were installed in the guesthouse (*casa de huespedes*) of the university. This was extremely convenient, first because it was paid for by the university but more importantly because it was on the university campus, nestled below the pine-covered hills that define the eastern edge of the city. The specific area is planted as a botanical garden, and the paths are open to the public during the day, all of which makes for a very pleasant setting. In winter, things are a bit less attractive, but for us, we could get from the guesthouse to the teaching rooms in ten minutes of easy strolling. This was ideal for our academic needs. Patricia was now located in a room in a large, wooden, gray house two blocks up from the university campus in a residential neighborhood. This was very convenient for us to meet frequently, and it was also a quiet neighborhood that had its own security person on duty day and night. The reason for this was not because of worries about robberies and other simple crimes; it was due to terrorism in the country.

Since the coup of 1973,[5] President Augusto Pinochet had taken strong control of all aspects of the government, suspending congress, manipulating the justice system, and censoring the press. In universities, some academic administrators were replaced by military personnel. This book is not the place to get into these many developments, but the fact remained that with the strict control that Pinochet had imposed, some Chileans were determined to fight back, and they did this through acts of sabotage, fires, and bombings of buildings, electric stations, roads, bridges—anything that might disrupt Pinochet and cause other citizens to join against the military occupation. Chile had experienced fifty years of democratic rule prior to Pinochet, and many educated persons were not happy with what was happening. In some ways, being in public parks, shopping districts, and even on university campuses was perhaps safer than it had ever been, due to the abundance of armed military personnel walking nearly everywhere. The bank entrances had two machine-gun–toting sentries on either side of the door, and inside were stationed two more for additional security.

For some citizens, this show of force was welcome, but in a university setting that seeks open and independent thinking, it was not appreciated, but complaints were not expressed openly for fear of losing one's job or worse.

As an example of the tensions of this period, an evening in Concepción is illustrative. Patricia, Dan, and I had decided to have a nice dinner in the restaurant Parque Ecuador (now converted into the Restaurante Colacho), appropriately enough nestled against the slope of one of the pine-covered hills at the back of the quite large public park of this same name, with its tennis courts, playground, and snack kiosks. Funds for this park had been given by the Ecuadorean government for reconstruction after damage done from the devastating earthquake of 1939, and it was now a prominent gathering place for couples and families. Dan and I always enjoyed this restaurant, having visited it frequently during our trips to Concepción. The specialty was fish, which Dan loved. I am more of a meat person, but the *corvina alcaparra* (white fish with capers) was really good, accompanied by *papas puré* (mashed potatoes) and asparagus, all washed down with a quality white wine. We began with the usual pisco sour,[6] a good start to the meal, with some fresh bread. For salad, we often ordered an ensalada chilena of tomatoes and blanched onions, sometimes with avocado added for better measure. Dessert was often a chocolate cake at that restaurant, or even chestnuts in honey, topped with cream, decadent but irresistible.

While we were enjoying the main course in the Parque Ecuador, suddenly there was a very loud noise coming from the center of the city, a strong boom, followed by a dimming of the lights, a flickering, and then complete outage. The waiter, apparently ready for these sorts of emergencies in the time of Pinochet, had a small flashlight and was armed with matches and a candle, which he lit and placed on our table, without missing a beat. It was romantic, really, but in an emergency context. After five minutes the lights of the restaurant came on, due to their having their own electrical generator. News finally came to us that a bomb had indeed gone off at the central electrical facility in Concepción, and this had knocked out part of the electricity of the town, including our area. We were able to enjoy our dinner all the same, but the plan afterward had been to attend a piano and soprano recital at the Goethe Institut very near the university, in fact, within walking distance from the restaurant, and this was one reason we selected it for that evening. We suspected that due to the bombing, there was little likelihood that the event would go on as planned, but then again, we knew that Germans are not easily deterred, so we decided after dinner to walk there and simply find out.

The streets of Concepción were totally dark as were most of the houses. Through windows facing the street one could see candles lighting the interiors, again, quite romantic but with realization that things were not quite right in the city. We arrived at the door of the Goethe Institut, and surprisingly, it was open. Furthermore, it had lights on, no doubt from an emergency generator. We entered this rather older building, with parquet floor and dark wooden doors, and immediately a well-dressed gentleman came to greet us. We inquired about the concert, and he assured us that it would go on as advertised, and he showed us into the salon, a small room with about one hundred chairs. We were the only people there. After about ten minutes, another fellow came into the room and took his seat. As the hour approached, a man and woman entered the stage, the man in formal attire taking a seat at the piano, and the lady setting up a music stand to her proper height and position and fussing with the sheet music. At exactly 8:00, the hour at which the concert was to start, the gentleman who had welcomed us at the entryway came upon the stage and introduced the artists, each one having had a long training and experience in numerous international events. This was a serious affair, and we were the sole audience, just the four of us. The concert began, very enthusiastic music, very well done. I confess to not remembering much about the pieces played and sung, but after each was over, we four applauded as loudly and enthusiastically as we could, trying to make up for the lack of the other ninety-six people who might have ordinarily been in the salon. It was essentially a private concert, and in fact, it would have been if the other gentleman had not showed up. Upon termination of the concert and our profusive thanks to the performers, who presumably got paid one way or the other, we exited for walking back to the university area, again down very dark streets. We had no fear, because on nearly every corner was another military man with his machine gun, so as long as we kept to our business, we felt quite secure.[7]

The month in Concepción with Patricia did one important thing—it cemented our relationship and made me question my resolve about relationships ten miles from home. Well, as any man (and woman, for that matter) knows, when matters of the heart start taking hold, logic tends to go out the window, flying in swooping curves up and over the trees and not easily returning to roost. I am a scientist, about as logical a human being as exists, all evidence-based, and usually calculating about most aspects of life, except those of the heart. I am, I confess, a romantic, who sees women as a center of warmth in a cold environment of life, a place to feel secure and to be open about emotions, feelings, attitudes, to share in an uncompromising way all the positive and negative aspects of life. This may coincide with why music has such a strong emotional hold over me. I confess to not relating in this way to most classical music, to some extent to Chopin's sonatas, some of Debussy's scores, but put me listening to Oscar Peterson and Art Tatum jazz piano, with their sweeps up and down and around, chords changing subtly, arpeggios brilliantly executed, timing perfect, and I disappear into a different world far from this present one.

Well, you can imagine what comes next. There was a visit from Patricia to Columbus to see if she might adapt to living in the States. This took place from September through November of 1987, and again, there had been discussions in her house on the advisability of letting this happen. Once again, Tessa supported Patricia in this adventure, and Don Juan was against it. As often happens in marriages, especially in matters of the heart, the women win out against the men, and the visit was set. I, being the perfect gentleman, sent Don Juan a formal letter promising, in effect, good behavior during her visit. My intentions were completely honorable, of course, but ahem . . . sometimes resolves have a way of fading a bit depending upon circumstances (as you may have noticed!). I rented an apartment for her within walking distance from my little house, which I had purchased the year before, so the structural aspects of the situation were well in place. Furthermore, the residential area where I lived was on the western edge of the university campus, which meant that Patricia could get to know it while I taught classes and carried out other duties.

We did a lot together during this time in Columbus. There were some downs along with the ups, but very few of the former. Patricia (I was now calling her Patty) seemed to be adjusting well enough to Columbus. She was pre-adapted for life in the United States because in Santiago she had attended Santiago College, a K–12 private school based on the American system and conducted in English. Her language skills were excellent, and she had also studied American history, giving her a cultural background surpassing that of most young Americans. Columbus is a very comfortable city in which to live. People are friendly, as in many parts of the Midwest, and this was very different from the custom in Chile, where people

1 San Juan Bautista, with the peak Damajuana on the right

2 Bizarre endemic species of *Dendroseris* on Robinson Crusoe Island. Left, *D. litoralis* (with a villager); right, *D. pinnata* (with Roger Sanders).

3 *Robinsonia* in the Robinson Crusoe Archipelago. Left, *R. gayana*; right, *R. evenia* (female flowering heads on the left side and male heads on the right).

4 Endemic tree ferns in the Robinson Crusoe Archipelago. Left, *Dicksonia externa* on Alejandro Selkirk Island; right, *Lomariocycas cycadifolia* on Robinson Crusoe Island.

5 Unusual endemic plants of the Robinson Crusoe Archipelago. Left, *Lactoris fernandeziana*; right, *Thyrsopteris elegans*.

6 *Gunnera* in the Robinson Crusoe Archipelago. Left, *G. masafuerae* on Alejandro Selkirk Island; right, cross-section of the base of the leaf stalk (petiole) of *G. peltata* on Robinson Crusoe Island, showing colonies of the cyanobacterium *Nostoc* embedded toward the perimeter.

7 Endemic flowering plants of the Robinson Crusoe Archipelago. Top left, *Pernettya (Gaultheria) rigida*; top right, *Ochagavia elegans*; bottom left, *Wahlenbergia fernandeziana*; bottom right, *Rhaphithamnus venustus*.

8 The endemic shrub *Chenopodium sanctae-clarae*. Left, habit; right, only known locality of this species on the top of the small islet Morro Spartan, alongside Santa Clara Island.

9 Alejandro Selkirk Island. Top, as seen from the distance aboard *Río Baker*; bottom, the settlement of Las Casas, occupied by fishermen eight months of the year (October through May).

10 "Alpine" vegetation zone (*Lophosoria quadripinnata* fern community) at the top of Alejandro Selkirk Island (Plano de la Mona)

11 From near the shore at Puerto Francés looking toward La Pascua

12 Panorama of the southern edge of Robinson Crusoe Island as seen from La Pascua; Santa Clara Island in distance

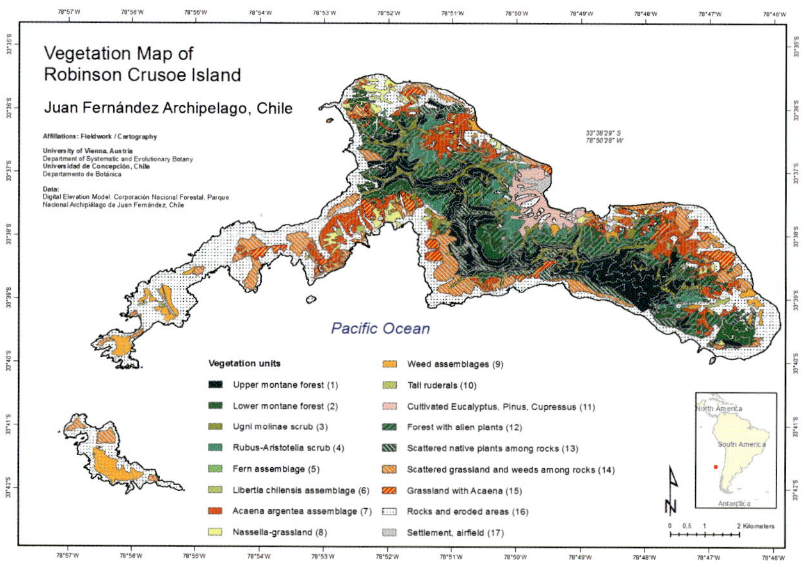

13 Vegetation map of Robinson Crusoe Island, published by Josef Greimler and colleagues in 2002

14 Southwestern side of Robinson Crusoe Island

15 Serious invasive plant species in the Robinson Crusoe Archipelago, especially on Robinson Crusoe Island. Top left, zarzamora (*Rubus ulmifolius*); top right, maqui (*Aristotelia chilensis*); bottom left, murtilla (*Ugni molinae*); bottom right, trun (*Acaena argentea*).

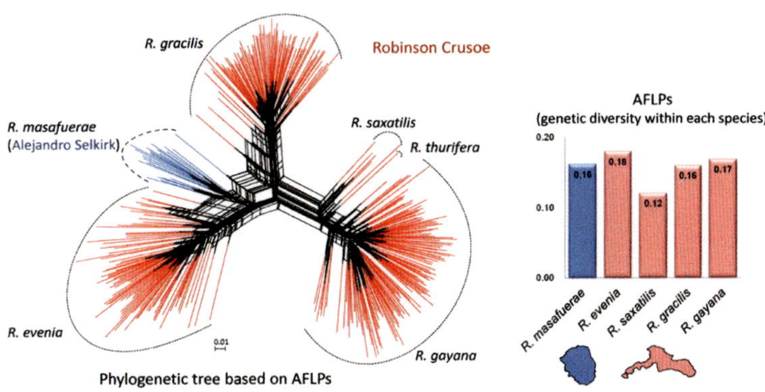

16 Phylogenetic tree and graph showing genetic diversity among species of *Robinsonia*, based on amplified fragment length polymorphisms (AFLPs). Lines represent individual plants.

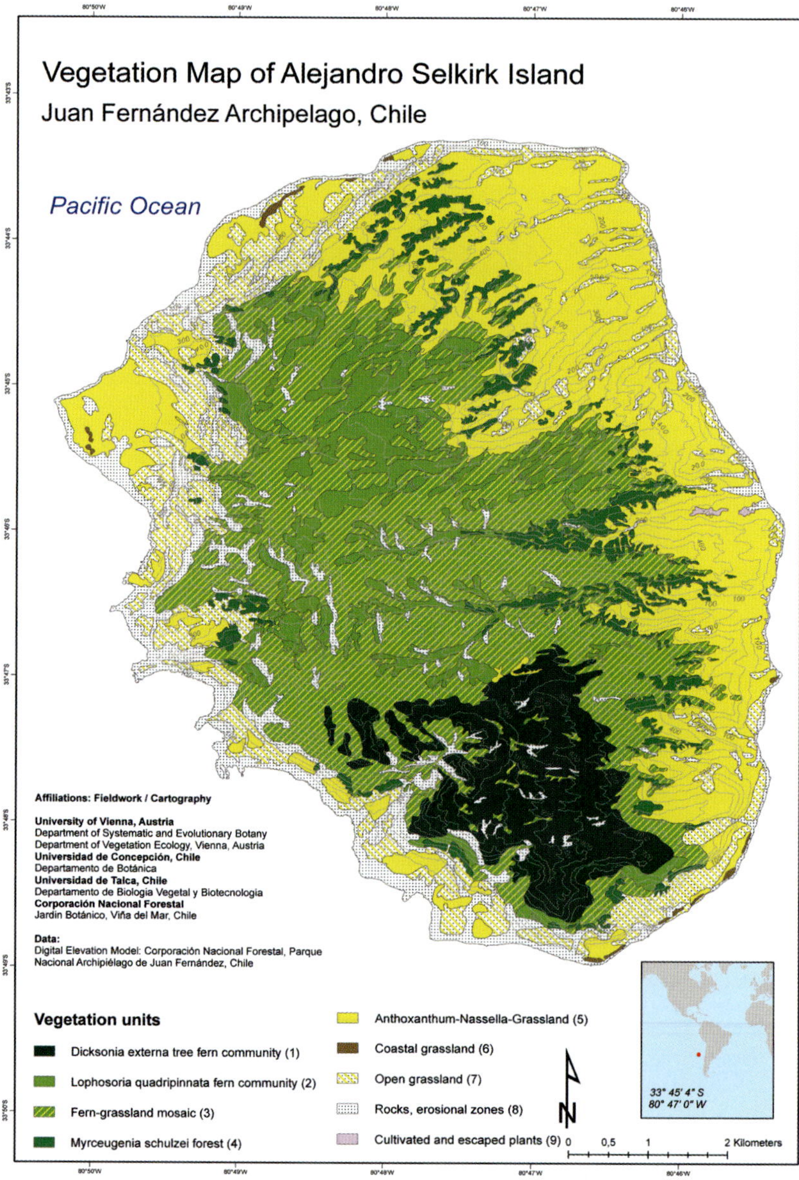

17 Vegetation map of Alejandro Selkirk Island, published by Josef Greimler and colleagues in 2013

are more reserved. In fact, sometimes when I walked in a public park in Chile and approached someone very near, I would offer a greeting and smile just to see what they would do. Most often, they either didn't look at me at all, or if our eyes met, they uttered not a word, probably thinking: "What does this fellow want?" On rare occasions I got a friendly response. In Columbus people are extremely friendly and expect to exchange greetings, and Patty liked this informality; this helped add social oil to the experience. Columbus is also an inexpensive city and easy to get around. Despite two million inhabitants in the city and suburbs, it is spread out with a good system of freeways, so mobility is simple. Access to the airport from our house is twenty to thirty minutes depending on time of day.

At the end of the visit, Patty headed back to Santiago, and I returned to my routine at the University. We agreed to continue the relationship, although here I was again, not this time 1,864 miles from my girlfriend, but instead 5,138 miles! As luck would have it, I was again asked to come to Concepción, Chile, this time for discussions on the *Flora of Chile* project, and this afforded me a chance to see her again in Santiago. At lunch, we discussed the relationship and she said she was willing to wait for me, but that there would be a limit to this. I would eventually have to come to some decision on what lies ahead. You see, this is the independent Patty putting it on the table, which in some ways I liked, but on the other hand, I am independent enough that I didn't want to be pushed. Still, I got the message.

Marriage of Patricia Ghivarello and Tod Stuessy, 16 January 1988, in Santiago (with Juan Ghivarello in background)

Yes, and you also get the message, I think. I soon afterward proposed to her, and we were married on 16 January of 1988 in Santiago. It was a beautiful wedding in the home of her aunt, which had a nice, covered terrace and broad garden behind, sufficient for a reception afterward with lots of food, drink, and dancing. After marriage, we both returned to Columbus and have lived here permanently, except for two years in Los Angeles, California, and sixteen years in Vienna, Austria, these experiences resulting in other interesting tales but not in context of this present book. We have now two boys, Daniel and Martin, ages thirty and twenty-eight, who live near London in Milton Keynes, location of the Enigma project at Bletchley Park headed by Alan Turing during WWII.[8] As always, there are so many unanticipated twists and turns in life that it is impossible to know where one will end up. In my case, meeting Mario Silva back in 1977 and developing the Robinson Crusoe research project has led to becoming married again, having two new children, and with family ties in Latin America. None of this could I possibly have imagined at the beginning of our work.

CHAPTER 10

The Call

Our expedition in February of 1999 to Robinson Crusoe Island began innocently enough. This was another round of observations on the vegetation of the island, this time with the expressed purpose of producing a modern vegetation map. In charge of the scientific work was Josef Greimler (Sepp) from the Department of Botany and Biodiversity Research at the University of Vienna. Sepp did his Ph.D. on vegetation analyses of high-mountain catchment areas of the Vienna headwaters as well as of a mountain national park in Austria,[1] which made him very proficient with methods of vegetation sampling and analysis. It had become obvious to us that we could not continue detailed evolutionary studies in the archipelago until we had modern comparative vegetation maps of both islands. Previous maps done by Federico Johow of Chile in 1896[2] and by the Swedish botanist Carl Skottsberg in 1953[3] were not very precise. There had been several other attempts since,[4] but none was what we needed for our own studies. This trip, with support from the National Geographic Society,[5] was to plot vegetation on Robinson Crusoe Island; a future trip would deal with Alejandro Selkirk Island (actually done in 2011 and published in 2013[6]; see chapter 11). The plan was for Sepp and I to fly on Friday, 29 January 1999, from Vienna to Chile, first to Santiago (arriving on Saturday morning) and then flying quickly on to Concepción, about 434 kilometers farther south. By this time, flight connections had become routine between the two cities. In Concepción we were to join with Patricio López-Sepúlveda (Pato), who was a Master's student in the Depart-

Participants on the expedition of 1999 to Robinson Crusoe Island for vegetation analyses. Left, Josef (Sepp) Greimler; right, Patricio (Pato) López-Sepúlveda.

ment of Botany at the University of Concepción. Pato was experienced in the islands, having accompanied us on three previous expeditions.

In Concepción, we organized the equipment for the expedition, which was immeasurably easier than with the larger previous trips. We were only three persons: myself, Sepp, and Pato, and our objectives this time were not to collect plant materials, but instead to make observations and photographic documentation of the vegetation. We didn't need to bring food, either, as we were going to stay in a tourist cabin and purchase all food in the village. Things had evolved by this time such that life had become much easier, especially for a group of only three scientists. We organized what we needed, socialized a bit with friends in the department at the university, and then boarded an all-night express bus (Turbus) to Santiago. These buses are very nice, and the seats recline quite far back, so that it is possible to get some good sleep. The steward also passes out a blanket under which to snuggle, and the lights are turned out. The idea is that the bus travels all night, making very few stops until arriving in the capital, Santiago, early the next morning. This is one of the cheapest and most convenient ways to travel to the capital city, ideal for a businessman to make a morning meeting. The cost was 9,000 pesos, which at the exchange rate of the time of 470 pesos/one dollar, came to a little less than twenty dollars each. I can sleep in these buses, but some folks find it difficult.

There is a rhythm of the coach, the hum of the motor, and all passengers respect the need to keep quiet and at least get some rest.

My in-laws, Teresa (Tessa) Pesce and Juan Ghivarello

We arrived punctually at the central bus terminal in Santiago at 7:00 in the morning, and after getting our bags retrieved below the bus, we walked to the taxi stand, climbed into a black-and-yellow cab, and drove on a small and quiet street in the suburb of Las Condes to the home of my in-laws, Juan and Teresa (Tessa) Ghivarello. She had prepared a breakfast of ham and cheese sandwiches, juice, and coffee. Tessa and Juan were always very interested in our expeditions, and they always offered help with lodging, a place to stage equipment, transport to another location, or whatever else was needed. On this occasion Tessa gave us a bag of lemons harvested from the tree in their garden, which proved to be perfect for making pisco sours on the island. We appreciated their kindness, and on this day Juan insisted on taking us to the airport for our morning trip.

Flights to Robinson Crusoe Island now all left from the old airport, Los Cerrillos, which was reserved for military and short commercial jaunts. Regular passenger traffic now used the modern Arturo Merino Benítez International Airport located fifteen kilometers northwest of Santiago. When we arrived at Los Cerrillos Airport, it was virtually deserted. We had no problem getting our bags and boxes up into the terminal (only one small building) and deposited by some chairs to wait for someone from our airline, Transportes Aéreos Isla Robinson Crusoe. We said goodbye to Don Juan,[7] thanking him profusely for all his kind help. He had visited the island many years before on the Chilean Navy training ship *Esmeralda*, and that is one reason he was particularly interested in our research. Furthermore, he was basically a scientist, having received his professional degree in chemical engineering. We waited about an hour, ambling around the small terminal, chatting, and hoping we were still given the green light for the flight. Eventually the representative from the airline showed up, Sra. María Sepúlveda, the wife of the pilot Santiago Figueroa, and we started to get things weighed and tickets purchased.

At that time, the way to pay for these passages to the island was in cash at the air terminal, so it was important to always have plenty on hand (Chilean pesos only). All costs on the islands also had to be paid in cash, all of which meant that we always were carrying lots of money on our person, usually in some type of money belt. Euros or dollars that were brought into Chile were changed into pesos before heading to the islands. This always took some time, as few houses (*cambios*) could handle the large amounts that we needed to change. On this trip we had exchanged Euros for pesos in Concepción at the Hotel Araucano (the largest hotel in the city), but often we did this in Santiago, depending upon our schedules. At first carrying so much money around made me very nervous, but after a while one gets used to it. But—it was always important to keep eyes open, avoid crowds or suspicious circumstances, and not store money in outside pockets, instead keeping it always hidden inside our clothing.

Years earlier after the military coup in Chile in 1973 and with Pinochet as the new president, it became illegal to buy or sell U.S. dollars. The government did not want Chileans to hoard dollars, but of course with fear of inflation, that is exactly what everyone started doing. Individuals were more than happy to buy our U.S. dollars and pay a good exchange rate. During this time, one could go to a certain street in Santiago, and there were men on the sidewalks asking if we wanted to sell dollars, which of course we did. The national police had to be aware of this operation, and perhaps they simply elected not to get involved, or they were paid off to stay away. On one occasion we followed a fellow and ducked into a gray, tall building and walked up a creaking wooden stairway to the second floor and into a room, where a man was sitting at an old wooden desk. We told him how many dollars we wanted to exchange, and he told us how many pesos that would buy, which was about twice the governmental rate. We laid our money on the table, about $3,000; he counted it and then pushed aside a picture of a typical Chilean landscape on the wall behind him, revealing a hole into which he handed the money. At this point, it would have been so easy to just tell us: "So long, goodbye, thanks for the contribution," perhaps emphasizing the point by placing a pistol on the table, but fortunately, this was a legitimate business operation (well, sort of, considering that the whole operation was technically illegal). Within five minutes, huge quantities of pesos came back through the hole, and they were counted out for us on the table—all correct. We divided this up among us, stashed the bills deep within our backpacks, and headed back down the stairs and out into the street. If ever a person could be a sitting duck for a robbery, it was us at that moment, but we never had a problem.

Back at Los Cerrillos Airport, it was time to pay for the tickets to the island, and Sra. Sepúlveda got us checked off the list and then collected the money. The tickets cost, round trip, 200,000 pesos and also 8,000 pesos for prepayment for the boat ride from the airstrip around to the village. All told, this was 208,000 pesos or approximately $443 each. Because of the number of extra boxes with supplies and equipment, we also had to pay sixty-two kilos overweight, or 80,000 additional pesos (about $170). The cost for the round-trip tickets had been going up over the years; in 1980 it had cost about $290 and in 2011 approximately $650. One of the reasons that few Chilean tourists traveled to the islands, a national park promoted within the country, was simply the high cost in contrast to taking a bus to some other location on the continent. Furthermore, there have been at least two plane crashes in recent decades that have received high attention from the press.[8] Most of the islanders come and go by boat because it is more economical, and we have also done this, especially on our first expeditions when there were virtually no supplies on either island (see chapter 6). On the negative side, the schedule is somewhat uncertain, with ships sailing only once or twice per month. One can take much more weight, which is the only way to bring large quantities of food, building materials, and domesticated animals.

This time in 1999 our trip to Robinson Crusoe Island was in the Nomad,[9] a larger two-propeller plane that could take nine people. On this day, there were six, the pilot, Sr. Figueroa, his son, the three of us (and our cargo), and a Chilean fellow who intended to write a book about the islands, their history, natural resources, and economy. Taxiing to the central runway was no problem, as we were the only plane active at the moment, which allowed us to accelerate down the very long runway and lift quickly off at exactly 10:45 in the morning. The trip begins from Santiago, nestled on the flanks of the Andes, down westward into the central valley and then over the much lower coastal range, passing just south of

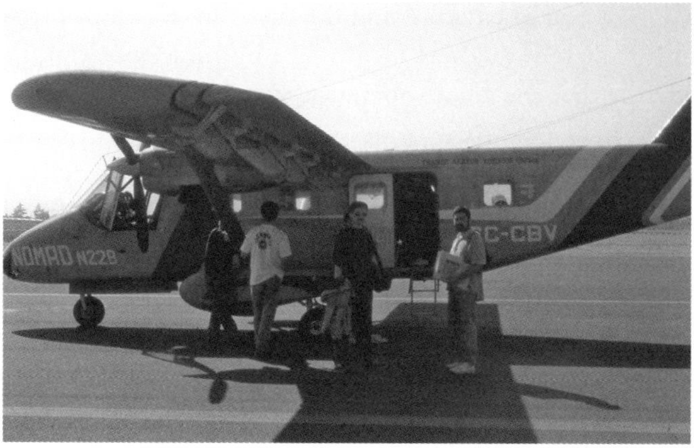

The Nomad airplane used on the Santiago–Robinson Crusoe Island route

Valparaíso, and then away from the coast and out into the open, blue sea. As one gains elevation, the waves can be seen below, silvery-capped flashes being reflected by the sun.

As always, and in part due to the length of the flight, a sack lunch was provided, with the old standbys of a ham and cheese sandwich, peanuts, chocolate-covered small cookies, peach juice, three small hard candies (Ambrosoli), and even coffee (with sugar already in the thermos). Again, as this plane had no bathroom, we drank sparingly. Due to the very loud noise of the engines, we were given blue headsets, but they were a bit uncomfortable, and we mostly didn't use them. As we neared the island after almost three hours, it was covered by clouds, so that we couldn't see below, but as we reached the western part of the island, and we dropped just below cloud cover, the landing strip became clear. We descended, slid back and forth on the air currents, and arrived on the airstrip at exactly 1:44 in the afternoon, definitely on schedule. The Nomad travels a little more slowly than the smaller planes, which can make the trip in two-and-a-half hours. The airstrip had been improved somewhat, as it was now compacted soil permeated with some sort of hardener—not asphalt, but more water resistant to provide safer takeoffs and landings when wet. The bulldozed, zig-zag jeep road down to Bahía Padre had also been widened for safety.

The airport at that time had only a small wooden building, sitting in a flat area with virtually no vegetation, only a few introduced weeds that became green during the spring rains. It is a stark landscape, attractive in

Western end of Robinson Crusoe Island. Top, dry landscape in the airport region, with Santa Clara Island in the distance; bottom, the pier at Bahía Padre.

its own right, but this is a zone of subsidence and erosion that has formed over the four million years of existence of the island. We waited patiently for the Nomad to be unloaded, moved our bags and boxes to the waiting dilapidated jeep, and then hung on for the short trip down to the ocean. We unloaded the contents of the jeep on the rocks by the sea, and it then returned to the airport to pick up the rest of our baggage. Bahía Padre is a nearly circular calm bay that serves well as a loading and unloading spot for the launch. In the 1980s this boat had to anchor offshore, and a small rowboat was used to ferry people and materials. Improvements had been made, and now a pier had been constructed of a chicken wire–type mold into which rocks had been filled and all topped by a boardwalk. This struck me as a very sensible and economical way to make a pier, using beach stones and obviating the need for shipment of concrete from the continent. The municipal launch was waiting to carry us on the one-and-a-half-hour trip along the northern side of the island to the village of San Juan Bautista, the only permanent settlement in the archipelago. Painted blue and white, the launch was tethered directly to the pier, and loading was relatively efficient. Everyone pitched in with the work, as no one could leave until everything was loaded aboard. We waited for the jeep to bring the other boxes of supplies destined for the village, which took about thirty minutes more, but there was no rush. It was, after all, summer in the Southern Hemisphere, and we had just come from cold winter in Vienna, Austria. Soaking up the sun felt extremely welcome.

The trip around the island can be very pleasant, but if the sea is angry, it can be difficult. Tall waves can splash water over the deck and soak all passengers, which is one reason why we always carry rain gear in our backpacks when flying into the island. Fortunately, this day was beautiful—the sky was clear with a very calm sea. As the launch left Bahía Padre we passed a colony of the Juan Fernández fur seal, a species that had been brought nearly to extinction at the beginning of the nineteenth century.[10] Under strict protection by CONAF, it has made a good comeback, and approximately 20,000 individuals now exist on this island and Alejandro Selkirk Island. As we left the protection of the bay, the sea became slightly more active, rolling and tossing slightly as we maneuvered along. Due to erosion over the millions of years, steep precipices line most of the perimeter of the island, and one can see the layers of lava from the initial volcanic eruptions, with occasional bands of intrusive lava and basalt. The ocean, because it is deep around the island that drops off quickly, was a dark lapiz lazuli blue, accompanied by white-tipped waves that struck the lava walls and threw spray up into the air. On this day the trip was superb, offering good photographic opportunities and generating thoughts about how the vegetation research would be conducted. As the trip progressed, we passed the starkly eroded valley, La Vaquería, and farther on another broad valley, Puerto Inglés, which has a cave near the shore that is believed to have been inhabited by Alexander Selkirk, who was the stimulus for the famous novel *Robinson Crusoe* by Daniel Defoe in 1719.

Northwestern side of Robinson Crusoe Island on trip in the municipal launch from Bahía Padre to San Juan Bautista

We rounded the bend with the small white-and-red–striped beacon light announcing our arrival to the village of San Juan Bautista that now spread out in front of us. In 1999 about 600 persons lived in the village,

The village of San Juan Bautista on Cumberland Bay (Bahía Cumberland), Robinson Crusoe Island, prior to the destructive tsunami of 2010

having risen to about 900 by 2020. One sees small houses, a gymnasium, a church, warehouses, and rows of lobster boats on the shore, as well as other boats tethered with buoys in the bay. The colors of paint are various, without uniformity, which provides a charm to the setting. Sadly, most of the structures near the water were destroyed in February of 2010 from an earthquake originating off the Chilean coast that produced a tsunami, which hours later smashed into the island and swept many buildings out to sea, killing eleven persons.

The launch maneuvered close to the cement pier and closed up tight with one of the stairways leading to the surface. After it was made fast, we grabbed our backpacks and some of our bags and climbed up, where Don Iván Leiva Silva, the chief of the island national park, was waiting for us. He had been informed by CONAF that we would be arriving for research work, but he was not happy because we had not communicated directly with him. He was right, but we had obtained a permit to work a specified time in the island from the CONAF office in Valparaíso, and we assumed he would be told about it. He had, in fact, received notification, but he also would have appreciated a call or letter from us directly. We agreed that this was an error on our part and apologized, which seemed to settle the matter. He is a very good park manager and always has been helpful to us in our research, so the last thing we wanted to do was create an unnecessary problem with him. Without the support of CONAF in the islands, no research gets done.

Don Iván had, however, investigated lodging for us, and the proposition was to stay in the house of Sra. Yola, who was the mother of one of the CONAF guides (Bernardo López), which we happily accepted. Prices had been rising in the village due to more ecotourism, and our accommodations summed to 20,000 pesos per day for the three of us (forty-five dollars). This certainly was modest by any international standard, and we were pleased to agree. We also were delighted to accept an offer from Juanita López, Bernardo's wife (he was during this period stationed on Alejandro Selkirk Island), to prepare us dinner, 9,000 pesos per day (twenty dollars), another reasonable expense. The idea was for us to prepare our own breakfast in the house, packing a lunch for the field, and then relaxing with a good dinner with Juanita in the evening at about 8:00. Getting our bags to the house took several trips of walking back and forth from the pier, but since our equipment was relatively light on this trip, the work went quickly.

Having settled in, and because it was now late afternoon, we decided to shop in the small general stores in the village and purchase food and beverages for the next few days. The village is very tiny, so one just walks around, storing purchases in empty backpacks. There were three small markets in the village at this time, each one having a different selection of items, such that it always took several stops to find what we wanted, which for breakfast was bread (*pan amasado*), marmalade, margarine, liver spread (*paté de ternura*), instant Nescafé, tea, sugar, condensed milk (my ultimate favorite!), and orange juice (in small cartons). For lunch we selected cans of sardines, tuna, and mussels, more bread, bars of chocolate, and *aji chileno* (hot sauce). For snacks we also picked out some cookies and chips. Drinks included Coca-Cola in large two-liter bottles, beer in cans, wine in bottles, and two bottles of pisco. That, mixed with some juice or even Coca-Cola (making a piscola), tastes pretty good after a hard day on the trail. We had to remember that there was no refrigeration in the house, so everything had to be non-perishable. Sometimes cheese is available, and this adds considerably to the breakfast. We used the refrigerator to store food items when opened, but this was only to protect from flies, not for cooling. After shopping we returned to our cabins, got better organized, and decided to celebrate arrival with a beer and some snacks. At 8:00 in the evening we walked up the long street, La Pólvora, to Juanita's home where I had visited on several previous occasions. As usual, she had prepared an excellent dinner, lobster salad followed by rice and fish. All her food was tasty, and there was always plenty of it. We purchased a bottle of wine from her to help wash down the dinner, which kept the mood festive.

The next four days were a combination of getting coordinated better with CONAF and the guides and doing initial vegetation analyses. Sepp was key to this expedition, as the principal objective was to produce a vegetation map of just this one island, plus also Santa Clara, which lies 1.6 kilometers off the southwest corner of Robinson Crusoe Island. Alejandro Selkirk Island would have to be done on another trip. Our discussions with Don Iván were very amicable, as usual, but he made a strong point that our permit did not allow *any* collecting, only documentation of the vegetation. To me this seemed an overreaction, and I was unsure as to what might have caused him to be so adamant on this point. As the years have progressed, CONAF has thankfully become more conservation oriented, which is necessary for proper maintenance of the national park. But sometimes this can be overdone. We had no permission to collect anything, not even introduced species, and with this in general we had no problem, but occasionally with vegetation surveys, one finds an uncertain species. This was not likely to happen with any native or endemic species, because Pato, Sepp, and I had a very good understanding of these island plants, but occasionally an introduced species is encountered (from who knows what country) that needs to be investigated further. Nonetheless, Don Iván made it very clear that we could only photograph these and make notes without taking specimens. We got the message and agreed to cooperate fully.

The plan was for us to carry out trips into the vegetation, one excursion each day as weather permitted. Previously we had always taken a day to work up collections after each field trip, but since this one had no collecting, we had more flexibility of keeping going to obtain as much field data as possible. The limiting factor was our stamina, but fortunately the three of us were in reasonably good physical shape. No one was overweight, Pato was young, Sepp was a former mountain climber and kept in shape by hiking throughout European mountains, and I went to the fitness gym every other day. None of us was close to the condition of the CONAF guides, but we were starting from a decent point. The plan was to make the first easy trip on Monday, without a guide, up the main trail to the Mirador de Selkirk (Portezuelo), so that we could do vegetation observations together, and Sepp could instruct us on methods. On Tuesday we visited Cordón Salsipuedes ("Get Out If You Can"), a rather foreboding title, but it was a ridge on the western side of the central valley, and it provided additional vegetational conditions. Furthermore, it was easy to access, being so close to the village. Wednesday was a more strenuous trek up peaks to 700 meters at the back of Puerto Inglés, and this time with a French worker, Philippe Danton, who was preparing a conservation-oriented book on the flora and fauna (finally published in 2020[11]), and Thursday it was to be in the lower regions of this same valley. I enjoyed seeing some of the CONAF guides again, especially Ramón Schiller and Oscar Chamorro, who had accompanied us on so many previous trips dating back to 1980.

Cordón Salsipuedes, situated along the northwestern side of San Juan Bautista

As I have mentioned before, without the help of these expert guides, we would have been unable to achieve very much in the islands. On any oceanic island it is nearly impossible to get completely lost, but on Robinson Crusoe Island there are no marked trails, except for the principal touristic one up to the Portezuelo (Mirador de Selkirk, or Selkirk's Lookout). The others are sort of paths, and only the guides know when to go this way or that. If alone, it is very easy to take a wrong turn that can then require trying to find the proper route, and much time is lost. Also, one can simply try to descend a valley from the forest to reach the sea, but some areas are full of the invasive blackberry, *Rubus ulmifolius*, or the maqui, *Aristotelia chilensis*, which now grows so densely that there is simply no passage. A guide has to go with you, and furthermore, some of the areas are dangerous, with precipitous ridges, and knowing how to trans-

Oscar Chamorro (left) and Ramón Schiller, two of the outstanding CONAF guides in the Robinson Crusoe Islands national park

verse these safely is based on experience. Even slopes can be very difficult, requiring zig-zagging along and watching footing so as not to slip and fall far down a considerable distance. For me with good Spanish skills, it was also fun to converse with the guides who have lived in the islands to learn what they think, and what sorts of tales they have about the islands and inhabitants. And likewise, they were curious about our lifestyle. Now that television had arrived in the island, which was not the case in 1980, everyone had an idea of what life in developed countries was like, so that mystery had dissipated, but still, chatting cross-culturally was always fascinating.

On Friday, 5 February, we arose at 7:00 for our usual breakfast of coffee (Nescafé, the water heating up quickly in our electric teakettle), followed by bread with margarine, marmalade, and liver paté, which had become our usual fare early in the morning. We had already prepared our backpacks the previous evening, so we only needed to add a can of sardines, bread, chocolate bar, and water bottle for lunch and head down to the dock. We were supposed to be there by 8:30 and we arrived by 8:35, but no matter because the two CONAF guides didn't arrive until 9:00. Along with them were the three French workers, Philippe Danton, plus his two colleagues, Michel Baffray and Emmanuel Breteau, all of whom were staying in the CONAF guesthouse up the hill by park headquarters. These investigators were interested in conservation of the flora, especially as documented by high-quality photographs. The trip today was being

The Zodiac rubber boat on the shore at Puerto Francés, Robinson Crusoe Island

conducted in the Zodiac, a sturdy rubber boat with outboard motor that had the advantage of coming nearly to shore, and it could easily be pulled up further for safekeeping during the day. All eight of us had lifejackets, and the day was a good one, very calm seas, which made the thirty-minute trip toward the eastern end of the island, Puerto Francés, possible without incident. The Zodiac was able to motor very close to land, and we could hop out onto the slippery rocks and struggle on shore before the next wave arrived. All having made it to shore safely, we tugged on the rope and pulled the Zodiac far up on land for safety during the rest of the day.

The plan was to divide into two groups, the French contingent with the two guides were to head westward into the valley and then up toward La Piña, a site next to upper montane forest, and from where impressive panoramas could be photographed. We were allowed to head up and farther east to La Pascua, but without a guide. This was permitted because I had done the trip several times, and I knew the way. The lower route up to La Pascua is easy enough, an open, very eroded area (see color insert, **11**). Few trees are left, except some dead trunks still in place on the lower slopes. A few scattered alien herbs (e.g., *Rumex acetosella*) occur interspersed with dry remnants of annual herbs and patches of trun (*Acaena argentea*). The soil is reddish-tan, giving the impression of a Martian landscape. This was an area that was originally forested to the ocean, as shown by historical documents from the seventeenth and eighteenth centuries. Little by little the forest was cleared for lumber to repair ships, for firewood, and for construction needs in the village. This part of the island is so deteriorated that it will likely never be reforested with native trees. It might be possible to import aggressive invasives, such as eucalyptus or pines, but encouraging these invaders creates another worry, should they disperse to other sectors and outcompete the native flora.

By lunchtime, we had arrived to the edge of the *Nothomyrcia* forest, which began the native vegetation, and which on the western side dropped off in steep precipices down to the sea, but the views were spectacular (see color insert, **12**). You could see all along the southern side of the island past the highest peak, El Yunque, at 915 meters, then farther toward the landing strip, and finally in the distance the small island, Santa Clara, which lies 1.6 kilometers south of the main island. Does one ever get tired of seeing beauty? I suppose you begin to take it for granted if it surrounds you all the time, becoming less significant in comparison to new stimuli, but seeing these incredible vistas never bored me. After a few photos, we selected a lunch spot under *Nothomyrcia* trees that had been sheared down (krummholz) by the wind that comes rushing up the slopes, and there were also ferns scattered about, a pleasant, grotto-type location to open a can of sardines in olive oil. Sometimes I brought a can of tuna or *machas*,[12] small, very tasty mussels, or cholgas, another larger mussel, but these were usually in water, not olive oil, and I love olive oil. I always ate all the oil, dribbling the remnants over the bread to get the rest from the can. Sometimes the oil dribbled down on hands, or got on pants, but this did not matter—we were sitting, after all, in the dirt. My companions ate similar fare, and we always carried an extra lunch for each of the guides, which they appreciated. They were to bring their own water bottle, but we supplied the food. A lunch in the field is really a treat, so that you can relax, recline, and let the muscles get a break from climbing. If a person is a real athlete, these excursions should be no difficulty, but although we were in pretty good shape, the exertion did take its toll. Lunch never lasted more than about thirty minutes, because if you rested too long, it was harder to get moving again.

After lunch, our job was to make relevés, inventories of the vegetation. One of the principal objectives of this trip was not only to observe and photograph the vegetation in selected different locations, but also to make quantitative determinations of species of plants in plots about 20 × 20 meters. Sepp was proficient in this technique, and he instructed us on how it was to be done. One counts the number of individuals of each species, listing each by name, and then also providing an estimate of cover for each one, e.g., 5 percent, 15 percent, etc. This gives an idea of what is there and also which species are dominant. The plots are selected when the vegetation changes, so that it is a deliberate approach to maximize data on different vegetation types. GPS locations are taken, but for security, we always carried a paper topographic map of the island, and we marked exactly on this map where each relevé had been made. This relevé technique was pioneered by the French vegetation scientist Braun-Blanquet back in the 1920s,[13] and it is still being used today, especially in Europe. I was extremely skeptical about this method at first, because it seemed too subjective. How do you decide where to make the plots? Are you sure you have included all the species? As the days rolled on, I became a convert to the method. You have to be experienced with this, and Sepp had that experience, including knowing not only the native and endemic species of the island but also European introduced plants, which he learned years ago from studies in Austria and Europe.

Day by day the information on the vegetation began to accumulate, and we could relate the patterns to what could be seen in the aerial photographs taken by the Chilean Air Force back in 1980 (a copy of which we had brought with us). Back in the laboratory in Vienna the real work would be to relate the data from the relevés with the aerial photos, ground observations, and our photos and sketches. Computer software is used to ordinate the data and develop a synthesis of vegetation zones.[14] Finally, these zones are plotted on a map that has to be corrected for distortions from the aerial photos. It is extremely complicated and requires lots of experience, and again, Sepp, along with help from colleagues in the Department of Vegetation Ecology of the University of Vienna, did an amazing job with this effort, based on more than one hundred relevés from this island alone. The map of Robinson Crusoe Island was finally published in the journal *Pacific Science* in 2002 (see color insert, **13**).

We continued in the afternoon working along the ridge westward and making relevés whenever the vegetation seemed different. From this vantage point, we had an excellent view of El Yunque ("The Anvil"), the tallest point on the island, rising to 915 meters. Although this is an insignificant hill on the world's surface in comparison to major mountains of the world in the Andes or Himalayas, such as Mt. Everest at 8,850 meters, El Yunque is challenging because it is extremely difficult to climb. The only possible access route is a basaltic ridge that, over time, has weathered to

El Yunque ("The Anvil"), the tallest peak on Robinson Crusoe Island. Top, as viewed in the distance from Cumberland Bay (the access ridge is on the left side); bottom, at the base of the access ridge.

an unstable condition. The attraction from a botanical perspective is that only at the very top of this peak, in a small, secluded, moist valley, is found the endemic genus *Yunquea* (with the sole species *Y. tenzii*), belonging to Asteraceae. The first naturalist/botanist to make the climb to the top of El Yunque was Otto Tenz, from Hamburg, Germany, who ascended with the islander Uberlindo Andauer in February of 1922. Tenz made a small collection of plants from the summit, and it was from the large leaves brought back that Carl Skottsberg described a new genus and species.[15] So far, no complete specimen has been collected, which makes the need for better material critical.

Over the centuries, several non-botanists have successfully climbed El Yunque. In 1795 the governor of Robinson Crusoe Island, Fernando Amador de Amaya, wanted an inventory of the top of this tall peak. Accordingly, he offered freedom and a cash bonus to any convict who might successfully make the climb. Ramón Negrete and Francisco Clavel responded, and on 19 November they made the first ascent, followed by another successful effort five days later, this time accompanied by two soldiers. In recent decades several islanders have made the climb, taken as a sign of macho status in the village. The botanist Philippe Danton in February of 1999 made the ascent with CONAF guides Ramón Schiller and his son Rodrigo, Alfonso Andauer, and Oscar Chamorro, and a detailed report of observations was published,[16] but again, no collections were brought back. *Yunquea tenzii* was located and photographed, but the plant lacked flowers or fruits. In 2015 two naturalists and CONAF guides also made the trip,[17] but no collections were brought back (just photographs of *Yunquea* showing only leaves).

These efforts stimulated me to attempt an ascent of El Yunque. You might think this sounds like insincerity or deception on my part, or even insanity, based on my vertigo explained in chapter 1. The idea was to employ a professional rock climber to make the ascent and then use a chair harness for me, blindfolded, to be winched to the top. Yes, this really was the plan, but a bit more realistic was plan B. If I lost my nerve as the expedition loomed close, a substitute botanist (probably a sacrificial graduate student!) might have taken on the task. It would be similar to the approach of Governor Amador with the convicts, something like: "If you want me to approve your Ph.D. thesis, you had better do this job for me." (Relax—just kidding, just kidding.) In any event, I submitted a proposal in 1983 to the National Geographic Society to obtain funding for the effort. The objectives were not only to complete the inventory of the summit of El Yunque, but also to collect more rock samples on the island for additional radiometric dating analyses, and to prepare a quantitative description of plant community structure, in collaboration with Assistant Professor Ralph Boerner from our Department of Botany at Ohio State.

The climber on the project was Timothy H. Jefferson, a Ph.D. in geology and paleobotany from Cambridge University, but more importantly he had technical climbing experience. But to my horror—about one month after the proposal was submitted, the shocking news came that Tim had been killed on 12 September 1983 in an avalanche in Peru while on another expedition with the Institute of Polar Studies from Ohio State University.[18] This blow on me was profound. He was only twenty-seven years old and just starting his academic career (and entire life). For a brief period I worked to find a substitute climber, and an excellent candidate did emerge, but somehow this tragedy cast an ominous shroud over the project, and I never resubmitted the proposal. Considering my vertigo and the weak structure of the volcanic rock on the passage to the summit of El Yunque, perhaps cancellation was the appropriate decision.

Having made many more relevés at La Pascua, it was now 4:30 in the afternoon and time to start heading down the quebradas back to the shore. We had been told to be back by 6:00, and this time the guides meant punctually, because it was the end of the day and time to go back to the village and their homes. Dropping down into the quebradas is usually not difficult, because under the *Nothomyrcia* trees, it is dark and few other species are growing, except in occasional clearings due to tree fall. In this area we did two more relevés. As one gets about halfway down toward the sea, the forest stops and the open, eroded gullies get more complex, often covered with the ankle-grabbing runners of trun (*Acaena argentea*) that interfere with easy walking. More photographs and notes documented this vegetation area. The afternoon was very pretty, with the sun starting to lower, casting longer shadows along the ridges and gullies. The temperature was just perfect, and we had done a good job with the work that had been planned that day. Step by step we continued down at a good, but safe pace, until finally reaching where the Zodiac had been beached in the morning. Ramón Schiller and the French workers were waiting for us, and we exchanged information on what had transpired during the day. As we did not speak French and they no English, we chatted in Spanish, which was better anyway, so that the guides could follow what we were discussing.

Getting into the Zodiac is much simpler than into a lobster boat, because it can be placed on the rocks very near the shore after a wave recedes. The trick was for everyone to very quickly get into the boat before the next wave came in, which lifted and moved it quickly twenty meters out to sea. The motor was revved up, and the trip back started. It was a very smooth return trip, with calm seas, a slight breeze, and the sun now disappearing behind the ridge above, casting a yellow glow over the edges of the few clouds hanging low around the island. Because everyone was tired, there was very little conversation, each person mulling over personal thoughts. All was normal and perfect, but then came the call.

As we were rounding the bend of Pangal, which is on the eastern side of the main bay (Bahía Cumberland), a radio call came to Ramón in the Zodiac from CONAF headquarters in the village. He reported to us: "There is an urgent message for Professor Stuessy." Well, this was very unexpected and disturbing, as I had never received a call from the mainland and never while in the field. I probed to learn what the message might be about, but Ramón gave no clues, only instructing me to go immediately upon arrival to the CONAF administration building. I don't know if he knew what was going on or not, but in any event, I was not given any further information. This, of course, made me very anxious, and it could only be bad news. If it had been good news, some medal or prize (in my dreams—highly unlikely, of course), then everyone would be informed, and we would be celebrating. No, this was something serious and negative. My wife and two young boys were on ski holiday in Davos, Switzerland, with our Swiss cousin Martha Geier, who visits the resort each winter. I could only fear that there had been some ski accident, perhaps an avalanche. Frequently when I was on an expedition, something bad happened, usually to the house, such as a flooded basement, or backed-up drain, or some financial confusion with the bank accounts, so these options flashed through my head. Or it could be that my stepmother suffered some mishap, she now being eighty years old, but she had seemed in excellent health. Perhaps some terrible accident had happened to my older children from my first marriage, Mary and Alan, who lived in the New York and Baltimore areas.

When we reached the dock, Sepp, Pato, and I walked directly and briskly up the hill to CONAF headquarters to find out what had happened. The park chief, Don Iván, and his wife, Sylvia, were waiting for me. While Sepp and Pato stayed outside, Iván asked me to come into his office, and he then closed the door for privacy, which heightened my anxiety. Yes, he said, the message was very urgent: "We received notification that your mother-in-law, Tessa, has had a cerebral stroke and is in critical condition." This was hard to believe, as she had prepared a wonderful breakfast for us only six days ago, and she seemed in excellent health, only sixty-nine years of age. My wife's sister, Verónica, had called the carabineros, who relayed the message to the station (*retén*) on the island, and they brought the news to CONAF, after which Don Iván called the boat. He had the phone number of the hospital in Santiago (Clínica Alemana) where Tessa had been taken, and Sylvia placed a call directly from CONAF to the mainland hospital, and I got Verónica on the phone. Back in the 1980s there was no direct telephone service between island and continent, but in 1999 things had changed markedly. Verónica had been trying to contact my wife, Patty, in Vienna, but to no avail as she was in Switzerland, so I was able to give her this new phone number. We agreed to talk further after I had checked into my situation.

There was no getting around this bad news—I had to return to the continent as quickly as possible, and Sepp and Pato would have to continue the vegetation research. This was not as terrible as it might have been, because Sepp was the vegetation expert and Pato had experience from other expeditions and knew lots of people in CONAF and in the village. It was agreed that I should return to Santiago immediately, which meant walking down to the airline office on La Pólvora and seeing what the flight schedules for the next several days might be. The young Argentinean couple who worked for all the three airlines, including Transportes Aéreos Robinson Crusoe, was very sympathetic and informed us that there would be space for me on the flight tomorrow (Saturday), but Sunday was completely booked. I had to leave tomorrow, then, and this would be the best alternative because Patty and the kids were probably going to get a flight from Zurich, Switzerland, which would arrive early Sunday morning in Santiago. As it turned out, bad weather hit the island on Sunday, and no flights left until the following Wednesday. This is always a difficulty in the island. At one point in past years, some Japanese tourists were forced to wait in the island two additional weeks due to very bad weather! The airline couple very kindly offered to let me call Davos, Switzerland, to talk to Patty, but there was no answer, and I tried several more times with the same result, finally giving up the effort.

From the airline office, we headed back to our house, took showers, and got ready for dinner at Juanita's home. Walking around the village takes some energy, especially heading up La Pólvora, which starts nearly at sea level and then goes right up the steep slope. We arrived by 9:00, which is a bit late for dinner (we normally ate at 8:00), but considering the circumstances it was acceptable. She had heard the news (it travels fast in the village) and understood there might be some delays. On a hunch that my mother (actually stepmother) in Lake Bluff, Illinois (a suburb of Chicago), might be able to contact Patty more easily, I called her from Juanita's home, and I was able to get through. She had not heard the news but agreed to get a hold of Patty in Davos. The instructions were to tell Patty to: (1) call her sister Verónica in Santiago; and (2) get on a flight with the kids from Zurich to Santiago as soon as possible, probably late Saturday afternoon or early evening. I learned later that mother was able to call Martha's daughter, Eva, in Zurich, who then called Martha in Davos and got the word to Patty. Mother knew Tessa from our wedding in 1988 and also from a meeting in Columbus when both women visited us after Daniel was born. Mother couldn't speak a word of Spanish and Tessa knew only some English, but they got along very well, taking care of the children, chatting (very limited), and enjoying shopping together.

Saturday morning was going to be an important day, so we were up at 7:00. Unsurprisingly, I hadn't slept too well, thinking about Tessa, Patty, and the kids, and the expedition. Both problems weighed on me, because once an expedition starts, there are lots of expectations for good science to come from it. Having to abort it is not a good feeling. And I was worried about Tessa. I always liked to talk to her, because she was empathetic and kind, someone with whom you could really communicate. My hope for the day was to fly back to the mainland, and that depended on the weather. From the door of our cabin, the weather looked favorable, but that can change very quickly on the island. Still, it was not raining, and the cloud cover did not look too dense. I started packing up before breakfast, leaving some items for Sepp and Pato and also giving them all of the expedition money that I had been carrying, 350,000 pesos for Sepp ($745) and 100,000 for Pato ($213), not a huge sum, but they already had two thirds of the money, and expenses in the island were not high. The return plane fares were already prepaid for all of us. In the island, we have always kept money and tickets tucked away in some part of our suitcases rather than taking these on the trail. This would not be possible in a hotel in Santiago, but here in the island there was virtually no crime. After all, should the money disappear, the carabineros would be notified and an investigation would ensue. There is no easy escape from the island, so the thief would have difficulty avoiding capture.

At 8:00, I headed over to the airline office, and fortunately it had just opened, which signaled that there would probably be flights today. Valeria, the agent, said that it did look as if there would be a flight today, and I was on the list; I had to be down at the dock by 11:00. Actually, there were going to be two airlines flying today, ours (Transportes Aéreos Isla Robinson Crusoe) and Lassa, as long as the weather would hold. Asking permission, I again tried to call Martha in Davos, but the line again registered busy; something just wasn't working right with the connection. I got through to my mother in the United States, and she provided some relief by informing me that Patty was traveling to Zurich for a flight out to Chile that evening. This was very good news—the information chain across continents had worked. Incredible, really. In the 1980s the only way to communicate with the continent was by radio from the Naval station, and from there an operator in Valparaíso had to call internationally, all very complicated and very uncertain. In 1999, the world was much more connected.

I returned to our cabin at 8:30 and had breakfast with Sepp and Pato, our usual bread with paté, margarine, marmalade, and coffee. We discussed the research to be conducted in my absence, and all things seemed clear enough. Because this was vegetational survey and analysis, there was no tedious hauling of bags during the day, the pressing of plants, and then getting them dried in the electric generating plant. All of this takes considerable effort, and with one man short it would have been a real burden for them.

At 9:30 I went to the CONAF headquarters to say goodbye to Don Iván and whatever guides might be on duty on Saturday. Don Iván and Sylvia were in the administrative building, and he indicated that CONAF would continue to support our vegetation research, which was very encouraging. Iván is a quality person, trained as a biologist and with a strong conservation background, and someone you can rely on. If he says he will do something, it means he will do it. No discussion. After these farewells, I walked up the path to the CONAF guesthouse where the French workers were staying, and I said goodbyes to Philippe and the photographers. Our relationship with Philippe Danton had always been cordial, but it also became competitive when he started a flora project in parallel with ours. They were having a picture exhibition in a building in the village that day at 5:00, so I excused myself for not being able to attend. Furthermore, I had invited them to dinner with us at Juanita's at 8:00, and I insisted that the invitation was still open to dine with Sepp and Pato after the exhibition. We settled these costs with Juanita on a weekly basis, so adding a few persons now and then was not a burden and it always helped keep good communications with other persons working in the island.

Having said goodbye, and realizing that it was now 10:30, I walked briskly back to the cabin, picked up my backpack and green duffle bag, and headed down to the dock, accompanied by Sepp and Pato. Soon we were on the pier, and while we waited, workers packed lobsters into boxes for the trip in the municipal launch around the island to Bahía Padre and from there to the landing strip. The lobsters[19] can live for considerable time out of water, and they do not have sharp pincers, which makes for easy packing in cardboard or Styrofoam boxes. There were a few folks on the pier to say goodbye, but not many because there were only two passengers leaving the island, I and one other, and I had already said my goodbyes. When everything was ready, I embraced Sepp and Pato and wished them good luck with the research. As they had met Tessa in Santiago before coming to the island, they sent their sincere best wishes for her recovery. It felt very odd to be leaving before the expedition was over; I was always the leader (or co-leader) and the last one to leave. Again, these mixed emotions flooded over me, but now the continent was making a stronger call for serious family obligations. We waved enthusiastically as the launch pushed away from the pier and started out of the bay.

I knew that Sepp and Pato would do an outstanding job, but I wanted to be a part of that experience. That is one of the main reasons I have always loved to come to the islands for research purposes. There is a clear goal, but there are also lots of adventures and fun. For a scientist, this is probably the closest one can come to a real guilt-free vacation. You are out of the typical academic daily routine with meetings, classes, conferences with students, service obligations, and manuscripts. The kind of vacation that is difficult to stomach is one that most people relish—sitting on the beach doing nothing. This drives me to distraction. I can handle this situation for about two hours each day, but that is all, and this includes reading a book during the two hours! In the islands even daily plans can change due to weather and progress of research. It is dynamic, challenging, and simply fun. It is not a normal vacation at all and terribly hard work, but for the scientist involved with the particular research questions, it is extremely seductive.

On the trip around to Bahía Padre, I chatted with the other passenger, an American named Chris Barrett, a young freelance writer, doing a travel story for *The New York Times*. He received his B.A. from Pomona College in southern California and was now living in Philadelphia. One of the reasons he was in the island was to check out the claim of buried treasure in Puerto Inglés, but this has so far not yet yielded any fruit, despite considerable investment from the wealthy American businessman, Bernard Keiser.[20] The excavation is controversial. Even though it is being done in a part of the island that has no native vegetation left, this is a national park and therefore ought to be treated with care, according to critics.[21] To me the likelihood of finding buried pirate treasure is so low that the precedent of disturbing the environment is too high a price to pay. This exposure, however, has attracted more tourists to the island, which helps the local economy. As we neared Bahía Padre, suddenly the motor gave off a loud thud, followed by a clankity-clank, and then it stopped. The launch drifted while the two men in charge worked to find the problem and provide a solution. I was anxious to get to shore and up the hill to await the airplane, but I presumed that a solution would be found somehow in short order, which in fact, did happen after about fifteen minutes. The jeep was already waiting for us by the bay. Little time was lost in hopping up onto the pier with our bags and heading to the jeep, followed by chugging up the zig-zag road to the airstrip, where we waited.

After about thirty minutes we heard a plane coming down. We watched it intently as it touched the airstrip and taxied to the small airport. Three passengers got out, plus the pilot, and then boxes of supplies for the village were unloaded. This took about fifteen minutes, and then it was our turn to load up and return to the continent. Chris Barrett was scheduled to leave on the Lassa plane that had not yet arrived, but a lady had been waiting at the airport, presumably having been dropped off earlier, so she and I boarded the Nomad followed by boxes of fish and lobster until, as usual, little room was left.

Everything inside, the pilot started the engines, and I waved goodbye to Chris who was waiting by the airport until his plane would arrive, presumably soon. It took about thirty seconds to taxi into position on the airstrip, and Sr. Figueroa then gunned the engines as a check, and at 1:58 p.m. he let the brake off and we shot down the runway. I never felt insecure about departures, but the runway is short, which means the plane actually drops down slightly off the island and then rises with updrafts and increasing speed that produces more lift. We were successfully off the island, heading upward into the sky, and turning toward the continent for the return trip. There was a super view of the island on the southern side as we climbed, and although I had seen this many times before, I never lost the thrill of such a bird's-eye panorama. I didn't attempt to chat with the lady, which in any event would have been nearly impossible due to the engine roar. On we went, up and away from the island, as my thoughts now turned to what I was going to find in Santiago. How bad a stroke did Tessa have? What was the prognosis for recovery? Did Patty and the boys get on the flight to Santiago? How was Don Juan, Tessa's husband, holding up under this situation?

After two hours and fifteen minutes, we approached the Chilean coastline, the blue of the ocean changing to the beige of the land, and this really increased my concern for family matters. The objective was to get to their family home as quickly as possible. I was unsure who might be there, and who was already at the hospital. I needed to get integrated into the family and be as supportive as possible. After landing in Santiago at 4:40 p.m. at Los Cerrillos Airport, because I had such little luggage, I grabbed my two bags, thanked the pilot, and asked how I could get a taxi to the city. He handed me off to an assistant who walked with me through the nearly deserted terminal and called on a phone requesting a van to come and pick me up, which made predetermined stops at different parts of the city. In fifteen minutes it arrived, and through discussion, he agreed to take me to the Plaza Italia, which got me to the edge of the suburb of Las Condes, and from there I transferred to a taxi to Patty's parents' home. This all worked without difficulty, and within an hour I was standing before their house ringing the bell to see who might come out. With luck, my brother-in-law Jorge Serani (the husband of Patty's sister, Verónica) came down the steps of the front entrance and greeted me at the black iron gate. After this was opened, creaking as it always did, we greeted each other very warmly, embracing, sharing a moment together when a family has suffered a setback. After walking up the stairs to the second floor, I dropped off my things and came back for a chat with Jorge. What exactly had happened? How did things stand? How is Tessa doing, and how long will she be in the hospital?

The story, somewhat unsettling, is the following. Tessa had left Santiago and was vacationing with her daughter Verónica and family at a beach house on the coast near Viña del Mar. In the morning after a normal breakfast, she began to feel a little tired and lay down on the bed to rest. In short order, she passed out. Verónica saw what had happened and called an ambulance, which took off on a two-hour frantic ride back to Santiago. By the time they arrived at the hospital, Tessa was in a coma, totally unresponsive, but breathing normally. The prognosis was not good—she had suffered a stroke, blood pouring from a broken artery at the base of the brain that was pressuring the spinal cord, producing paralysis of the entire body below the head. According to initial tests, her brain appeared to be functioning fine, but the rest of her body was no longer under her control. The autonomous nervous system was working satisfactorily, but she had little authority over her muscular movements. All of this was preliminary, but this ultimately did prove to be the case. It was a most tragic situation—clear brain function but a complete body paralysis—she was trapped in her own body.[22] This meant she could not eat by herself, and her speech was also affected. Initially the attending nurse was able to understand her, but we had difficulty, and it worsened over the years. Her condition was to last for seven years, she bedridden, fortunately kept at home with twenty-four-hour care, but sadly, slowly becoming more and more isolated from other people. All the family was devastated, as she had been its center.

When Patty arrived early on Sunday morning with the two boys, Jorge and I were at the airport to greet them, and we went immediately to the hospital to see her mother. We filled Patty in on everything as best we could, and she had many questions. It was wonderful to be together again, especially for me also to see the boys, Daniel and Martin, who at that point were six and four years old. At the hospital I was prepared that Patty might be extremely emotional and difficult, but this was not the case. She was very mentally organized and dealt with the incoming medical information in an outstanding way. We were all learning, and slowly the gravity of the situation became clearer. It was likely that her mother would never walk again and would be confined to her bed. Thus, a family drama played out that started with a call, a simple call, across 687 kilometers of ocean, which then changed research plans and altered the structure of a family. This was true life in action, simply that, nothing more and nothing less.

CHAPTER 11

Isolation Revisited

The roar of two propeller engines was pretty loud—it was very difficult to talk among the six passengers. One passenger, Josef Greimler (Sepp) was in the co-pilot seat, and five of us were in the back, two seats facing forward (Daniela Bacho and me) and two backward (Eduardo Ruiz and Marcelo Baeza), and one sidewise (Patricio Novoa). Also with us were boxes and bags of equipment and supplies. No quick exit from this plane! All people and cargo had been carefully weighed back in the terminal in Santiago to be certain the plane could handle the people and supplies—all was calculated to the kilo. This was always a time for some joking, as each person stepped up on the scale and everyone could see the weight. The cargo weight also affects the rate of burning up fuel, and there has to be enough for the round trip, as there is no refueling possible on the island.

The day was perfect—sunny, clear, without any clouds—just the right sort of day to fly two and a half hours at 240 kilometers/hour at 2,347-meter elevation westward from Santiago, Chile, crossing Valparaíso, and then out into the vast expanse of the Pacific Ocean. Despite my vertigo, I am never fearful of being in a plane or looking out the window. I feel protected inside the cabin with a seat belt holding me fast—if the plane goes down, well, that's the way it is, we go down. If the pilot can't do anything about it, there is no way I could possibly do any better. Also, a small propeller plane would glide down to the ocean and slide into the water. A small consolation, perhaps, but better than being on a jet liner that just plummets directly downward and smashes to pieces as it hits the sea.

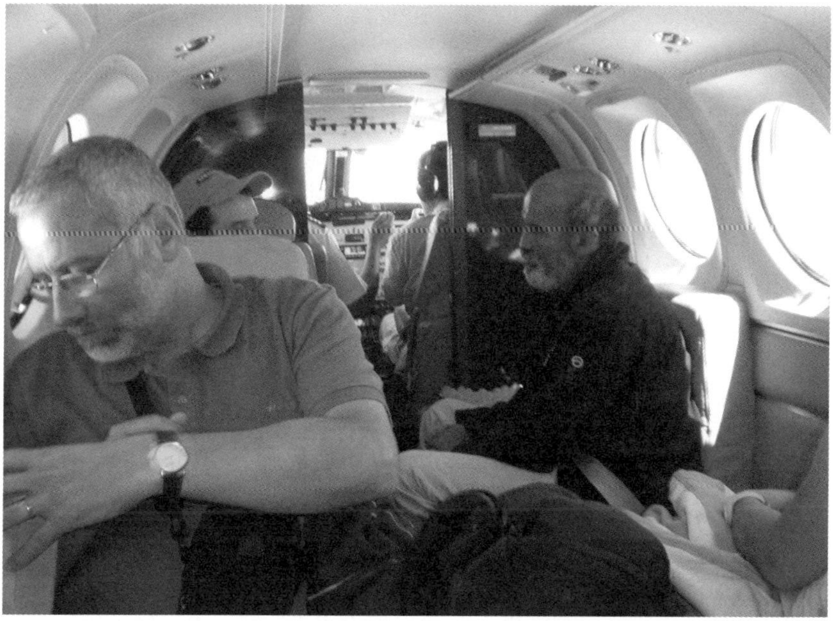

Airplane flight from Santiago to Robinson Crusoe Island, 2011 expedition. Left to right: Eduardo Ruiz, Marcelo Baeza, Josef Greimler (in co-pilot's seat), Patricio Novoa, and Daniela Bacho (arm only).

But these things were not on my mind that day, 29 January 2011. This was my twelfth expedition to the islands, and hence, nothing was completely new anymore. We were on a trip focusing on the far island this time, Alejandro Selkirk Island, a complex journey requiring first a flight to Robinson Crusoe Island, which has the only landing strip in the archipelago. We had to fly to this island, get further organized, and then go by municipal boat to the far island. All a bit complicated. But it was a very nice day, bright sun, blue sky, and deep blue water below. It always amazed me how the pilot could head out across the broad ocean without any landmarks and find the small island. I know, I know, there are instruments to guide him, but it would not be hard to miss the island, because it is only forty-eight square kilometers and 915 meters tall—not a huge land mass by any means. And if you miss the island by very much, even after finding it later, there would not be enough fuel for the plane to return to the mainland. Another plane with extra fuel would have to make the trip, or barrels would have to be shipped by boat from Valparaíso. But this never really concerned me either, more of a curiosity than a worry, because the pilot had done this hundreds of times, and he wanted to live to see another day as much as I did.

After a while the loud hum of the engine induced sleep, and some of us drifted off to a nap. We also had the usual snacks, courtesy of the airline, consisting of a ham and cheese sandwich on white bread, an *alfajor* (a type of cookie with dulce de leche inside and chocolate on the outside—terribly unhealthy and very tasty!), some Ambrosoli candies, and a small lunch-size carton of apple juice. No one wanted to tank up on food or drink too much, because the plane had no bathroom. Too much coffee in the morning before takeoff would result in a delicate situation, especially on this trip with mixed company. We were all men except for Daniela Bacho, who was a former nurse but now a medical student at San Sebastián University in Concepción. Since we were going to work in the far island, which has no medical attention or supplies of any kind, we had to take someone along with medical experience and a good set of supplies. She proved to be indispensable. Three other men were also going on the expedition, Alejandro Gatica and Patricio Peñailillo from the University of Talca (Chile), and Patricio López-Sepúlveda from Vienna (originally from Concepción), but they had taken another flight.

As we approached two hours and fifteen minutes in the air, Sepp in the co-pilot's seat turned back to us and shouted: "The island is up ahead—we are getting close!" The rest of us could not yet see it, but this put everyone on excited alert to catch a glimpse from our side oval windows. Minutes later the island came into view (see color insert, **14**) as the plane flew westward toward the tip where the landing strip lies. The island is striking, with reddish-brown lava precipices, and brilliant white wave patterns ringing the shore like a decorative hemline. The plane roared along, and now it started to dip right and left because of the updrafts around the island. The plane slowly descended, getting closer to the level of the backside of the island, and suddenly it banked sharply left and dropped even more, revealing a landing strip below. Down we went, faster now, swerving left and right until a sharp bump and we were on the island, similar, I would imagine, to the experience of having landed on an aircraft carrier in the middle of the ocean. We all applauded, delighted to be down safely, the first leg of our journey. There have been infrequent accidents in the past, the most recent being an attempted landing in 2011 of an airplane from the Chilean Air Force, which got caught in unpredictable winds and plunged into the ocean, killing all aboard. But today we had landed safely. The plane

Arrival in 2011 at landing strip on Robinson Crusoe Island. The plastic tubs hold silica gel for drying plant tissues for DNA investigations.

taxied off the side to the small building, which is the "airport," and we began unloading. The airport has over the years greatly improved. In 1980, the landing strip was only bulldozed earth, then came a packed, oiled surface, followed by asphalt; the building now has a bathroom, whereas previously the only personal relief was obtained by hiking over the hill behind the building.

The expedition of 2011 was very different from that of 1986, in part because I was now sixty-seven years old rather than forty-three. Plant collecting in continental areas along highways or field stations can be done comfortably by persons of all ages. But here in these islands, especially on Alejandro Selkirk Island, collecting is an exhausting and challenging physical business. For the earlier trips I had always done an exercise regime three months prior to leaving, which involved weights, jogging, and leg exercises. If weather permitted, I even ran the steps of Ohio Stadium, up and down around the horseshoe, providing a super exercise mimicking what one encounters in the islands. Now in 2011 I had not done such a thorough preparation. I still went to the fitness center in the hotel adjacent to the Institute of Botany in Vienna, where I fast-walked the treadmill and used the step machine, plus calisthenics and weights, but it was much less than before. Also, I was just older, in good health, and ready for the trip, but it was clear that my role on this expedition was going to be coordination and leadership from below, while the younger crew headed to the hills for the difficult hiking and collecting. I didn't particularly like placing myself in this situation, which I regarded as a lesser role, but it worked out fine. Someone had to do the planning, processing of specimens, and keeping track of data, and that I could easily do. I went on some of the excursions, but not the ones to 1,000 meters or more, as I had done enthusiastically in 1986. In some ways, going on this trip at all might be viewed as a little dangerous for a person of my age; Dan Crawford, one year older than me, declined to go, with my complete agreement, because he had already had two heart attacks, each time in Columbus where he quickly got to the hospital and was stented. If this happened on Alejandro Selkirk Island, it would be the end. Dan hated not going, and I wanted him to go, because it always made for a more enjoyable trip, but we both knew not this time—it was too isolated and risky.

Getting to the village of San Juan Bautista from the airstrip was the usual routine of riding the jeep down the winding road (now also of asphalt) to the waiting municipal launch, followed by a one-and-a-half-hour trip to the other end of the island. Our work in the village consisted of making proper contact with the CONAF authorities and guides, checking on our supplies that were earlier sent by boat from Valparaíso, and getting everything loaded on the municipal boat the *Tío Lalo* that we had contracted for the trip to Alejandro Selkirk Island. But the voyage was to

Municipal boat for trip from Robinson Crusoe Island to Alejandro Selkirk Island

leave in the morning, and CONAF had kindly arranged for us to sleep in houses in the village for this one night. No one slept particularly soundly that evening because of the excitement of knowing we were shipping out to the far island in the morning.

The voyage from Robinson Crusoe Island to Alejandro Selkirk Island on the municipal boat left Cumberland Bay early the next morning. It seemed a very good day to be making the trip—the sea was calm, the sun was bright, and so far, at least, there had been no hitches or problems of any kind. The goods that had been shipped to us from Valparaíso had been checked thoroughly, and everything was in good condition; nothing was missing. Pato and Marcelo had done outstanding jobs of purchasing needed foodstuffs and camping equipment for the expedition and arranging for these to be sent beforehand to Robinson Crusoe Island. With everyone and everything safely on board, we separated from the pier and slowly headed out of the bay to open water. The route stays close to shore along the northern side of Robinson Crusoe Island, where we retraced our route back toward Bahía Padre near the airstrip, and then continued into the open ocean toward the far island. All went well as we chugged consistently westward through the sea, the boat rocking gently from side to side. This was not an exciting part of the expedition, as it was largely monotonous. However, the small boat did have a correspondingly small, but adequate, bathroom, and there was a diminutive galley, in which a decent pasta lunch was prepared. We rested, chatted, slept, and let the time pass.

In late afternoon we arrived within ten miles of Alejandro Selkirk Island, now arising out of the ocean like a loaf of bread that had been deeply cut horizontally several times. As we traveled closer, suddenly smoke began spewing from the small engine room below. The cook grabbed the fire extinguisher and rushed down the few steps to smother whatever was burning, and this action was followed by the motor stopping, leaving us adrift. Should the fire not be successfully quenched, I contemplated with some bravado the possibility of having to abandon ship and swim to the island, however unrealistic such an attempt would have been. Further, we were carrying a drum of gasoline, which if the boat had really caught fire, would have exploded just like in an action movie! All passengers were concerned, but after ten minutes the cook reported that the problem seemed minor and repairable, which gave us some confidence. In fact, this did prove to be the case, and in thirty minutes we were again heading directly toward the island.

Boxes of food and supplies stacked safely on dry ground after arrival at the settlement of Las Casas on Alejandro Selkirk Island. Left to right: Patricio López-Sepúlveda, Marcelo Baeza, Eduardo Ruiz, Alejandro Gatica, Josef Greimler, Patricio Novoa, and Patricio Peñailillo.

Our arrival to the shore of the settlement of Las Casas on Alejandro Selkirk Island this time was uneventful. The municipal boat was much smaller than the *Río Baker* upon which we had arrived in 1986, which made unloading into the lobster boats a simpler exercise. Several trips were needed to bring all boxes and baggage to land, following exactly the same system as in 1986. The only change was that now a mechanized winch was used to pull the lobster boats up over the horizontal tree trunks, which served as protection from the lava rocks. With all our expedition participants working together, we were able in quick order to heft equipment and boxes of food and drink out of the boats and stash them on dry land. From this staging location, personal items and scientific supplies were moved into our new house, and the food and beverages were placed in a vacant shed that could be locked; I kept the key.

A number of changes had taken place in the settlement since our visit in 1986, some twenty-five years earlier. Basically, Las Casas looked much the same, with a few more small houses for the fishermen. Most impor-

The CONAF house used for visiting researchers in the settlement of Las Casas on Alejandro Selkirk Island

tant had been the construction of a relatively large house for visiting researchers. This was a huge improvement over having to use tents in 1986. Furthermore, the installation of a diesel-powered electrical generator provided electricity from 8:00 in the morning until 10:00 at night. This innovation made possible satellite television with programs of news and entertainment. Also, because of recent satellite coverage, and with the proper lithium-ion telephone unit, one could call anywhere on the planet. We had brought one of these telephones along mainly for emergency needs, recalling our hassles on the expedition of 1986 (see chapter 8). It was expensive to make a call, but during the trip, each of us had a chance to call home once, and in my case that meant to our home in Vienna, Austria—direct from Alejandro Selkirk Island. Incredible! It was calming for all of us to chat briefly with our loved ones. Another very significant change from 1986 was that because of the diesel generator that constantly was blowing off hot air, we could use this to dry our plant presses and not have to fool with kerosene stoves that were messy and potentially dangerous. We had been advised from CONAF of this change and hence didn't bring a drum of kerosene with us.

In addition to our expedition crew of eight, we had a visitor who accompanied us to Alejandro Selkirk Island, but he stayed only a week. His name was Jonathan Franzen. When we were making preparations for the trip in Vienna, I had received an email from the *New Yorker* magazine, asking about the possibility of someone going along with us to Alejandro Selkirk Island. To me this would be fine, perhaps they might even do some reporting on our expedition, which would be good public relations. We agreed to a fee for the passage round trip from the near to the far island, because we had contracted with the municipal boat on Robinson Crusoe Island to take our group to Alejandro Selkirk Island, leave us for a month, and then come and get us again. We could certainly add a person to this trip, but he would have to make arrangements and payment for his own return trip as we would be working for several weeks more. When we first arrived by plane on Robinson Crusoe Island, I met him walking in the village. Because there are normally no other Americans there, except for occasional Mormon missionaries, it is natural for two Americans to greet and chat. Jonathan was a thin, tall fellow, with dark glasses and brown hair, friendly enough, and enthusiastic about going with us to the far island. His job was to write an article about the experience for the *New Yorker* magazine, which indicated to me that he was a very skilled, if not well-known, author. I invited him to attend a dinner that evening with the CONAF chief (Don Iván) and his wife, plus two of the guides who would be accompanying us and our entire crew on the expedition. We would be paying the guides extra for this service, because they were doing it during official vacation time. Jonathan apparently knew very little Spanish, so it was obvious we would be helping him. He was sociable and seemed like a pleasant enough travel companion.

Jonathan Franzen, writer from New York, who joined us for part of our 2011 expedition to Alejandro Selkirk Island

Jonathan proved to be an interesting person in several ways. After we had made it successfully to Alejandro Selkirk Island, we integrated him into our food regime because it cost us virtually nothing to add another mouth, and this simplified his life considerably. He stayed in the CONAF house with the guide Danilo, who was in charge of the settlement at this particular time. After we had gotten installed in the island, I was chatting with Jonathan about what sort of things he wrote, and he replied, "Social commentary." Well, that could have meant just about anything, so I pursued with more specific questions, and he responded, "You may have seen me on the cover of *Time* magazine last year."[1] Hmm . . . well, no, I had not seen that, but it did get my attention—this was apparently no schlock writer of supermarket romance novels, but someone of reputation. He worked in New York, and he said that he went to his own office each day for writing. Now, if someone can afford to have an apartment in the city and also an office, he or she is not a struggling author eating hand to mouth. In fact, I later learned that he had published his novel, *The Corrections*,[2] to considerable critical acclaim, and he had been hailed by *Time* as a great American novelist. To me, he was just someone who needed help in this island, and this meant mostly being an interpreter between him and the island CONAF supervisor Danilo (and his son, Danilito).

Jonathan wanted to be taken to the top of the island, 1,100 meters, where CONAF now had a small hut, and be left there for two full days alone. He wanted to experience this isolation as had Alexander Selkirk centuries ago, when he was surviving alone on Robinson Crusoe Island from 1704 to 1709. I had some reservations about Jonathan doing this, because when the fog covers the top of the island, you just cannot see anything. Because of the deep ravines that start at that level, there are serious dangers awaiting anyone who walks the wrong direction. I told him about this, but he insisted he would stay close to the hut if it got foggy and be cautious. He seemed to be level-headed, and in any event, it was really none of my business—he was here to accomplish what he had planned. On the trip over we had learned that Jonathan was a serious bird watcher, as he pulled out binoculars to identify the sea birds and report their names.[3] To me a bird is a bird is a bird, but I had enough sense to not communicate my profound ignorance of the avian world to him. He was eager to see the endemic birds on Alejandro Selkirk Island, particularly the fardelas that burrow in the soil among the tree ferns at higher elevations.[4]

We realized months later, after reading Jonathan's piece in the *New Yorker*,[5] published months after we returned to Vienna, more of what he had experienced in the highlands of Alejandro Selkirk Island. He had carried the ashes of his very good friend, David Foster Wallace,[6] up into the hills and scattered them as his friend had wished. We had no idea of this plan, nor did it really matter, but it gave a slightly different sense to the objectives of his trip. It was not just to write an article, but also to fulfill the wishes of a departed friend. The article has been reprinted in a compilation of several of his short stories.[7]

There were two principal objectives for us on this expedition to Alejandro Selkirk Island. The first was to produce a modern vegetation map of the island, as a companion to the one that had been published for Robinson Crusoe Island,[8] using the same methods. This was essential so that our evolutionary questions could be examined more precisely. Sepp Greimler was in charge of this subproject, and doing relevés was again on the schedule for the fieldwork. In addition to obtaining relevés, Sepp would take photographs of the landscape for ground truth when comparing patterns seen in the aerial photographs taken by the Chilean Air Force in 1980. We did have permission to collect on this expedition, which made it possible to bring back unknown plants for later identification.

The second objective was to examine patterns of genetic variation within and among populations of selected groups of endemic plants. We were particularly interested in measuring the degree of genetic divergence among populations in species on this young island in comparison to what we had already documented on the older island.[9] We expected lower degrees of populational divergence in the younger island because of less geological time for such divergence to develop, and this was the hypothesis we intended to test. A key group was the genus *Erigeron*, also of Asteraceae, which had five endemic species on this young island. Some of these species were easy to access, but others grew only in the higher elevations above 1,000 meters.

As in 1986, the key to success on this trip was forming teams of collectors who would go simultaneously to different parts of the island. Work in the hills near La Cuchara would be facilitated because there now was a small cabin, making it possible to cook on a small propane stove away from the wind and to allow a few persons to sleep inside. The other team members would have to sleep in tents. To lighten the burden of transporting these supplies and equipment from the settlement to this elevation,

Researchers' cabin at 1,100 meters (La Cuchara). Left, cabin in fern-grassland mosaic landscape; right, interior of cabin.

two mules were used. Other crew members would stay below in Las Casas, hike along the coast and up into the many quebradas (ravines), collecting plants. They also were to process plant materials for the population genetic studies. This involves the placing of leaves from individual plants into empty tea bags and then into Ziploc plastic bags with silica gel, a time-consuming endeavor.

To keep our genetic work focused and coordinated, we prepared a large wall chart that listed the species of interest and the numbers of populations and individual plants that had been sampled. The genera and species were *Erigeron* (five species), *Rhaphithamnus venustus*, *Myrceugenia schulzei*, *Robinsonia masafuerae*, and *Gunnera masafuerae*. These nine species were all endemic to Alejandro Selkirk Island, and we wanted to compare levels of genetic variation within and among populations with species on the older island. We sampled leaves from thirty different plants scattered throughout each population, so that our data would be statistically valid, but in some cases there were fewer individuals available. Placed next to the chart was a large satellite photograph of the island, on which we plotted locations of the sampled populations. These visuals allowed us to see progress with our sampling and gave purpose to the field effort.

As I have mentioned previously (see chapter 7), we were very lucky during our expeditions not to have serious medical issues. However, there were some close calls. On this trip we were accompanied by the medical student and former nurse Daniela Bacho. Her expertise came in very handy one afternoon. Most of us were working in the seaside house preserving leaf material in silica gel, trying to catch up on accumulated collections from the previous day. Sepp was working with his data on the vegetation,

Plotting of locations of vegetation samples (relevés) on topographic map of Alejandro Selkirk Island.[10] From left to right: Patricio Peñailillo, Patricio López-Sepúlveda, and Josef Greimler.

making a draft map as the trip continued, which he hung on the wall for all to see. I was updating our list of numbers of populations collected for the species of interest, plus updating the topographic map showing all the daily excursions.

While these activities were taking place, Bernardo was outside burning the garbage. CONAF had set up an old oil drum away from the house toward the shore, with holes punched out at the bottom for air. There is no trash collection on this island, and each person was responsible for burning his or her own garbage. This was something that Bernardo had done hundreds of times, so if anyone were expert, it was him. But, suddenly he came hurriedly into the house holding his arms up in the air—a plastic aerosol can had exploded in the fire and small pieces of plastic had shot out and embedded into his arms. Fortunately, Daniela was in her room and quickly came out to investigate. She sat Bernardo down, with his arms propped up on the table, and we could all see the burned area of his arms and the small pieces of plastic embedded in his skin. She got to work immediately, first washing the affected areas, uncomfortable for

Daniela Bacho, our nurse, treating the burned arm of Bernardo López, CONAF guide, on Alejandro Selkirk Island

Bernardo, and then with a tweezers picking out pieces of plastic, each about 5-mm diameter. This procedure took thirty minutes because so many pieces had entered his skin, but eventually they were all removed. She then applied an antibiotic salve on the affected regions and added some gauze and tape for protection. Bernardo felt much better about having received this very quick attention to the problem, and he now was able to relax a bit.

I was relieved that this crisis had been solved, at least for the time being, but I began to think about the next day, as Bernardo was scheduled to take the group up the hills to the small hut at about 1,100 meters, where they would stay for two days. Later in the day, I asked Daniela whether he would be able to handle this responsibility, and initially she felt it would be inadvisable. We chatted about the options for quite a while, and eventually she suggested that if his bandages could be changed each day, and if he were careful, it might be possible for him to go. For his part, Bernardo wanted to go, but he looked to Daniela for guidance. Eduardo Ruiz was also going along on this excursion, and he offered to change the bandages at the end of each day, which Bernardo himself could not do. I felt better, because no one else except Bernardo had the experience and leadership qualities to handle this type of more difficult excursion. The next morning, we had another discussion about his health, and he and Daniela concluded it was possible for him to go. He did, in fact, make the trip, and with the medical attention from Eduardo, everything worked out fine.

There were several days in which all personnel went out collecting except for me—I stayed back in the house. My usual early morning routine was to arise at 6:00, occupy the bathroom, and start making breakfast. I became skilled at toasting bread by holding it with a fork over the propane gas burner, heating up water for coffee or tea, opening cans of liver paté and boxes of crackers, and slicing up hard salami. Because of the early hour, candles were needed in the kitchen and on the serving table, which made for a soft light to start the day. During the day, my tasks were to work with the plant presses, process more plants in silica gel, tidy up the house, and take care of the garbage.

Having seen Bernardo burn the garbage, I was confident to handle the task. Most of the refuse is paper based, so most of it disappears in the flames. I brought out the garbage in plastic bags, shook the contents into the drum, and followed this by a few matches. A fire quickly started. I was especially careful to avoid any explosive objects. The fire got going very well with flames leaping upward, and it was clear that this technique would do the trick, getting rid of our garbage by the time the crew returned. There was also a good breeze that blew the smoke, what little there was of it, into the air. As the fire continued, it leapt higher from the drum, still well contained within it, but embers now were being thrown upward, too. The breeze began to get stronger, and now the embers were being tossed quite a distance into the air. Some of these were fairly large pieces, and landed back on the ground, in some cases among very dry grasses. In fact, as I looked around the settlement, I realized that all grasses surrounding the houses were very dry, just ripe for some of them to catch fire. No, this was not good, not at all recommended, but how to stop the fire that now was roaring from the mouth of the drum? The fire with the garbage was doing fine, but I did not at all like watching large embers flinging upward and then sinking down again. At this point I was on high alert—there was no way that I was going to allow a fire to break out in the settlement, especially in the afternoon when the fishermen were away, and burn down the village. I could see the headlines: "U.S. Professor Allows Fire to Burn Down Settlement of Alejandro Selkirk Island. Three Children Injured." A few embers landed right in a patch of brown grass, and I jumped in and stamped them out, just to be sure. This continued for a good thirty minutes until the fire had died down, and the size and number of embers dissipated considerably. But I had been watching hundreds of embers on their voyages up on the wind and down into the dry grass, all the while tense, worried. I still had another bag of garbage to burn, but I decided to wait for another day with much less wind, or even to leave the task to Bernardo, who always seemed to know how to handle these island chores. I felt as if I had narrowly averted a disaster, and I was left exhausted by the experience.

On a very isolated island and after weeks of strenuous physical effort, thoughts naturally turn to a fiesta. A really good affair took place during this trip in 2011, toward the end, with everything so far having been very successful. There were a few days more to go, but we needed to take advantage of the Saturday night for some fun. This was also our way of thanking the CONAF guides and the fishermen for all the help they had been providing. We also had given a seminar (in Spanish) for CONAF and the fishermen on our research and its significance. For the fiesta, we knew how many bottles of wine were left and how many we could open for this party, and it was substantial. We placed tables outside our CONAF house, and we lined up the bottles, some already opened. We had several bottles of pisco, too, but these were kept in reserve for later. All the families were invited, and they pitched in, some bringing large two-liter bottles of Coca-Cola or Fanta, the two most popular soft drinks. They also brought potato salad, potatoes being easily kept in the island, with added olives, some oil, some spices—really excellent. The grill was started, again using a drum cut in half with a grate on top, which allowed for a very good fire and development of lots of hot coals. The first course would be fish, and a fisherman began cutting open the fresh fish into fillets, followed by adding

Fiesta on Alejandro Selkirk Island during our 2011 expedition. Left, Danilo Arredondo starting fish on the grill; right, villagers relaxing.

spices for flavoring, and then immediately onto the grill. As the smell of the grilled fish started rolling out across the settlement, others came along. We had brought some paper plates to use for fiestas, and this allowed each person to just take one and help themselves to whatever was on the table. As far as I could tell, there was no one coordinating this, but food was eaten and more food just kept coming. As the wine was poured, even more folks appeared, and then we all enjoyed the fish and grilled goat, the latter of which, although tough, has an excellent taste. As time moved on, we brought out our cassette player and started the music going, adding more fun to the mix. The weather was perfect, the sun now completely below the tall hills to the west of the settlement, with only a light yellow glow edging the silhouetted rocks.

As the fiesta continued, I took a glass of red wine and walked away just a bit toward the sea. The stars had now appeared, and they were bright, very bright, with the Milky Way strewn out across the sky, as if a ladle had been dipped into a bucket of stars and thrown from one side of the sky to the other. I could hear the music playing, the Latin rhythms, and people standing close to each other and chatting, some now even dancing, and I could also hear the sound of the waves, now soft and comforting, a calm sea this evening. In looking at the people, the dark heavens with so many millions of bright stars, and listening to the ocean, I realized how incredibly lucky we were to be having this experience at this one moment in universal time. We were about as isolated as we could possibly be from the rest of humanity on this very small island, and we were the only people here, but now enjoying companionship, fellowship, and trust in each other. It felt very good to be alive, a part of this world, and I was so very thankful for being in this very spot at this very time.

Upon my return to the group, persons were starting to dance more, either with partners, or just alone, and the older fishermen, each of whom was well known, had a dance style that elicited lots of good-natured commentary. The group then said, "Profe Stuessy, Profe Stuessy," and it was my turn to attempt fancy footwork while not spilling the rest of my wine. When the electricity cut off at 10:00, a propane lantern was provided by one of the fishermen, which scattered enough light to keep the party going in high gear. As the evening wore on, two young men brought their guitars and began singing folk tunes of the islands, tales of the sea, and hardships, and everyone filled up their glasses once more, smiled more broadly, and felt contented. I went to bed about 1:30, but the fiesta didn't finish up until after 4:00.

Our trip to Alejandro Selkirk Island had one very emotional component, and that related to the tsunami that hit Robinson Crusoe Island on 27 February 2010. We had been in San Juan Bautista about a week before the event. A submarine earthquake had occurred three kilometers off the coast of Chile between Valparaíso and Concepción at magnitude 8.8, causing serious damage to coastal regions. A tsunami was also created that moved westward, finally arriving in San Juan Bautista four hours later, hitting early in the morning. This wave destroyed the lower part of the village and left eleven persons dead and several missing, never to be found. Two of the CONAF guides, with whom we had worked on numerous occasions, suffered serious personal losses. Jorge Angulo lost his home and restaurant, totally destroyed and without insurance (it is doubtful if anyone in the village had insurance), and Danilo lost his home and daughter (never found). We had made a good collection of money to donate to each of these friends, but the most poignant moment came on Alejandro Selkirk Island with Danilo. We were having all meals with him and his family, his wife and another lady doing the cooking, and we provided the foodstuffs and wine. At one dinner we presented him with an envelope with several thousand dollars, to show that we were there by his side. He was visibly moved, having to leave the room for a few minutes.

When Danilo returned, he told us about that horrible night. He woke up in his house in the lower part of San Juan Bautista on Robinson Crusoe Island with his room filling up with seawater, followed by the house ripping apart, with him being swept out into the bay. He held onto a large piece of wood and managed to swim to the cement pier, just before another wave hit, while he held on for survival. After this wave passed, he climbed up onto the pier and ran across the debris of the village to higher ground before another wave, now smaller, swept across the lower region. He was very lucky, but his daughter did not survive. During this expedition of 2011, I talked to him about his loss, and I could share my own feelings, because my daughter, Mary, had died from a heart attack at age thirty-one. In five minutes she was gone, just like that—no warning, no preparation, nothing. Dead. This is another story for some other time, but the point is that I could relate to what Danilo was going through. It was even worse for him, because his daughter's body was never recovered; it just disappeared, no trace, gone. One never completely gets over the loss of a child because it is not the proper order of things. The parents are supposed to die before the children, who are the next generation, and the reverse just feels wrong. I still have a lonely, restless longing inside for my daughter, to hold her, to hug her, to share her experiences in life, to relive my own experiences by sharing hers, to look into her eyes and tell her how much I love her. But we must move on, because we cannot change the past—we can only adjust to the future and deal with it in a positive way. Mary, after all, would hardly wish that my life should be miserable due to her death. I understood the pain on Danilo's face when he talked about his daughter, and it just takes time to be able to talk about her openly again, smiling, remembering experiences, and celebrating accomplishments of her short life. Leaving the island this last time had special poignancy.

Departing Alejandro Selkirk Island in 2011 for the final time

CHAPTER 12

The Heavy Hand of Humans

I don't know how you feel about it, but the older I get, the more I worry about the planet and its condition. Our generation has not done a great job of managing resources to protect our environment. Within this context, it is worthwhile to examine the impacts from humans on the ecosystems of the islands of the Juan Fernández Archipelago. It is a good case study because there never were any indigenous peoples prior to discovery of Robinson Crusoe Island in 1574 by Juan Fernández. Only after that point did Europeans come to this island and eventually to Alejandro Selkirk Island and begin to alter the landscapes. Because of the strategic importance of these islands, many visitors stopped to rest, take on fresh water, eat fresh goat meat and fish, consume vegetables, repair boats, and then continue westward across the Pacific or harass Spanish ships and coastal colonies in South America. Many of these voyagers, especially captains or educated officers, kept logs or wrote diaries of their trips, and some of these were published in European cities upon their return.[1] The European public was thirsty for information about the New World, and these publications, mostly books, provided an additional way for the authors to profit from these arduous travels.

It is very important to stress that the human impacts on the ecosystems of the Juan Fernández Archipelago occurred only during the past four centuries, whereas the natural impacts on the archipelago had taken place during the past four million years. Once oceanic islands arise from a hot spot beneath the sea floor, they usually undergo additional episodes of volcanic activity, eventually stabilizing as the submarine tectonic plate moves toward the zone of subduction. The Juan Fernández Islands are attached deep on the Nazca Plate that subducts under the South American continent. As the plate moves eastward, it slowly subsides, taking the islands downward as it goes. At the same time, erosion by waves along the coast and by wind and rain on the exposed landscape causes the islands to change in size and shape. These impacts after four million years have resulted in much smaller islands, much lower than when formed, and with a greatly modified geomorphology.[2] It has been estimated that up to 95 percent of the surface of Robinson Crusoe Island might have been lost through subsidence and erosion during its ontogeny.[3] With the younger Alejandro Selkirk Island at only one million years of age, these forces have had less time to work their inexorable modifications, and perhaps only 28 percent of the land has been removed. Overall, these natural impacts on the flora and vegetation of the archipelago have been substantial, much more dramatic than any impacts from humans during the past four centuries. These types of modifications have been happening everywhere on Earth since it was first formed 4.5 billion years ago[4]; this is the natural progression of events in nature. We obviously do not have a written record of these natural impacts, only hypotheses based on logical inferences. With human activity, we do have documents that reveal when people visited the archipelago, and what they did when they got there. This is the substance of information in the present chapter.

From the period of discovery into the middle of the eighteenth century, ships from many different European countries visited the Juan Fernández Archipelago, especially Robinson Crusoe Island. Although no one disputed ownership by Spain, nothing had been done to fortify the islands so as to control who came to the islands. Privateers loved using Bahía Cumberland as a location to recover from the arduous trip around the tip of South America, and also for staging attacks against Spanish ships and colonies along the western side of the continent. The situation got progressively worse, until Spain finally made the decision to take control, which they did in 1750 by founding a permanent colony, San Juan Bautista, beside Bahía Cumberland, and building a fort with cannon to defend the bay.[5] This did solve the problem, although no such attention was given to Alejandro Selkirk Island, largely because fewer persons went there and because there was no protected harbor. The level of population in San Juan Bautista has varied from a few persons to more than 900 today. There is a limit to the available resources in the archipelago for maintaining the colony.

Several factors combined to influence attitudes toward San Juan Bautista from the Spanish and later Chilean governments. First, European alliances and conflicts were constantly changing, with wars being a familiar part of foreign policy of most nations. It was a burden, therefore, for administrators of San Juan Bautista to be certain who was regarded as friend or foe, as it took months for news to trickle from Europe to South America and then to the islands. Second, the isolation of the village made it challenging for governments to provide proper supplies on a regular basis. There is always enough food from the sea to keep people alive, but for building materials, foodstuffs such as sugar and flour, luxuries such as tea or coffee, and alcoholic spirits, all had to be imported from the continent or purchased or bartered from visiting friendly ships. Third, due to the isolation of both islands, beginning in the late eighteenth century, they were used as a penal colony, housing ordinary criminals (some quite hardened, i.e., murderers and robbers) as well as political prisoners, particularly during the epoch of Chilean independence in the early nineteenth century. This continued on Alejandro Selkirk Island up to 1930, finally being abandoned for good. It was completely unrealistic to expect decent colonists to enjoy living in the islands when accompanied by criminals who routinely rioted and threatened families. Keeping prisoners also required having food and clothing. Being administrator of San Juan Bautista was not a desirable responsibility in the eyes of most qualified persons. It was too isolated from continental authority, and the military forces needed to contain the prisoners were never sufficient to handle the job securely. Even worse, sometimes the military mutinied against the administration, which added more complexity to the job.

There are, to be sure, economic potentials in the Juan Fernández Archipelago, but they have rarely been developed efficiently and sustainably. The main products available traditionally have been lumber, fish, and lobster. Soon after discovery, native trees were felled for processing into boards for houses and repairs of ships, and also for use as firewood. Ships also purchased wood to use to fire up their galleys, and when steamships were invented, these sought wood to keep the boilers going. Fish were abundant in staggering quantities and of great variety, and the lobsters (*Jasus frontalis*) were very good eating and super abundant up into the twentieth century. The lobster industry has declined in recent decades, being overfished, but the government has now placed strict control that prohibits fishing during the reproductive season (15 May–30 September) and allows keeping only larger individuals (with more than a 115-millimeter carapace), all to maintain a sustainable industry. There is now also prohibition against granting any new lobster fishing concessions until 2024. In the meantime, initiatives to develop ecotourism have sprung up, in part due to the acceptance that there are limits on income from the lobster and fish industries. At the end of the eighteenth and beginning of the nineteenth centuries, many profits were made from harvesting the fur seal on Alejandro Selkirk Island, and this species was brought nearly to extinction.[6] Somehow it miraculously survived, and it now forms a stable population on each island under strict protection by the park service. During the past decade ecotourism has evolved as an alternative income source, which has great potential due to the richness of the natural history of the archipelago.

In addition to human impact, San Juan Bautista has periodically suffered from tidal waves that have destroyed part or all of the village. One spectacular devastation occurred in 1835, which killed few but reduced the village to rubble. This event was chronicled by Thomas Sutcliffe, the governor at the time.[7] Recently, in February of 2010 an earthquake off the coast of continental Chile, which caused much damage to southern coastal towns, yielded a tidal wave that slammed into San Juan Bautista in early morning, killing eleven persons and destroying about half of the buildings, including the municipality, office of the captain of the port, post office, school, gymnasium, discoteque, one church, small restaurants and pensions, and many private houses.[8] Reconstruction of buildings is now resulting in an attractive lower village, but the quaint historical character of the older structures is gone, altered forever.

After Europeans discovered Robinson Crusoe Island, and after word spread back to Europe about the archipelago so useful for navigation purposes, many visitors came. Most preferentially arrived to Bahía Cumberland on the northeastern side of the island, which provided the best protection from wind and waves; ships sometimes also anchored off Puerto Francés, Puerto Inglés, and La Vaquería. There was constant need in San Juan Bautista for constructing new buildings and also for firewood. Early attempts at colonization began at the end of the sixteenth century, but these failed in short order. Most ships were interested in obtaining fresh water, giving the crew a chance to rest and recuperate, and cutting lumber

Drawing of Lord Anson's campsite in the central valley, Robinson Crusoe Island, in 1741

for repair of ships. To appreciate the impacts on the forest in this part of the island, we need to look first at a modern vegetation map (see color insert, **13**). This excellent illustration, done by Josef Greimler from the University of Vienna, clearly shows the extent of erosion and lack of native vegetation in the regions of the four anchorage points.

The historical documents from previous voyagers to Robinson Crusoe Island show that originally the northeastern part of the island was covered by forests nearly to the ocean. George Anson visited the island for several months in 1741, and he published detailed textual comments as well as several maps and drawings that also reveal trees in this region.[9] For example, he illustrated his campsite for the officers near the shore with the "laurel trees" (*Nothomyrcia fernandeziana*) covering most of the valley, part of which was likely cleared by earlier visitors. He also included a map of the entire island that shows tree symbols over the northeastern region.

By the nineteenth century, major loss of tree cover had taken place in the regions of the four principal valleys on the northeastern side of Robinson Crusoe Island. Perhaps the most dramatic example of this is a plate from Claudio Gay, sketched during his visit in 1832 and published in 1854.[10] It shows the very small village of San Juan Bautista, with houses, some people, and domesticated animals. More importantly, areas adjacent

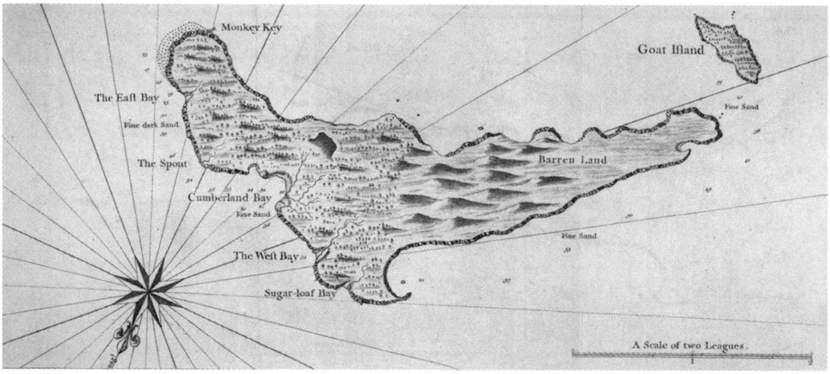

Map of Robinson Crusoe Island published in 1748 by Lord George Anson

to the village were now without plant cover, with many trees having been removed. The situation was no better in Puerto Francés, Puerto Inglés, or La Vaquería, the forests having been cut substantially, and with erosion continuing to the present day.

In the twentieth century, efforts began to reforest the area around San Juan Bautista with introduced tree species to help hold the soil and to provide wood for lumber and firewood. Eucalyptus, cedar, and pine were

Drawing by Claudio Gay of the village of San Juan Bautista along Cumberland Bay, published in 1854

those that took hold and still grow in this region today. Historical documents reveal a slow growth of these plantations,[1] such that over the decades a green zone has been reconstituted around the village, but now of exotic species rather than the original endemic ones. The park service has encouraged these efforts, because use of wood from the plantations eases pressure on native trees, and at the same time they carefully monitor spread to avoid these species becoming uncontrollable pests in the flora.

As the forests were being cut, fire was needed for cooking, keeping warm during the chilly winter months, making charcoal, and other miscellaneous needs. Sometimes fire spread unintentionally, burning areas in and around the village. Occasionally fishermen would make fires on land during a rest period, such as lunch, and these could spread through strong winds, casting embers into dry grasses. This is, in fact, exactly what happened in February of 1996 on the western side of Alejandro Selkirk Island, and seventy hectares of native forest burned.[11] Historical reports indicate that large fires had burned in Valle Inglés (started by visiting California 49ers), and also in Puerto Francés.[12] While the long-term consequences on the vegetation of these areas are unclear, the fire surely did not help maintenance of the native species. After such disturbances, introduced invasive plant species would have been provided opportunities to colonize these areas.

In addition to cutting trees for lumber and firewood, colonists and visitors over the centuries have harvested other plants. One species that suffered great loss is the endemic palm *Juania australis*. This attractive species, with its bright orange berries, was sought for its edible growth apex (apical meristem). How and who first discovered that this two-kilogram tissue was edible is unknown, but when cut from the tree and boiled, it tasted somewhat like cabbage, hence the common name of the plant, cabbage tree. The problem with harvesting this resource was that the entire tree had to be sacrificed by chopping it down and cutting out the apex. This, combined with grilled fresh goat meat, made a spectacular meal for sailors who for months had been eating hard bread and dried meat. It is not known how many plants of this palm were taken, but at the present time between 1,000 and 2,000 individuals remain,[13] mostly on the higher ridges and largely safe from further molestation by people. It is now illegal to cut down any of the native plants without permit, a rule strongly enforced by the park service, which provides good protection for the cabbage tree (also known as the *chonta*). The species is very distinctive mor-

Endemic trees harvested on Robinson Crusoe Island. Left, the palm, *Juania australis* (the "cabbage tree," *chonta*); right, the extinct sandalwood, *Santalum fernandezianum* (drawing by B. Krüssel).

phologically, and that is why it is considered an endemic genus, as well as species. Some molecular phylogenetic studies have been conducted on relationships, but so far we only know that it belongs to the Ceroxyleae taxonomic tribe, perhaps near *Ceroxylon*.[14]

A more tragic story deals with the sandalwood, which has now become extinct. *Santalum fernandezianum*, based on historical documents, was once relatively common on Robinson Crusoe Island. The problem was that during the nineteenth century, a large industry developed throughout the Pacific region for harvesting species of sandalwood for sale to China,[15] where it was used as incense, burial objects, and expensive furniture. As with the cabbage trees, to harvest the sandalwood, one had to cut down the entire tree, because the strong scent was found only in the wood in the lower stem and upper portion of the roots. The upper stems, which could easily be pruned off, did not contain enough fragrance to be commercially profitable. Many species of the genus *Santalum* were cut for this industry, and some, thankfully, such as in the Hawaiian Islands, were protected by King Kamehameha to provide a sustainable business for him and the royal family.[16] In Robinson Crusoe Island, there was no such control, with the last tree remaining in weak condition in Valle Inglés in 1908.[17] By 1916 it

Animals endemic to the Robinson Crusoe (and Desventuradas) Islands. Left, the Juan Fernández lobster (*Jasus frontalis*), with Eduardo Paredes (Don Uto), one of the exceptional CONAF guides; right, the Juan Fernández fur seal (*Arctocephalus philippii*).

had disappeared, the wood having been cut up for sale to tourists. This is the most conspicuous case of extinction of a species in the archipelago by deliberate overharvesting.

In addition to harvesting of plant species in the archipelago, people have placed pressures on populations of the marine animals, especially fish, lobsters, fur seals, and sea lions. The fish have always been incredibly abundant around both islands of the Juan Fernández Archipelago. The diversity was high and the total abundance was astounding, easily providing food for crews on passing ships as well as colonists. There still is a large variety of species,[18] but none is incredibly abundant as in the sixteenth to nineteenth centuries. In part this has been due to fishing industries that became active at the end of the nineteenth and early decades of the twentieth centuries, with tons of cans of fish products being exported to the continent.

Another important marine resource has been the lobster, *Jasus frontalis*. This very tasty food was appreciated by early visitors to the archipelago. Even now, lobsters are exported to restaurants on the Chilean continent, and they are regarded as a real delicacy, commanding high prices. During the same period of the fish canneries, lobster was also captured and processed, which led to declines in production into the middle of the twentieth century.[19] It was finally realized that to maintain this resource, the principal income stream for the islanders, conservation of populations was essential. Now there is a minimum size required for keeping a catch, and harvesting lobster is now illegal during their reproductive season (mid-March through September).

A sadder tale, perhaps, is the fate of the populations of the Juan Fernández fur seal (*Arctocephalus philippii*) and the southern elephant seal (*Mirounga leonina*). Visitors to the archipelago during the seventeenth and eighteenth centuries were amazed by the large quantities of these animals on both islands, where they formed dense colonies in low areas, such as in Bahía Cumberland. Because the fish were abundant, this provided adequate food for these colonies. Early travelers commented that it was very difficult to get on shore because of so many animals lying about. Most of the animals were fur seals, which are endemic to the Juan Fernández Archipelago and the Desventuradas Islands, 824 kilometers north (also owned by Chile).[6] Initially they were eaten for food, but the meat had a very fishy taste (unsurprisingly). The elephant seals were less abundant, but they were killed and boiled up for their copious oil, which was then used for cooking. They were also killed for sport, whereby six or more sailors with long pikes would surround a male animal and prick it to death, which usually took about two hours. Such a cruel treatment is difficult for us to comprehend today, but at that time, there was little ecological sensitivity.

A market in China developed for the pelts of the fur seal, which led to a major harvesting industry. The activity took place primarily on Alejandro Selkirk Island for several reasons. Most of the animals had already been killed on the northeastern side of Robinson Crusoe Island, where ships had anchored. Spain had also taken firm control of this island, and ships from other countries were less welcome, especially to harvest natural resources. Ships went instead to Alejandro Selkirk Island, where several low areas provided ample space for massive seal colonies. Most of the sealers were from England and New England (in the United States), particularly between 1798 and 1808. There were an estimated three million animals, and since they had no fear of humans, they could be approached and killed easily with one blow of a club. It was not unusual for a ship to stay several months and carry off many thousands of pelts on the route to the Orient. This slaughter was so effective that by the second decade of the nineteenth century, the level of fur seal population had dropped so low that the industry was no longer profitable.[6] Some harvesting continued later, but only sporadically, until in the middle of the twentieth century it was believed that the species had been extirpated from the archipelago.

Feral animals in the Robinson Crusoe Archipelago. Left, the goat (*Capra aegagrus hircus*), common on Alejandro Selkirk Island; right, the European rabbit (*Oryctolagus cuniculus*), abundant on Robinson Crusoe Island.

There was still a population on the Desventuradas Islands, but that, too, had been harvested for profit. In 1965 some fur seals were found again on Alejandro Selkirk Island,[20] and now under protection by the park service, a sizeable and stable population has been reestablished. There are also some fur seals on Robinson Crusoe Island, fewer, but these, too, are recovering under protection, and in fact, are now valued for attracting ecotourists.

Along with direct human disturbances to the ecosystem of the Juan Fernández Archipelago have come introduced animals and plants. It is impossible to say which has been more injurious to the native vegetation, but both have created serious problems. Without doubt, the animal that has caused the most damage, and which was deliberately left on the islands by early visitors, has been the goat. Goats are very adaptable creatures and can survive by eating nearly any plants, including endemic species. Furthermore, their sure-footedness allows them to take refuge in steep, mountainous terrain, making hunting them by men or dogs a dangerous proposition. At the present time the highest concentration of goats is on Alejandro Selkirk Island, perhaps as many as 4,000 animals.[21] On Robinson Crusoe Island the goat population is much less, perhaps fewer than one hundred, and in fact, I have never seen one in the forest during our expeditions to the archipelago. On the younger island, one can easily see these animals and also hear them bleating. A high conservation priority for the flora of the archipelago would be the complete elimination of the feral goat. This is a tall order because of the rugged terrain on Alejandro Selkirk Island and the improbability of ever hunting all individuals, which over time would simply allow regeneration of a large population.

Perhaps a better approach might be the development of a male-sterile line that would be set free on the island to mate with existing females, hence yielding no progeny. I might add that we have had many opportunities to sample grilled goat meat, and it is excellent. The meat from these very fit animals is not fatty, and the flavor is very good. One imagines how this must have tasted to sailors who had been eating hard bread and beef jerky for months.

The other animal that was intentionally brought to the Juan Fernández Archipelago and has now become feral with negative consequences is the European rabbit. Brought to San Juan Bautista in the twentieth century, probably as a domesticated food source, it escaped and has now multiplied into the thousands on Robinson Crusoe Island.[22] Rabbits eat vegetation and they also burrow, which loosens the soil and hastens erosion. As with the goats, the rabbits are so numerous that it is difficult to imagine a program that would lead to their successful extermination, although this has been accomplished by hunting on the small Santa Clara Island. Studies involving biological control (with viruses) have been considered,[23] but so far no action has been taken. It is important to keep in mind that goats and rabbits also eat the invasive plant species, so in effect an artificial equilibrium has been established, but pressure against the native flora still remains.

Sheep and cattle have long been housed in the archipelago, particularly in and around San Juan Bautista. These animals have rarely been kept in pens or corrals, being left to forage in the valleys. The geomorphology of the valleys, such as the central valley, Valle Inglés, Puerto Francés, and La Vaquería, is such that the sides are steep enough to keep most animals from leaving the valley in which they are placed. There they sit eating away at the native vegetation, and in all of these valleys today there are virtually no native species from the shore to 350 meters, the plants in this zone now being invasive weeds. During our trips in the 1980s, we would occasionally find a cow or steer deep in the forest munching on some native plant. This now is much better controlled and is no longer permitted. Sheep have been equally destructive to the native plants. Thousands were imported to the village during the early decades of the twentieth century,[1] with the idea of producing wool and meat for sale to passing ships and in the Chilean continent. This was somewhat successful, but the concept was completely opposed to that of the national park, which was established in 1935.

Other animals have also come to the archipelago, some intentionally and some inadvertently. The usual traveler with ships is the rat, and this has arrived to both islands, although we have only personally seen rats on Robinson Crusoe Island. In addition to creating potential health difficulties, rats also eat tender shoots of vegetation, native or introduced. To stem the rats, cats also have been introduced, and this has been reasonably successful, although some of these have become feral. Cats love to eat birds, and the native bird populations have therefore suffered. Dogs were brought to Robinson Crusoe in the seventeenth century to run down the goats, in the hopes of removing a food source for ships arriving that were at war with Spain. Dogs cannot follow goats to their steep hiding places, so this initiative was only partially successful. Coati (*Nasua nasua*) were also brought to the island, and having become feral, they now prey on eggs of native birds.

In addition to the several animals that have pressured the native flora, a number of introduced plants have also outcompeted native and endemic species. The four worst offenders in this regard are *Rubus ulmifolius* (*zarzamora*), *Aristotelia chilensis* (*maqui*), *Acaena argentea* (*trun*), and *Ugni molinae* (*murtilla*). The zarzamora (blackberry; see color insert, **15**), introduced from Europe, might seem benign enough, having been brought from continental Chile about 1920[24] as a food source and natural fence to contain livestock. The problem with this species is that it grows so densely that no other species can live underneath it. It also is a prolific reproducer, easily increasing asexually from sprouts at the base, and sexually from the edible and tasty berries. The birds also love these fruits, and they disperse seeds over the entire island.[25] Huge patches of this species exist, especially in regions around the village, and it is impossible to get into and through them except by tedious hacking with a machete. In disturbed areas after landslides or washouts during spring rains, this species is super competitive and can invade newly opened areas quickly, keeping the native flora from recovering. Cutting the stems only briefly benefits the environment, because soon more new shoots sprout from the base, and the plant comes back more robust than ever.

Another very competitive invasive plant species is *Aristotelia chilensis* (maqui), a large shrub five to nine meters tall (see color insert, **15**). The species prefers moist quebrada bottoms and fills these so densely that no other species can grow underneath, not even ferns. The maqui can also be cut down, but it also sprouts from the base and maintains its claim on an area. The species is native to continental Chile, but natural factors in the environment there constrain its growth, never allowing it to become a serious pest. The maqui was apparently brought to San Juan Bautista deliberately, to serve as a natural fence row, and the fruits, although small, are edible and tasty, suitable for making marmalade. As with the zarzamora, if maqui has claimed an area, there is no getting through—it is just too dense. The only realistic way to control both of these species would be by poisoning the cut-over stems, but this could have ecological consequences for species in the native flora. Investigations would need to be conducted before initiation of any large-scale herbicidal efforts.

A smaller shrubby invasive species, *Ugni molinae* (murtilla), also native to continental Chile, is about one to two meters tall (see color insert, **15**) and grows on exposed ridges or open slopes. In some ways this species is more problematic than the others because it can grow in the higher elevations of Robinson Crusoe Island, and to a lesser extent on Alejandro Selkirk Island, and it fits well into the landscape. That is, it doesn't appear out of place in the vegetation, but there is an endemic relative, *U. selkirkii*, that grows at about 700-meter elevation on Robinson Crusoe Island, and which now is being pressured by the invasive relative. Some scientists have advised that it may be more practicable to live with invasives instead of trying to remove them entirely,[26] and the murtilla may be a candidate for this perspective. Some species can grow beneath it, but not very many.[27] One can also walk through a stand of the murtilla, with some difficulty, but this is not possible with the zarzamora or the maqui.

The last of the four most aggressive invasive plant species is *Acaena argentea* (trun; see color insert, **15**). This herbaceous species reproduces prolifically both asexually and sexually. It probably was introduced inadvertently along with packages and/or foodstuffs coming to the village from the continent. It is also native to continental Chile. It is a problematic species from a conservation perspective because it does have one positive attribute: it can take over a completely bare area and become established. It will become so dense as to impede growth of nearly all other species, but it holds the soil against further erosion. It reproduces asexually via runners over the ground, which is why it can take over an area so quickly. At the same time, it is a prolific sexual reproducer, putting forth roundish "heads" of small, dried fruits, each possessing a hook (like with Velcro fasteners) that attaches very easily to the fur of animals or clothing. After walking through a patch of trun, there are usually large clusters of entire heads stuck to the bottoms of pant legs and shoelaces. Should one bend down too far, these fruit clusters adhere to sweaters with a vengeance. We have sometimes been on the trail with dogs from the village, and the trun adheres all over the poor animals, requiring a good cleaning at the end of the day. One can imagine how effective sheep would have been as dispersal agents when there were thousands of them on Robinson Crusoe Island earlier in the twentieth century.

Many other introduced species in the flora of the Juan Fernández Archipelago are of serious conservation concern. More than half of the existing flora consists of introduced species,[28] and many of these are serious competitors. What is needed for better management is phytosanitary control. At the present time, packages come from the continent by plane and boat and are simply received into San Juan Bautista. There is no inspection station whereby supplies are examined and pests removed and destroyed. Such a program would be useful not only for the direct benefit of keeping out undesirable species, but also to serve as an educational program for the islanders. In this way they would come to understand what species are permitted to enter the park and which ones are prohibited. Resources are needed to support such an important phytosanitary program.

CHAPTER 13

Scientific Accomplishments

To counter the possibility that the reader may finish this book believing that all I have done in these forty years is to have adventures, get "vacations" away from routine university tasks, find a wife, and to indulge my own interests in language, cultures, and evolutionary questions, the purpose of this chapter is to explain some of our accomplishments during these years of research efforts. And I say "our" because this project has been a group effort from the very beginning. During more than four decades, thirty-eight participants from eleven different academic institutions have been involved with the fieldwork (see Appendix 1) and dozens more with laboratory investigations, and eight people have been on three or more expeditions, believing enough in the project to return for more hardships, adventures, and scientific insights.

Prior to our research, the basic inventory of the vascular plants (ferns and flowering plants) had already been done by previous workers, mostly Bertero, Gay, Philippi, Hemsley, Johow, and Skottsberg (see chapter 4). We, and others, have described some new species in more recent decades,[1] but these have only added frosting to the cake, so to speak, rather than baking a completely new one. This realization, in fact, was one of the reasons we chose to focus on evolutionary patterns and processes in the archipelago. If one does not have a good inventory of the organisms in a region, it is nearly impossible to make convincing statements regarding evolutionary affinities. It is like trying to interpret relationships among children in a family when one of the parents is unknown. Missing organisms create gaps in the analyses and weaken conclusions. Previous work

Colleagues who have participated in the Robinson Crusoe expeditions three or more times. Top, left to right: Gregory Anderson, Marcelo Baeza, Daniel Crawford, and Josef Greimler; bottom, left to right: Patricio López-Sepúlveda, Patricio Peñailillo, Eduardo Ruiz, and Hugo Valdebenito.

had shown convincingly that more than 60 percent of the species in the archipelago were endemic, that is, they only occurred on these young islands, which meant that a progenitor arrived from some continental source area and evolved over millions of years. Understanding processes of evolution in the islands, therefore, involves focusing on these endemic groups. Biogeography in the islands has a similar requirement. One must have a good inventory of what exists in the archipelago, as well as knowing what the close continental relatives are, so that geographic origins of the island species as well as their modes of dispersal can be inferred. And finally, a complete inventory is requisite for evaluation of the conservation status of the flora, which means knowing how many species (often in percentages) are endangered, threatened, or vulnerable. It would be impossible to set conservation priorities if a comprehensive inventory did not exist.

For evolutionary and biogeographic studies, we knew it was essential to have geological ages established for each island. Based on K-Ar radiometric dating from rocks that we collected and brought back in 1980 for analysis at Ohio State University by our colleague Kenneth Foland in the Department of Geology, we learned that Robinson Crusoe Island is about four million years old and Alejandro Selkirk Island approximately one million.[2] This makes sense, because the islands have originated from a hotspot from beneath the ocean floor, and as the tectonic Nazca Plate moves eastward, finally subducting under the South American continent, periodic eruptions of new magma led to the production of new islands, some of which grew large enough to emerge above the sea. Robinson Crusoe Island first appeared four million years ago, and after three million years, and much eastward movement of the plate, Alejandro Selkirk Island formed in the same original location above the hotspot. The plate keeps moving, which means that the original hotspot (which is fixed) is located approximately one hundred kilometers west of where Alejandro Selkirk Island now occurs. These results have powerful evolutionary and biogeographic implications.

Armed with dates for the archipelago, we directed our attention to evolutionary (phylogenetic) relationships among the endemic species of the flora. What is related to what and by how much? This is the eternal quest of botanical systematics, and here in the islands we had particular interest in *Dendroseris* and *Robinsonia* and all the other complexes that contained two or more endemic species. Morphology, the external features of organisms, serves to give a good approximation of affinities, and this is always the first line of examination, but in some groups it helps to have more insight, and here is where DNA data can be extremely helpful. It must be remembered, however, that no single type of data is a panacea, because each plant group has had a very different evolutionary history.[3]

Once relationships among endemic species of the flora have been sorted out, biogeographic comparisons can be made with suspected close relatives in source areas. The challenge is to identify from where the progenitors have come and how they arrived. The earlier botanists, especially Hemsley, Johow, and Skottsberg, speculated on such origins, but these were only beginnings. More recently Gabriel Bernardello from the University of Córdoba, Argentina, and Gregory Anderson from the University of Connecticut, have put special attention on biogeographic origins, and they have shown that most ancestors lie in southern South America, some from Chile, others from the Neotropics of South America, and also some from the western Pacific.[4] Birds transported most of the propagules, small seeds or fruits that had been ingested, stuck in feathers, or plastered on muddy feet. One does not know in a particular case exactly what transpired, but by examining the morphological features of the fruits and seeds, it is possible to suggest modes of dispersal.

Once propagules arrived to one of the islands of the Juan Fernández Archipelago, most often to Robinson Crusoe Island because it was closest to the mainland and existed three million years before Alejandro Selkirk Island was formed, evolutionary processes began to unfold over many generations within established populations. For proper interpretation of evolutionary events, we have stressed perspectives from historical ecology. All oceanic islands have an ontogeny from birth beneath the ocean, full development as an island, erosion and subsidence, and finally disappearance under the sea. On this metamorphosing stage is played the evolutionary drama within each plant group. It is imperative to infer, therefore, what the geomorphological dynamics of each island have been. With Alejandro Selkirk Island at only one million years of age, the present island would have been altered only slightly by subsidence and erosion during this time. In contrast, we might expect Robinson Crusoe Island to have lost considerable surface area (and ecological zones) over four million years of its existence,[5] such that the species we now observe did not evolve in the present condensed landscape. That is, the present spatial and ecological context for the endemic flora may reflect very little about the processes by which these species originated. One of the contributions from the Juan Fernández project has been a strong emphasis on understanding environmental history for evolutionary investigations in oceanic islands worldwide.[6]

In addition to inferring major natural changes to the landscape over the past four million years, it has become important to examine the historical impact of human activities during the past four centuries. Prior to discovery of the archipelago by Juan Fernández in 1574, there were no indigenous peoples, so the human impact on the ecosystems (especially the vegetation) began with the arrival of European visitors. Because of abundant historical documents from visitors to the islands, particularly those written by captains or other officers on board the ships, there exists a detailed chronicle of human activities when they arrived, especially to Robinson Crusoe Island with its protected bay, Bahía Cumberland. Most impact occurred on the northeastern side of the island, where forests were cut for lumber to repair ships, to construct buildings, and to make firewood. Plants and animals were also introduced, particularly the goat and rabbit, which became feral and pressured the native flora. The marine resources, such as the fur seal, fish, and lobsters, have been harvested extensively, such that these populations have been greatly reduced, but thankfully, not to the level of extinction. Chapter 12 provides more detail on all of these points.[7]

With geographical, geological, and historical ecological contexts having been established, we selected specific plant groups for intensive investigations. Those that contain the largest number of endemic species harbor the maximum amount of evolutionary information, and in the Juan Fernández Archipelago, that means *Dendroseris* (with eleven endemic species) and *Robinsonia* (with eight species), both in the sunflower family (Compositae or Asteraceae). This was convenient, because this was the family in which Clodomiro Marticorena, Dan Crawford, and I had the most previous research experience. The basic questions were always: (1) How are the species related to each other and to continental relatives (that is, what has been the *pattern* of evolution)?; (2) from where did the original immigrants come and when and how did they get there (that is, their biogeographic history)?; and (3) what have been the evolutionary processes that have produced these island lineages (that is, what have been the *processes* of speciation)? The strategy is to first get the patterns of relationships and then seek the processes that give rise to the patterns.

It is important to emphasize that oceanic islands are, indeed, natural laboratories for studying evolution.[8] Plant immigrants, which have the capacity and luck to disperse to a newly formed island, often change over millions of years. Some lineages change more rapidly and dramatically than others, but many groups speciate. This is evolution in action. If anyone seeks evidence for the existence of evolution on our planet, floras and faunas of oceanic islands offer excellent case studies. The theory of evolution, which was presented in a coherent and well-argued fashion by Charles Darwin in 1859,[9] has been tested thousands of times over more than 160 years and is now regarded as a *fact*, not just a theory. It makes no sense to talk about "believing" in evolution; it is a fact, it exists, and our task as scientists is now to learn more about how it has happened in different groups of organisms, such as in the endemic plants of the Juan Fernández Archipelago.

The only way to answer the evolutionary and biogeographic questions posed above is to gather comparative data from the organisms. To understand relationships among plants requires first examining their external features, their morphology. Evolution works on changes in the genetic makeup of organisms within populations, but natural selection filters the genetic effects through the environment, which results in the phenotype, which we see as the morphology. Anatomical, biochemical, and other aspects are also affected by the evolutionary process, but what we see most easily is the morphology. For our evolutionary studies on the Juan Fernández flora, we have always begun with hypotheses of relationships based on morphological data.

Related to morphology are the internal, or anatomical, features. Back in 1983, I had received a request from Professor Dietmar Behnke of Heidelberg University, Germany, for stem sections of *Lactoris*, one of the unusual plants of Robinson Crusoe Island discussed in chapter 5. Professor Behnke was conducting an investigation of the types of starchy or proteinaceous inclusions in the membrane-bound plastids of the sieve elements, sugar-conducting transport cells that occur in all families of flowering plants. Lactoridaceae had not yet been examined. For his research, he needed small pieces of freshly cut stem tissue placed in a special preservative (formaldehyde with glutaraldehyde), which meant this had to be done in the field on Robinson Crusoe Island. Professor Behnke had sent me, prior to our 1984 expedition, a vial with this special fixative, and I agreed to help if we encountered *Lactoris*. One day in the field, we did have such good luck, locating a few plants nestled in a high mountain secluded forest. We brought back our day's collections for processing, and in the evening I prioritized the preparation of *Lactoris* for Professor Behnke.

After an agreeable dinner, now 10:00, I set to work on *Lactoris*. I got out my pocketknife, sharpened it on my quartzite stone to make a fine edge, took out the sealed plastic bag with stems and leaves of *Lactoris*, and selected a particular stem for sectioning. The species has semi-succulent stems, which makes cutting quite easy. As there was no electricity in the village after 8:00 at night, I had to work by candlelight, which was sufficient for the task at hand. I was struck by the context of the situation: there I was, by flickering candlelight, slicing through this enigmatic ancient plant family on an isolated Pacific Island, far from the rest of the world. After our return to Ohio, this vial would be shipped to Professor Behnke in Germany for laboratory analysis, who would make extremely fine slices of the stem tissue with an ultramicrotome, cutting through the microscopic plastids to allow observations of their inclusions, and making these observations with the very high-powered transmission electron microscope (×20,000 magnification). The contrast between the rough physical fieldwork and the precision of high-tech lab work struck me as fascinating, beautiful, and also totally appropriate. It underscored the importance of collaboration between workers doing different types of scientific studies. The story ended well, with material being sent to Professor Behnke, followed by successful anatomical work, and in 1988 he published his observations, showing starchy inclusions in the plastids of *Lactoris*.[10]

Other types of characters can provide useful information, such as chromosome numbers, reproductive biology, flavonoids (secondary plant products), isoenzymes, and DNA sequences. Chromosome studies can be very informative for suggesting evolutionary affinities among plant groups. In plants of continental regions, some lineages have undergone remarkable changes in chromosome number over millions of years, involving multiplication of whole sets of chromosomes (polyploidy) or addition or loss of single chromosomes (dysploidy). For example, among the forty species of the genus *Melampodium* of the sunflower family (Asteraceae) centered in Mexico, which served as the topic of my Ph.D. thesis, the following haploid chromosome numbers have been reported: n = 9, 10, 11, 12, 14, 18, 20, 23, 24, 27, 28, 30, and 33.[11] These chromosome levels tell much about evolutionary relationships involving polyploidy, with one line containing species with n = 10, 20, and 30, a second with n = 9, 18, and 27, a third with n = 11 and 33, and a final lineage of n = 12 and 24. One of the important results of our chromosomal surveys among the endemic plants of the Juan Fernández Islands is that *no* such changes can be seen,[12] that is, whatever the chromosome level of the initial immigrant, this is what has remained during speciation in the archipelago. An explanation for this phenomenon is that directional selection in the new island environment is so stringent that any mutant individual with differing chromosomal levels would be selected against and lost from the population.

Another very important type of data obtained from the endemic plants of the Robinson Crusoe Island was from reproductive biology.[13] There are many dimensions of a plant's reproductive system that might be investigated, but most relevant for our interests were modes of pollination and breeding systems. Gregory Anderson and his colleague, Gabriel Bernardello, carried out intensive studies on many of the endemic species of Robinson Crusoe Island during the expeditions of 1991 (Gregory alone) and 1996 and 1997 (together). These results revealed the near absence of insect pollinators in the island.[14] Some plants had adapted to hummingbird pollination (e.g., *Dendroseris* and *Rhaphithamnus*), but others were pollinated by wind (e.g., *Lactoris*),[15] even though the flowers were not structurally of the type normally regarded as having a wind-pollinated syndrome (e.g., long, exserted stamens, small or inconspicuous floral parts). They postulated that the ancestors of many of the island endemics may have been insect pollinated, but in the absence of insects in the island, natural selection put a premium on evolution of alternative pollination mechanisms.[16] Their companion studies on breeding systems have revealed which species are normally self-compatible and which are out-crossing, with these data being fundamental for understanding patterns of genetic variation within and among populations.

Another set of data that we compared and contrasted has been the flavonoids,[17] and the lead workers in this area were Daniel Crawford and Mario Silva. These are compounds that serve many functions, different ones for different tasks, including association with basic metabolic pathways (e.g., photosynthesis), colorful pigments in flowers to attract pollinators, and defensive roles to protect plants from UV radiation and insect predators. The laboratory techniques involve extracting the compounds, usually from leaves, and separating them with different organic solvents on large sheets of filter paper. Back in the 1960s and 1970s, there was a huge interest in these compounds for assessing evolutionary relationships, and during our early work in the islands we examined a number of plant groups (e.g., *Dendroseris*, *Robinsonia*) for these features. Our data revealed that the compounds in most cases correlate with morphological

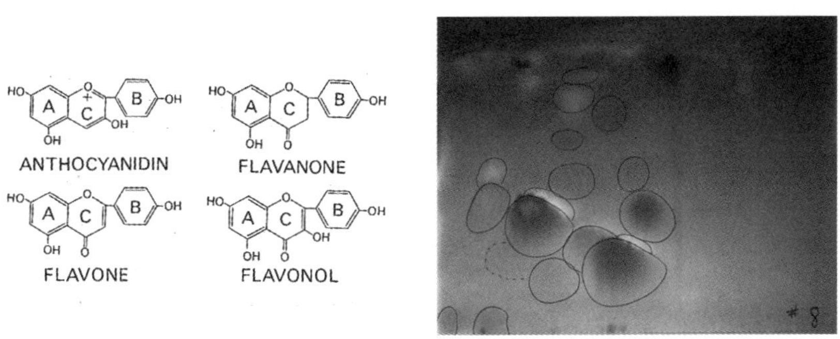

Flavonoids as characters for assessing plant relationships. Left, basic classes of compounds; right, representative paper chromatogram of compounds separated and viewed under ultraviolet light.

affinities and hence serve as good specific markers.[18] More importantly, complexes of compounds sometimes corresponded with subgeneric groupings, indicating that morphology and flavonoid chemistry have been modified in similar directions during evolution within the archipelago. What we did not see, however, is a dramatic loss of compounds in comparison with continental ancestors. The idea had been suggested that because flavonoids may have a defensive role, once in islands where former continental predators (e.g., phytophagous insects) were no longer present, the diversity and quantity of flavonoids should be expected to be less.[19] Our results suggest that flavonoids in these groups are performing other important functions and/or they may be linked genetically to other features that simply cannot change without serious negative results to the organisms.

During the 1970s, a new technique developed of using different forms of metabolic enzymes for assessing relationships and for determining genetic variation within and among populations.[20] The idea was that these enzymes, fundamental for many life processes, can occur in different allelic forms (allozymes), and gel electrophoresis was the method for separating them by electrical current and then staining for visualization and scoring. Workers seek to analyze twenty or more enzyme systems to generate enough data for statistically valid conclusions about patterns of genetic variation. These data can tell much about genetic relationships among closely related populations and species, as well as indicating levels of genetic variation within and among populations of the same species.

But I have to tell you a story about our experiences with isozymes on our expeditions to the islands. Science involves having good ideas, but it also requires perseverance, which is the basic theme of this story. On trips in 1986, 1990, and 1991, our main objectives were to obtain isozyme data, and at that time, the way to do this was to bring back seeds and germinate them in the laboratory, which then provided fresh tissue for grinding up, using buffers, and running electrophoretic separations. Fresh, young leaves collected in the field could be used, but because of the time involved with travel schedules from these isolated islands, getting the samples back to the laboratory at Ohio State in suitable condition would be difficult. With time, it was possible to use liquid nitrogen to preserve the tissues, but this created logistical problems.

On the trip to Robinson Crusoe Island in 1984, our crew collected literally thousands of seeds, and these were successfully brought back to

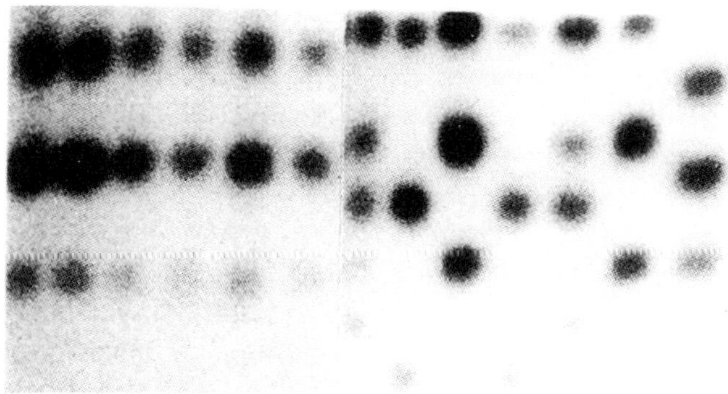

Isozyme patterns (zymograms) among individuals of the endemic *Chenopodium sanctae-clarae* from Morro Spartan (see color insert, **8**). Left, no genetic variation among six plants with the enzyme 6PGD (6-phosphogluconate dehydrogenase); right, considerable genetic variation among seven individuals with IDH (isocitrate dehydrogenase).

the United States at the end of the collecting efforts, arriving at the Miami International Airport, which has a phytosanitary inspection station. The U.S. Department of Agriculture (Plant Protection and Quarantine) inspects all incoming live plant materials, and there are lots of these botanicals, especially cut flowers from Colombia and Ecuador. This inspection is sound policy, because if insect (or other) pests get into the U.S. environment, they could seriously harm the citrus and horticultural plant industries in states such as Florida, Arizona, and California. We declared our collections, more than 10,000 seeds, and for this large volume, we were escorted to an adjacent inspection building, thankfully air-conditioned, where an official began examining every seed, some under the dissecting microscope. After several hours, it was obvious that we were going to miss our connecting flight to Columbus, and it was also evident that his work would not be finished quickly. The inspector then said, "Boys, I am going to need more time to look all of these over, so why don't you leave me to handle this and come back at 9:00 in the morning, when it will be finished, and you can continue your trip home. To be on the safe side, reschedule your flight after 12:00 noon tomorrow. If I find any seeds that are infected, I will have to destroy these, but so far, I have not seen too many of these, so don't get worried about losing your research materials." Well, when the boss in a situation so declares, one obeys. Simple. We took

Inspection of seeds from our Juan Fernández expedition of 1984 in the U.S. Department of Agriculture plant quarantine station at the Miami International Airport. With Hugo Valdebenito.

our leave, got our luggage together, and checked into the airport hotel for the evening. This was quite pleasant, really, getting a good shower, a great dinner, and also a good sleep after the all-night flight from Santiago to Miami. In the morning we visited the inspector again, and all seeds had been examined and repacked, along with a certificate of inspection. Done. We did make Columbus on time with our rescheduled flight, and the seeds thereafter were safe in the laboratory. Many of them did germinate, and Dan got very useful genetic data from the collections. Seeds of some genera, however, did not germinate under laboratory conditions within our timeframe, so this got us thinking about the next trip and how a more efficient system might be established.

Based on innovations with isozyme techniques in different labs around the country, Dan thought that using liquid nitrogen might be the best approach. Liquid nitrogen is not particularly expensive, is relatively easy to obtain in nearly all countries (including Chile) because it is used in medical procedures, and in the right container, it can last a very long time. Dan tested these aspects in the lab. Specialized two-handled carboys were purchased, and samples of local plants were made and preserved inside, each in a small, plastic tube coded by number and held in metal trays. Liquid nitrogen is kept at −195°C, which requires special handling and always wearing insulated gloves when inserting or removing samples. Dan wanted to be certain that tissues, after being in the chamber four weeks (the duration of a typical expedition), would provide high-quality isozyme activity and acceptable results. We also wanted to be certain that the liquid nitrogen would last that long, as once in the islands, there could be no replacement. After the month, tests showed excellent results, which convinced us that this would be the approach on the next expedition. We would have to box up a carboy for carrying as excess baggage on our flights, but without the liquid nitrogen, which we could purchase in Concepción after arrival.

We departed for Chile in January of 1990 for our next expedition to the Juan Fernández Islands, which proved eventful. One of the boxes we checked in at the Columbus airport was that containing the carboy for liquid nitrogen, which when empty, is not particularly heavy—no worse than a normal suitcase. We made it to Concepción without difficulty, and it was easy to get the carboy filled with liquid nitrogen from a company outside town. Again, simple, inexpensive. Much of our supplies had been sent by boat (*Afrodita III*) from Valparaíso to Robinson Crusoe Island accompanied by Pato López-Sepúlveda and Mauricio Rondanelli, during which they suffered through a powerful storm, but our supplies were not damaged thanks to the careful wrapping in plastic. For us, the trip to Robinson Crusoe Island was by small plane, and Mario had arranged for the planes to leave directly from Concepción, which meant a short hop up the coast to Los Cerrillos Airport in Santiago, refueling, and then heading immediately out to the island. The advantage of this was that there was no discussion about the dangers of hauling the liquid nitrogen on board. We were traveling with it, obviously, so this had been cleared previously with the airline (Taxpa), and no hassles ensued. Arriving to the island, it was an interesting contrast to watch Dan holding tight to the carboy on board the municipal launch from Bahía Padre to the village, a contrast between the simple life of the fishermen in the islands and our high-tech scientific world. The carboy got installed in the CONAF work area, and all went well for the collecting of samples for more than three weeks. Day after day, more and more young leaves were prepared and inserted down into the liquid nitrogen for examination later in the laboratory. At the end of the four weeks, we had thousands of good population samples of many endemic groups in the flora, all of which would tell us much about levels of genetic variation and also help inform on conservation priorities.

Use of liquid nitrogen for preserving plant tissues for isozyme studies. Top, Dan Crawford holding onto the packaged carboy on boat trip from Bahía Padre to San Juan Bautista; middle, preparing to insert the DNA samples into the liquid nitrogen carboy on Robinson Crusoe Island; bottom, pouring off the liquid nitrogen at the University of Concepción.

At the end of the expedition, we packed up all equipment, plus now the ultra-valuable carboy with liquid nitrogen containing thousands of leaf samples, and we headed down to the wharf for the boat trip around to the western side of the island and our flight back to Santiago and then Concepción. Everything worked as planned, and in short order we were back in Concepción. Once arriving at the university, we settled into the routine of separating voucher specimens, visiting with administrators, talking with students, the usual post-expedition activities. We then had to decide how to get the carboy with samples back to Columbus, as it still had liquid nitrogen inside. It would be impossible to show up at the airport window and just declare this for shipment. Dan assumed that we could decant the small plastic vials from the liquid nitrogen and repack them in regular ice for the flights back, which we could easily do in a hand-carried picnic-style small cooler. We poured out the remaining liquid nitrogen on campus, and all seemed to go as planned. We did, in fact, get these samples safely back to Columbus, and once there, they were placed in a deep freeze in Dan's laboratory.

The weeks rolled on, and Dan and I got back into the usual routine of classes, meetings, conferences, the typical challenging set of obligations that confronts all professors. After about a month, Dan came into my office and asked to chat with me. It was always a pleasure to see Dan and talk, so I said, "Sure, come on in and sit a spell. What's up?" "Well," said Dan, "there is a small problem with the isozymes." To which I responded, "Oh, what happened?" Dan is never hysterical, nor am I, so all this was pretty low key. "Well, you remember when we took the leaf samples from the liquid nitrogen and repacked them in regular ice? It turns out that such a drastic change in temperature affects the activity of some of the enzyme systems, such that we will not have as many as we had hoped. We will get some good data, but not as much as planned. I had not realized this, and I am really sorry. It just happened." At first hearing, this came as pretty bad news, especially considering all the very hard work by everyone on the expedition climbing up and down the ridges and valleys and collecting so many interesting samples across the endemic flora. But, I also knew that however bad I felt, Dan must have felt ten times worse. I just assured him that I understood the problem, and he could count on me to help with the interpretation and writing up of any results that he might be able to obtain. I couldn't offer more than that, as Dan was the real expert with isozymes and their techniques. Dan left, and I confess to feeling down about this news, but I also had confidence that if anyone could extract data from the samples, despite problems, it was Dan.

Dan was able to get good data from the damaged liquid nitrogen samples, but we resolved to do better on the next trip, which was also to use the same sampling methods. I checked into the possibilities of shipping liquid nitrogen from country to country, and yes, there were specialized companies that handled the permits and certifications. We checked with our Chilean colleagues to learn if such a company existed in Santiago, and they responded affirmatively. I obtained the telephone number of the company and called them about our needs, and yes, they could do the job for us. I sent them a fax with our planned field schedule, which they accepted. We only needed to call them six hours before flying from Santiago to the States, and they would come to our address, pick up the carboy, and handle the rest. We had to pay cash upon pickup. Based on this information, Dan and I agreed to do the liquid nitrogen program again. Many of the active labs were having very good success with this preservation method, and we knew how to get the carboy to Robinson Crusoe Island and back again.

This time in 1991 we convened our expedition team in Concepción. After the usual organization and discussions, we assembled in the bus depot (Turbus) in the city at about 10:00 for the all-night bus trip to Santiago. All participants were excited about the trip, some of whom had never visited the archipelago. Among our group were Gabriel Bernardello and Gregory Anderson, who were investigating reproductive biology in the flora of the islands (mentioned earlier). At the bus terminal, all went well enough until it was time to give the luggage to the porter who put the bags and boxes in the storage area under the bus. He hefted our personal bags with no comment, and when coming to the equipment boxes, he asked what each contained. There was no difficulty until he got to the box containing the carboy with liquid nitrogen, which we had filled up again outside Concepción. At this he balked, and he stressed that it was impossible to send "this dangerous chemical" by bus. We assured him that if any of the liquid nitrogen evaporated, it was not flammable, nor poisonous, and so represented no threat to the bus or passengers. He resisted repeatedly, such that I was beginning to get concerned about not being able to travel that evening with the most important element of our equipment. I started to think about alternative plans, such as trying to find a truck or van that would take us to the airport by early morning in Santiago for the scheduled flight, but these sorts of last-minute arrangements I dislike—too much adrenaline and pressure. After more discussion, Gabriel managed to convince the bus conductor that we ourselves would not want to travel on the bus if anything were dangerous. This seemed to win the argument, and he finally put the box inside the bus. We waited until the door of the storage area was closed and secured, with our items safely inside, then following the conductor into the bus, being polite to let him enter first, so that there could be no last-minute removal of our carboy with the precious liquid.

We did get all equipment to Los Cerrillos Airport in Santiago, and on to Robinson Crusoe Island. All material and expedition members were finally safely in the village of San Juan Bautista, with the carboy nestled in the storage room at CONAF headquarters that we intended to use as a laboratory. We carried out the same routines as on the previous trip, collecting young leaf samples of many of the endemic species, and especially of *Dendroseris* and *Robinsonia*. After sampling, we mashed the fresh leaves in a mortar and pestle with buffer, decanting the slurry into very small plastic vials, and then placing these into small trays and down into the liquid nitrogen. This had to be done for all samples, totaling many thousands by the end of the trip. There were no crises on this outing, and at the end of the several weeks, we got ready to depart, as usual, with the carboy filled with extensive samples for isozyme studies in the laboratory in Columbus. And so, as on previous trips, we headed back on the municipal launch to Bahía Padre, up the hill in the jeep to the landing strip, and finally winging once more eastward toward the Los Cerrillos Airport in Santiago.

The plan was to leave some luggage and equipment, including the carboy, at the home of my in-laws in Santiago, while we took the bus to Concepción and spent several days there sorting specimens, organizing our receipts, and talking to students. After this, Dan and I headed back to Santiago, again via overnight bus, finally arriving at 7:00 in the morning at the central bus terminal, and from there taking a taxi to my in-laws' house. All was well with the family, and Tessa, Patty's mother, prepared a nice breakfast for us, the usual toast, marmalade, sliced cheese, and coffee, quite sufficient to get us going for the day. We also took the opportunity to shower and shave, and assess our equipment and personal items for the return flight to the United States. Most important was putting through a call to the company that would dispatch the liquid nitrogen to Columbus;

they said they would come to the house at 4:30 that same afternoon. This worked very well with our schedule, because we were supposed to fly out that evening from Santiago at 10:00 to Miami, and then the last hop to Columbus the next day. We worked to repack all gear for the return flights, estimating size and weight of each item, until we were satisfied that all would be within the requirements of the airline. More importantly, at 4:45 p.m. a representative of the company arrived to the house, and after doing the necessary paperwork and making payment, he took the carboy away and with it all the samples for isozyme studies accumulated through very hard fieldwork over almost three weeks. Dan and I looked at each other and hoped that these samples would reach their destination as planned.

They did arrive safely. Two days after we had returned to Columbus, I received a call from customs at the Columbus airport, indicating that the liquid nitrogen carboy was there, it had passed customs, and I could come and claim it. Quite excitedly I immediately drove out to the airport, which in Columbus is only about thirty minutes away from the university. It took me a while to find the customs building, but having finally located it and parked, I entered the office and inquired about the shipment. I had brought my copies of the shipping agreement from the company in Santiago, just in case, to ensure proper identification, plus my own personal IDs. The official was satisfied, went in the back and brought out the carboy, without any packaging, and presented it to me. It looked a little beat up, but otherwise intact, and it still had some weight, which meant that there was still some liquid nitrogen inside, a very good sign. I nestled the carboy into the back seat of my car, making sure it would not tip over, and headed back to campus. Upon arrival, and getting the carboy into the building and up the elevator, I headed right to Dan's laboratory. He was there working away, as usual, and upon seeing the carboy, he broke into a broad smile. We had done it—the samples had arrived all the way from Robinson Crusoe Island to his lab in Columbus in good condition. Our plans had worked, everything seemed perfect, making all the efforts of the trip completely worthwhile.

We again got back into the routines of the normal academic life with teaching, dealing with students, attending meetings, and working on manuscripts. One of the delights of the academic world is the huge variety of things to be done all the time, and the professor has a lot of control of what to do when. There is maximum freedom, but there is also maximum responsibility. No one asks what you are doing for months at a time, but at the end of the year, if there is no productivity, questions start being asked and warnings are given by the administration. If that continues, a young professor without tenure will not be made permanent. For a professor already with tenure, low productivity will mean low raises, perhaps even with more committee work or increased teaching load in the following year.

Dan appeared in my office about a week after we had come back, and he asked to chat with me. We touched base frequently. Our offices were at different ends of a long hallway, but close enough to take a break and drop in for a discussion. Dan did not look quite right, his face tight. He began: "The good news is that the liquid nitrogen worked well, and all the samples arrived in good shape. But, there is bad news—we have taken out some of the samples and run them electrophoretically, and some work, and others do not. There seems to have been a problem with the buffer that we used in the field. It was not mixed properly by one of the students, which means that some enzymes will work and others won't." I could hardly believe what I was hearing—another massive collecting effort resulting in only partial success with the samples. This was the second time that this happened.

Dan did get some good data from the recently collected materials, and good publications resulted, but I was deflated about ever having success with isozymes in very isolated situations. But, as hope shines forever bright, we discussed preparing another isozyme grant for a more conservation-oriented project involving genetic variation within and among endemic species of Robinson Crusoe Island, and using these data as a basis for setting priorities for conservation. I said to Dan, "Look, we have had so many problems with using liquid nitrogen and trying to use this to get good samples and data in the islands, that I have no taste this time for repeating these efforts. There must be a better way." With time, progress had come to San Juan Bautista such that there was now electricity from 8:00 in the morning until 8:00 at night, thanks to having two large generators. This meant that, in theory at least, we could run electrophoretic gels during the day, if we could get all the lab equipment to the island. This was an exotic plan, but Dan was positive about it, and it represented a chance to actually get the genetic data immediately after collecting the materials. If this would work, it would mean making collections on one day and having genetic data from that population the next day. The gels would be sketched and photographed, so that the data would be captured immediately. These were encouraging thoughts, indeed.

By some miracle, we did get this new grant funded from the U.S. National Science Foundation (NSF). Whenever an investigator gets a grant funded, it seems like a miracle, because the competition is so intense (funding levels can be less than 10 percent of submitted proposals). We now had to get very serious about getting the isozyme equipment and materials assembled for shipment to the islands. One very important item needed to be constructed, and that was an electrical switching board that would connect the power sources and gel trays all together, making it easy to control the current to specific pieces of equipment. Several gels could be run at the same time, too, which would increase efficiency. While these thoughts were percolating in our heads, I mentioned it to my wife, Patty, and she suggested that I ask her father, Don Juan, if he could help us. He was a chemical engineer and sort of an inventor, always making things for the home and for his small factory where he produced commercial-grade detergent. I asked Patty to inquire if he would be interested, and would have the time, to put together such an apparatus. She did so, and he agreed to help, which made me feel bad in one sense, taking advantage of another family member to get professional items constructed, but on the other hand, I was delighted that he took on the challenge, and I had no doubt that he would do it well. Dan sketched a diagram, a schematic, showing the parts and how they needed to be connected, and I faxed this to my father-in-law. He modified the drawing a bit, asked some questions, and sent this back to Dan for further comment. These details were ironed out, and Don Juan started constructing the switchboard, finishing it well before our arrival in Santiago for the trip.

This time in January of 1996 Dan and I were not going to Concepción at the start of the trip but would do so afterward. Our Chilean friends were coming by bus from Concepción, and we would meet at the airport for the flight to Robinson Crusoe Island. We had agreed to organize this trip not only as a research expedition, but also as a formal course from the University of Concepción, which would give the students hands-on experience with isozyme methods as well as general collecting techniques and discussions about the evolutionary aspects of the project. Dan and I arrived at the Santiago airport, got our gear together, and piled everything into a taxi with a rack on the top, which allowed our boxes to be lashed securely, with our bags and us inside. We zoomed into the city and out

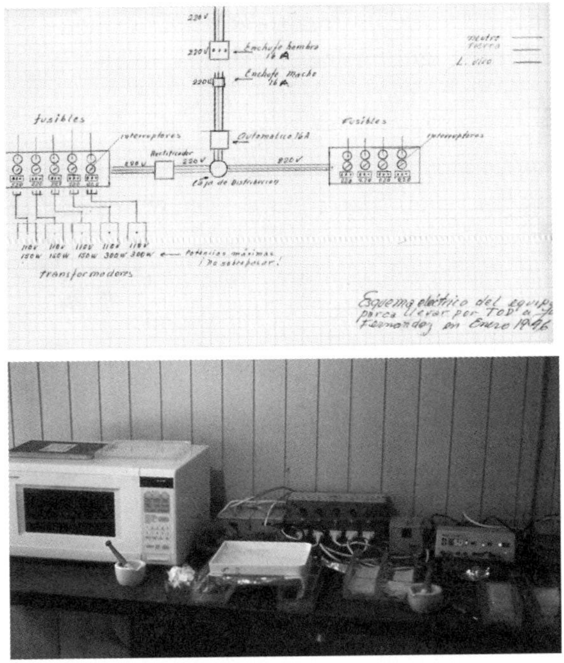

Electrophoretic equipment during the 1996 expedition on Robinson Crusoe Island. Top, schematic electrical diagram by Juan Ghivarello; bottom, switchboard and equipment installed in our island house.

toward Las Condes, the suburb where my in-laws lived, eventually stopping in front of their house, with the same old black iron gate in front. Familiar address, familiar territory. After unloading, making arrangements to store our boxes in the garage on the ground floor, having a nice breakfast, and catching up on family news in the process, our discussions turned to the switchboard that Don Juan crafted.

After breakfast, with stiff coffee to wake up from the overnight flight, we adjourned to the terrace where Don Juan had set up the switchboard on the ping-pong table under their grapevine trellis. There it was, the switchboard, done all in brass, and soldered together, very professional, very well crafted. We agreed that it made sense to get out the electrophoretic equipment, attach everything appropriately, and give it a small test. Don Juan brought an extension cord from the garage, from where it could be plugged in, and in the meantime Dan and I found the boxes with the pieces of equipment, which we opened up and arranged together on the ping-pong table. The current in South America is 240 V; Dan had brought a step-down transformer to get the current to 120 V, required for all his lab equipment brought from the States. There was also a voltage regulator that Don Juan had purchased, which would allow Dan to watch the current as the electrophoretic runs were progressing to avoid any surges that might occur in the island. Dan was completely satisfied, Don Juan was proud, and I was relieved that everything seemed in order. As the years rolled forward, this switchboard was used on three expeditions to the islands.

Our colleagues from Concepción did arrive punctually the next morning, and we all met at the Santiago Los Cerrillos Airport at 9:00. The trip over to Robinson Crusoe Island was completely routine. CONAF had arranged for us to stay in the village in one of the houses of the Servicio Aeronáutico, which is conveniently located on the old site of Fort Santa Bárbara. Two other modest houses are there for the officials, but this third one is used for visiting researchers. Important for our needs was that it had good electricity inside, and one "bedroom" (without beds) could be converted into the electrophoretic laboratory, with a few modifications, such as taking down the closet door panels and using them for tables on which to arrange the equipment. The house had seen considerable wear over the years, so we were not doing anything that was inappropriate. There was access to running water in the bathroom and kitchen, and Dan had brought a small refrigerator for the lab that could be plugged in, also with a step-down transformer. After everything was set up and connected, the very good news was that everything worked just fine. The collecting of samples soon began, and it was fun for all of us to collect specimens one day and get the report about genetic variation the next day. This was very exciting, educational, and the data were good, very clean, and accumulating daily. All had finally worked as planned—but it had taken three expeditions to get it that way. We used this equipment on another expedition in 1997, and Eduardo Ruiz used it in 1998 for his research on genetic variation in endemic plants of Alejandro Selkirk Island, as part of his doctoral dissertation from the Universidad de Concepción. This was an even more isolated location.

But science progresses and methods change. In the early 1990s, new techniques were being developed using DNA to assess relationships and also to more precisely determine levels of genetic variation within and among populations.[21] The very good news in terms of field collection is

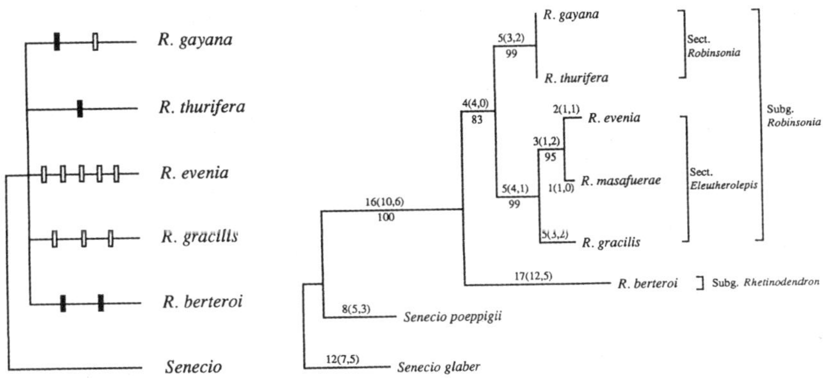

Phylogenetic trees of DNA relationships among species of *Robinsonia*. Left, based on restriction site data; right, based on sequences of ITS (a nuclear gene).

that leaves dry very quickly in silica gel, the same compound used to preserve flowers for dried arrangements. The cells dry very quickly, preserving many DNA sequences.[22] This is the case with DNA contained within the nucleus, and even more so with DNA contained within membrane-bound chloroplasts and mitochondria, the former providing better data for assessing plant relationships. Use of inert silica gel in the field was infinitely easier than working with liquid nitrogen. Dan again took the lead with these techniques, first working with restriction site data.[23] The DNA molecule is cut at specific sites, and the fragment profiles are compared between different species. As methods got easier (and cheaper), we were able to sequence certain DNA regions, several hundred base pairs long, and make comparisons based on these data.[24] Into the 1990s and later, fragment analyses of DNA at the populational level using amplified fragment length polymorphisms (AFLP)[25] allowed us to distinguish each individual plant based on its distinct fragment profile (see color insert, **16**). Into the 2010s, we investigated microsatellites, repeating DNA sequences that are highly reflective of genetic relationships.[26] And in a study just completed, the entire molecular structure of the chloroplasts of seven species of *Dendroseris* has been determined and inferences obtained on relationships within different parts of the genus (among the subgenera).[27] All these methods and analyses were done back in the laboratory, which meant the only challenge for the fieldwork was getting the silica gel to the island and back again with dried leaf samples. The only threat to this chemical was water, particularly seawater, which meant we had to be especially careful to pack samples in double plastic bags and then in strong plastic tubs, these roped up much like Christmas gifts. This allowed each tub to be easily picked up and passed to someone else, hence lowering the chance of falling into the ocean.

The last type of data generated during our expeditions dealt with the vegetation. There are lots of ways of dealing with plant associations, from the very simple description of what one sees to more quantitative and statistical approaches. The principal investigator for this line of research was Josef (Sepp) Greimler, a colleague and good friend in the Department of Botany and Biodiversity Research in Vienna. He did his Ph.D. dissertation with vegetation analyses in Austria[28] and was very qualified to lead these studies in the islands. In addition to using data from high-quality stereo aerial photographs taken by the Chilean Air Force in 1980, Sepp was skilled in the methods of a former Swiss worker, Braun-Blanquet, who had developed a way to inventory all species within sample plots, plus estimating their abundance and dominance (the relevé technique[29]). These copious data were analyzed with computer software to yield groups of similar plant associations that were used to plot different vegetation zones on the maps.[30]

These modern vegetation maps show several important points (see color insert, **17**). The vegetation of Alejandro Selkirk Island still remains in broad zones, which would be expected after only one million years. The older island, however, shows a very low and dry western portion, due to subsidence and erosion, and also large areas of disturbance centered around San Juan Bautista and the smaller bays where ships anchored during past centuries. In these areas many introduced invasive plants now grow, often mixed with native forest. Erosion has also been dramatic in some regions, particularly in La Vaquería and Puerto Francés. These natural and human pressures on the native flora have resulted in a compaction of the vegetation, such that the endemic species are now confined primarily above 350 meters, many at higher elevations. The most undisturbed areas are those of difficult access, where people have simply not often visited. The interpretation of evolutionary events in the islands must take these vegetational modifications into account.[6]

One of the central evolutionary questions in the archipelago has been: What mechanisms of speciation have taken place to produce the endemic species? We have expended considerable effort to answer this question. Armed with phylogenetic data about affinities, distributional data showing ecological tendencies of the species, information on landscape change, and knowledge of chromosome numbers, we have been able to infer many aspects of the evolutionary process.[31] From our chromosome studies that showed uniform numbers within species complexes, we were able to eliminate polyploidy (whole genome duplication) or gain or loss of chromosomes (dysploidy) as factors. For speciation to occur, reproductive isolation must take place, and these isolating mechanisms can be categorized for convenience into pre-zygotic and post-zygotic types.[32] The former include spatial, ecological, and phenological (flowering time) differences. The latter are genetic and developmental barriers that come into play after a hybrid cross has been made; seeds do not develop, or they develop but the seedling is weak; or hybrid plants survive but they are sterile. Populations that become separated (isolated) between the two islands of the Robinson Crusoe Archipelago often speciate, even when each still maintains the same ecological tendencies. The distance is 181 kilometers, more than adequate to interrupt gene flow in angiosperms (flowering plants) and allow divergence to occur over time. Such geographic speciation probably also happened within Robinson Crusoe Island when it was first formed, and when it was much larger in size and more ecologically diverse. In this case separation of species into different adaptive zones may have been common. This is harder to see today because the island has been reduced and modified so much over time, but there still are some ecological tendencies, including flowering time differences, among closely related species.[33] Groups of closely related endemic species have come about through the splitting of lineages, or cladogenesis,[34] plus adaptive radiation, a frequent mode of evolution in plants of oceanic islands.

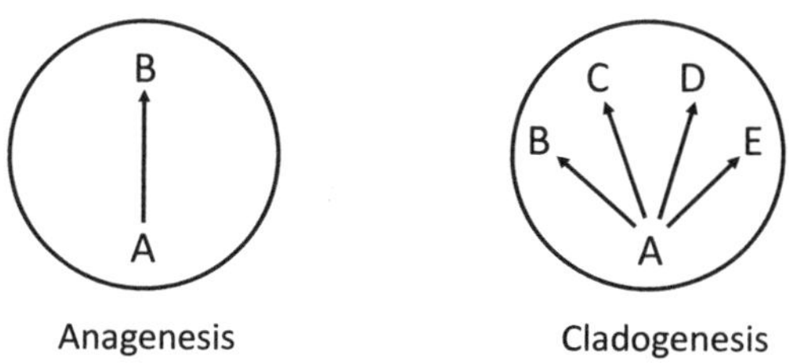

Generalized modes of speciation, cladogenesis, and anagenesis, in the origin of endemic plants of the Robinson Crusoe Archipelago. "A" represents the newly established immigrant (ancestral) population, and "B-E" are derivative species.

In addition to species complexes containing two or more endemic species, about half of the endemic flora of the archipelago consists of only single endemics.[35] In these cases, a colonist arrived and successfully established and changed through time, but it never split into more derivative endemics. This process of speciation has been called anagenesis, a concept developed in part through our studies on the flora of Ulleung Island, Korea,[36] where nearly all the endemic species have originated anagenetically.[37] We have now tested the concept more thoroughly among the Juan Fernández endemics, and results indicate that about half the species have originated this way.[38]

Hybridization can be a major evolutionary factor in floras of oceanic islands, such as has been demonstrated in the Canary Islands[39] and to a lesser extent in Hawaii.[40] In the Robinson Crusoe Archipelago, there is very little evidence of present or past hybridization, with a few exceptions. The most conspicuous example occurs between *Gunnera bracteata* and *G. peltata* on Robinson Crusoe Island.[41] These species are separated spatially and ecologically, with the latter occurring between 350 and 450 meters and the former higher than 500 meters. In one frequently visited valley, Villagra, a series of hybrids between these two species occurs along the trail, giving the impression that human disturbance may have been responsible for inducing hybridization. Because the species involved are quite different morphologically, it is quite easy to detect the hybrids, but a detailed molecular analysis would be worthwhile to understand exactly what sorts of hybrids have been produced (i.e., F_1s, backcrosses, advanced introgressants). Another example of hybridization is the relatively infrequent occurrence of the intergeneric hybrid between *Acaena argentea*, an introduced species, and *Margyricarpus digynus*, an endemic species, with the hybrid being given the name ×*Margyracaena skottsbergii*. We have seen only a few such plants in the field, but this rather unusual situation has been confirmed at the molecular level.[42] The final documented hybrid in the archipelago is the fern, ×*Pleopeltis cerro-altoensis*, a hybrid between the two native species, *P. macrocarpa* and *P. masafuerae*.[43] There are some concerns regarding the taxonomic affinities of these two species, but the main point is that this does seem to be a case of hybridization, this time between fern species.

Lastly, we have always turned a close eye to the problems of conservation of the native flora of the Juan Fernández Archipelago.[44] It is impossible for a biologist to work in these islands and not take interest in this very important dimension. Imagine sitting under a tree on Robinson Crusoe Island, perhaps eating a piece of bread with sardines on it, and realizing that this plant species we are leaning against occurs nowhere else on Earth, just here on this one island. One also realizes that if the ecosystem is not managed properly, this one species could disappear forever. Persons working in the archipelago need to make contributions to understanding this natural world in the way they are best able, but certainly conservation must always be in the background of all research. If the plants go extinct, there will not be much left to investigate. As part of our efforts, we have shared information about the endemic plants and their evolutionary significance with personnel of the park service. We have highlighted changes in the flora from Skottsberg's studies at the beginning of the twentieth century, plus pointing out the endangered status of many of the endemic species.[45] The sandalwood, *Santalum fernandezianum*, has gone extinct due to human overharvesting, so this is not just a hypothetical worry. Approximately four-fifths of the endemics are now categorized as being vulnerable to extremely endangered, which places conservation high on the list of priorities for future efforts. Our studies of genetic variation within the endemics have been very helpful, because these data have revealed which species have depauperate genetic profiles and are at higher risk for loss of populations and possible extinction.

CHAPTER 14

Perspectives

Biologists are special in their communion with other organisms on this planet.[1] I don't know why this should be so. Can we explain why someone becomes an actor? An auto mechanic? A banker? Chance, of course, plays an important role, as do early influences from parents, teachers, friends, and relatives. But I believe there is more here—I believe that biologists feel a closer affinity with other life forms than do most other people. Bankers like to know and understand our own species, and psychologists are dedicated to understanding our behavior. Historians try to educate the next generation about our past accomplishments and foibles, which so often have been at the expense of other organisms. But biologists are different: we *need to sense* the existence of other life forms and to study and interact with them[2]—hence our fascination with insects, snakes, frogs, mosses, fungi, flowering plants, and all other packages of DNA that inhabit this globe. Can you stand in front of a stately magnolia tree, marvel at its height and stature, be amazed that this creature lives in one spot its entire life (that it never travels to any other place on earth), that it seduces insects into transferring its pollen from one flower to another, that it has the miracle of making its own food from the sun—do you think about and *feel* the meaning of all these ideas? Biologists often do, especially those dedicated to field studies, such as ecologists and systematists, who get very close to other organisms in their own home territories. We have chosen to spend our lives studying other forms of life at a much more detailed level than most other people.

The opportunity to learn about life forms restricted to faraway places is a biologist's dream—a fantasy converted to reality. And this is what happened to me with the Juan Fernández Islands. A bit of background reading on a cold, snowy winter evening in Columbus filled me with ideas about islands, which when opportunity beckoned in Chile in 1977, led to a successful collaborative research project. We became extremely interested in the plants of this archipelago, especially those that had arrived by long-distance dispersal, survived, and evolved into new species. That the flora was reasonably small, and the number of islands few, both augured well for gaining insights, plus I was fortunate to have help from many excellent collaborators—professors, postdocs, and students, in Chile, the United States, and Europe. All of us have been excited about uncovering secrets of the endemic plants of this magic place, and all of us have also understood the importance our results would have for conservation efforts.

There is something special about the Juan Fernández Archipelago. Because of favorable geology and geography, it is particularly well suited for asking and attempting to answer scientific questions.[3] As we have said in so many publications over the past decades, this is a manageable system, a reasonable system, which offers the chance to understand what has happened over the past four million years. I have visited the Hawaiian Islands, the Canary Islands, and the Galápagos Islands, and all three are fascinating, each harboring many endemic plants that cry out for evolutionary explanations. But these archipelagos are also geologically and geographically more complex. The Hawaiian archipelago is linear, with the Big Island being the youngest and Kauai the oldest, but there are also the Emperor Seamounts, submerged islands extending to the northwest of the present islands, some of which were previously above the ocean. These potentially impart a much longer time and greater complexity for evolutionary events to have taken place.[4] The Galápagos Islands, so stimulating to the young Charles Darwin aboard the *Beagle*, harbor all kinds of interesting plants and animals, but there are eighteen major islands arranged in a complex mosaic pattern,[5] making it more difficult to infer hypotheses about origins and processes. And the Canary Islands, with seven major islands including snow-capped Tenerife (ca. 4,000 meters) with its bizarre plant forms, is especially fascinating, but some of the islands are near the African continent and much older (20.2 million years for Fuerteventura), which yields a complex mixture of islands with different geological histories.[6] All this is alluring and challenging, but also much more complicated. The Juan Fernández Archipelago consists of only two major islands, of different geological ages, arranged in an east–west line from the major source area (South America), with an excellent historical record of natural and human impacts on the vegetation, with modern vegetation maps, and with a well-known floral inventory. This is an ideal place to try to determine how life forms arrived and changed through time, and I was lucky enough to wallow in this scientific quest. It is a natural laboratory for revealing evolutionary processes, the same ones that have taken place elsewhere on the planet, which means that our investigations in Juan Fernández do have general implications for understanding these processes elsewhere.

When we began work in the Juan Fernández Archipelago, systematic and evolutionary botany was in the midst of conceptual and methodological change. My first postdoctoral fellow on the project, Roger Sanders, and I were both very interested in determination of branching patterns of evolution using quantitative methods, called cladistics. The idea was to take evolutionarily significant characters of the plants, code them numerically, and use these data to make comparisons among species to assess affinities, resulting in the building of evolutionary trees, or cladograms. This was just becoming popular in the 1970s in botany,[7] and we took advantage by applying these methods to the study of *Dendroseris* and *Robinsonia*, the two largest genera in the archipelago. These approaches were greatly aided into the 1990s by the availability of DNA data, at first sequences and then fragment patterns (AFLPs and microsatellites), integrated into the Juan Fernández project. Today the costs to obtain massive amounts of genetic data from organisms are low, which has shifted challenges toward the analysis of mountains of molecular (nucleotide) data instead of straining to obtain sufficient data to answer the questions.

For me, the Juan Fernández project has provided many satisfying academic challenges. There was the intellectual challenge of asking the right questions and getting answers, but this wasn't so terribly difficult. Academics are very skilled at this sort of thing, and these islands just begged for questions. The logistics were particularly challenging. These islands are far from the Chilean mainland, and there are no regular boat trips or daily flights. Weather has a big impact, too. I was amused when I first travelled to Hawaii, arriving by large jet to a modern airport on Oahu, where Honolulu lies. This is a big city with big tourism, although other islands are more isolated and harbor secluded forests and valleys. The level of specific endemism among the flowering plants is 89 percent,[8] extremely high and incredibly fascinating, but getting to the archipelago, at least to the major areas, is not a logistical challenge. One needs money and time, and any airline will get you there. Juan Fernández is another world by contrast. In 1980, even Robinson Crusoe Island didn't have enough food for our expedition—we had to take everything we needed. Even now, this is the case for Alejandro Selkirk Island, complicated, lots to worry about, somewhat dangerous.

Another challenge with working in the Juan Fernández Islands relates to difficulties of making collections. No roads exist where cars can take you to your collecting localities—all must be done on foot, and the landscape relief is steep, requiring climbing up hills and down into slippery valleys. Persons who have not been to the islands will never appreciate the physical efforts for our plant populational sampling over the landscape of both islands, and it is doubtful that such an effort will be duplicated anytime soon. Sometimes the conditions can be very hard—long hours, rain, heavy packs, illness—but through it all the publications make it worthwhile. The field hardships also breed comradeship, allowing us to chat about some of the agonies in subsequent months or years, always bringing a smile and satisfaction to those concerned. Certainly all thirty-eight persons who have participated in our Juan Fernández project on the twelve expeditions have this sense of fraternity.

Challenges for work in the islands extend also to laboratory analyses after the collections are brought back to campus. It is one thing to get samples, but it is quite another to push the work through to generate comparative data required to answer the questions posed. I have been very blessed to have collaborators, postdocs and students, especially Dan Crawford, who have always done their share of lab work after the trips. I never had to badger or cajole any member of our team to get data completed. It just happened because everyone wanted to get the publications finished and out. This is another related point. On our trips, all scientific personnel have become coauthors on publications from the expeditions, such that everyone profited professionally. This was a correct approach to distribute academic credit, and it was also a good motivator for encouraging exceptional efforts in the field as well as in the laboratory.

Another academic reward from the Juan Fernández project appeared after several years of participation with our Chilean colleagues in the Department of Botany at the University of Concepción. Dan and I were asked to conduct short courses on several occasions for the graduate program in botany. These were block courses of several intensive weeks, equaling in hours what would be encountered during a full normal semester. We gave lectures, had students prepare reports, wrote exams, and assigned grades. We also invited graduate students to study in our laboratories at Ohio State University and the University of Vienna, which led to four receiving their Ph.D.s on Juan Fernández topics (Patricio López-Sepúlveda, Patricia Pacheco, Eduardo Ruiz, Hugo Valdebenito). This project also opened the door for a memorandum of understanding between Ohio State University and the University of Concepción, whereby botanical interests were joined with medical education interests (coordinated by Gregory Trzebiatowski, Associate Dean of Medical Education at Ohio State). This provided funds at both institutions for other professors to travel and carry out research and/or teaching, which greatly expanded the opportunities for academic life at both ends.

But all of these academic satisfactions from working in the Juan Fernández Archipelago wouldn't have occurred without the origin of my obsession with the Spanish language and cultures of Latin America. My connection with Spanish began very indirectly, starting in 1961 when I was an undergraduate at DePauw University. This is a small liberal arts college in Greencastle, Indiana, which at that time had a foreign language course requirement of two years (four semesters; it is currently two semesters). As a freshman, I elected to take Russian, believing that it would be scientifically important in the future. My attitude in this first year was not exemplary, to say the least, and the grades in different courses reflected this (A through F). Russian dealt me a D– the first semester (the teacher told me later that the only reason she didn't flunk me was that I had signed up for the second semester!). At the end of the second semester, I earned an F. But in my defense, this Russian class couldn't have been more boring. The teacher, a lady whose name I have long ago repressed, was wedded to the textbook, asking student X to read paragraph A, and then asking student Y to continue with paragraph B, and so on. My poor attitude didn't help, but the presentation couldn't have been drier. Nonetheless, after my failure in the course, I still had to get two years of language, which confronted me squarely in the beginning of the second year.

Mercifully, several factors were beginning to combine to straighten out my attitude and performance. One was that I had decided to become a botany major, thanks to my revulsion at the horrors of the zoology laboratory (the pithing of the frogs was the last straw for me), and also the positive mentoring from Professor Preston Adams.[9] Adams had arrived from Harvard University with his Ph.D. at the same time I came as a freshman. I took his botany class in the spring semester of this first year and really loved it. Toward the end of the spring semester, Professor Adams asked me one day in the hallway of the science building, Harrison Hall, where he had his office, what my career goals were. I said, "Well, Dr. Adams, I would like to be a botany professor like you." To which he responded, "Oh, you would, would you?" And I said, "Yes sir, very much so." He turned directly to me, looked at me in the eyes, and stated clearly, obviously wanting to get an unequivocal message into my skull: "Well, if you don't get your grades up, boy, you will not get into graduate school and you will not become a professor of botany or any other subject. Think about that." Humph, what did he know, and I certainly did not need this sort of lecture, my rebellion against the system rushing forward, but I had enough sense not to rebut his claim openly, letting it go, preferring to let it slide off me, saying goodbye and moving on to my next class of the afternoon.

A number of factors combined to get my academic engines started for my sophomore year. One important aspect was this hallway mini-lecture from Professor Adams. I had huge respect for him, and his admonition was there somewhere in the recesses of my youthful, resistant brain, and it kept osmosing into my consciousness from time to time, sort of a nagging cerebral tutorial. Another influence was that I had applied for a position as assistant in the Botany Department at the Field Museum in Chicago for that summer. Our family lived in Lake Bluff, one of the northern suburbs, about thirty-four miles from the city, and my father commuted into the city on the Northwestern Railway (now belonging to Union Pacific), which was an hour each way. I would be able to follow this schedule with him and work at the Museum if they offered me a job—which they did! They really did! And so, that summer I learned about herbarium specimens and what a systematic botanist did for a living. Now, this was not a university environment, but it was certainly a research unit, with Ph.D. botanists working on different projects, especially in Costa Rica and Guatemala, among other regions of the world. My tasks were pretty menial, but I learned a great deal about how specimens were prepared, processed, and used in research.

At the start of my sophomore year at DePauw, I returned to campus and enrolled in the first semester of beginning Spanish. I figured that since

I had to get this language requirement taken care of, Spanish must be easier than Russian. So, there I was sitting in the old wooden classroom in the oldest building on campus (East College), and in walks Señor Randall Listerman, a young enthusiastic professor, dressed in a crisp business suit (typical for that time). "Buenos días, clase!" Students who obviously had some Spanish in high school responded, "Buenos días, Profesor!" Professor Listerman continued speaking with simple Spanish words, then calling roll: "Señor Abel?" The student responded, "Presente." Listerman moved on to "Señorita Becker?" And she replied, "Presente, Profesor." And finally he came alphabetically to: "Señor Stuessy?," to which I responded, "Presente," the first word I had learned and used in Spanish. Listerman continued that entire class speaking only in Spanish, repeating things, acting out words, writing a few expressions on the blackboard, but all in Spanish. I was overwhelmed with this new approach to language. What a contrast with the stuffy academic approach of Russian—this new class was about communication in another language, a puzzle to figure out, and a challenge to train the brain to connect words into concepts.

At the third class meeting on Friday of that first week, Señor Listerman was late, and my personal "code" of behavior was to wait only fifteen minutes for a late professor, and then I would leave. This didn't happen very often, but today it did, and so, I picked up my books and left. I learned later that he arrived about a minute after I had exited. On that day a quiz was administered, and of course, I missed it, not a good start to the semester. So on Monday morning after class, I asked to speak with him about it, and he invited me to his office in the same building, the dark brown floorboards creaking as I followed him to his very small office. I told him that I had waited fifteen minutes and then left. He apologized for being late and asked if I could take the quiz right then, to which I responded positively. I have no idea if this was the same quiz, but it didn't matter, because I got a B+ on it, the highest grade I had ever gotten in a language course! He complimented me and said he looked forward to seeing me in class on Wednesday. I was really overwhelmed by his positive attitude, his support of me personally, and his general enthusiasm for teaching Spanish. Oh, he was really good, coming into the classroom shouting Spanish phrases, pounding the table, dramatically emphasizing things, all creating a very lively learning environment, which I will never forget. When I became a professor, I never felt comfortable acting out this way so dramatically in the classroom, but it suited him, and I lapped it up.

Thanks to Professor Listerman, I was addicted, really addicted to learning Spanish. I immersed myself in the course, spending Friday evenings in the language lab listening to the tapes and trying to duplicate the sounds as best I could (it turns out that I have a good ear for tones, which is a real advantage in learning a language). Other students were going to movies, getting a beer, or whatever, but I was joyfully immersed in the language lab. I got an A in that first course, and to move my progress faster, along with the second course in spring semester, I took an individual studies course with Señor Listerman, reading Antonio Buero Vallejo's *En la Ardiente Oscuridad*.[10] The idea was to work more intensively on vocabulary and reading ability. I got As in these courses, and I managed a 3.6 grade average overall (4-point scale), which put me on the Dean's List and the right academic path. In addition to this progress, I signed up for a six-week summer program in Saltillo, Mexico, run by DePauw and other midwestern liberal arts universities.

I continued Spanish classes during my junior and senior years, in composition and conversation, the literature of the *Siglo del Oro* (sixteenth and early seventeenth centuries), and so forth, which caused me to wonder if I might better become a Spanish major. I had received only As in all these courses. However, at the same time I was very enthusiastic about my botany classes, getting As in these, too. I finally decided that for me the thrill of Spanish was being able to communicate with another set of words and phrases, which affects one's view of everything else, including personal behavior. I loved to talk in Spanish (still do), but I decided it would make sense to continue as a botanist but to work in Latin America. And that is exactly what transpired—and I do believe I made the right choice. For Ph.D. work I attended the University of Texas at Austin (Professor Adams pushed this alternative for me very much, based on my Latin American interests, even though I had fellowship offers from Cornell and Vanderbilt), and my thesis topic at Texas dealt with a group of plants (*Melampodium*) in the sunflower family that was widely distributed in Mexico and Central America. Perfect. It proved to be an excellent match for my interests and talents, and it started me on a career of botanical research in Latin America that eventually led to work in Chile and the Juan Fernández Archipelago.

And so, as this book is winding down toward its conclusion, there is still one more item that I would like to share with you. It deals with Roger Sanders, my first postdoctoral fellow. About fifteen years ago he sent me a letter in Vienna that he was coming through the city and would like to have breakfast together. Very good news! I always enjoyed his company and seeing him again would be a distinct pleasure. On the appointed day, I arrived at his hotel, and there was Roger in the lobby, same as always, with a broad smile and warm greeting. We were both happy to see each other again. After Ohio State, Roger had taken a position as research botanist at the Fairchild Tropical Botanic Garden in Miami, Florida, where he had worked for several years. He subsequently moved back to Arkansas, from where he had come originally, and where he had family ties.

As Roger and I ate breakfast and chatted, he dropped the bomb: "I very much wanted to see you on this trip so that I could talk to you personally." Hmm . . . when someone starts out like that, it means something pretty heavy is going to follow. He continued, "As you know, I have always been very religious [I knew that, but it never created any problems personally or professionally], and after much soul-searching, reading, and discussions with different people, I can now see clearly that much of evolutionary theory is just wrong. The world and the wonderful life forms have been specially created by God during only the past several thousand years." Roger had become a creationist! I was a bit stunned and speechless, having another sip of coffee and looking at Roger in the hopes that he was kidding somehow, perhaps testing me, pulling my leg. I responded, "Am I hearing this right, Roger, that you have become a creationist?" "Yes," he said. "That is correct." I hastened to add: "But Roger, remember the radiometric datings from the islands that you and I published in 1984,[11] showing Robinson Crusoe Island to be four million years old, and our phylogenetic analyses of the species of *Dendroseris* and *Robinsonia* that evolved during this time.[12] How do these data coincide with a creationist view of an extremely youthful origin of only a few thousand years?" He was pleasant but serious, and responded respectfully: "Well, I know it must be a surprise to you, but I have been doing a lot of reading and study of these radiometric dating techniques, and I believe them to be inaccurate and false. The Bible tells us what happened and how it happened, and that for me is the most reasonable explanation." I was still trying to wrap my head around how this Ph.D. in systematic and evolutionary botany from the University of Texas at Austin (from the same doctoral program where I had graduated), and who had published with me on the evolution of the flora of the Juan Fernández Islands, could now claim that what we had been doing all this time was simply wrong.

Roger was convinced that the belief system of his new religious community in Arkansas, using statements from the 2000-year-old historical text, the Bible, was a more reasonable explanation of how life forms originated than the facts of evolution confirmed over the past more than 160 years. To me this was an eye-opener on how a belief system can be powerful enough to overshadow hard scientific evidence.[13] It is dangerous if this attitude is applied to other aspects of our civilization, such that people reject science as a way of knowing about the world. One should not forget that the reasons we can sit at the computer and type up this book, open a refrigerator to get a snack, talk on a smart phone, turn on the lights, get in the car for a nice ride, or take a prescription pill, are due to scientific accomplishments. All of these aspects that make our civilization so agreeable and our life spans longer and more pleasant are due to progress in science. If belief systems take over society and are used to provide answers for our daily lives, we will be pushed back to Medieval times, a huge backslide in human progress.

In fairness to Roger, a few years after this breakfast bombshell, I attended in July of 2008 a meeting of botanists in Vancouver, Canada.[14] And guess who also attended—Roger! It was great to see him again, and we caught up on family news, but then I couldn't resist asking, "Why are you here listening to these talks dealing with population-level evolutionary phenomena?" He said, "Well, I am not against all evolution, not against micro-evolution at the population level. That is OK. What I cannot accept is that humans evolved from other primates. This I just do not believe happened." I resisted the temptation to start quizzing him about the other known extinct hominins, *Australopithecus*, Neanderthals, Denisovans, and the like,[15] simply because I judged that this would not get anywhere, and I still liked him very much as a friend. In a way, I was pleased that he had softened up a bit, and I was amazed that he had come to the scientific meeting, where there were no sessions dealing with creationism (or intelligent design). But I was again reminded that a strong belief system trumps all hard scientific evidence, no matter how strong it may be. This is why creationism cannot be taught in a science class. It can be discussed in a course in comparative religions, or worldviews, that sort of thing, but not as science. A belief system of creationism has no place being discussed along with the facts of evolution. This experience with Roger,[16] a very nice and extremely intelligent person, made me realize that scientists in society have a special role to be vigilant against persons who wish us to accept religious beliefs as substitutes for scientific facts.

Arriving now to the end of this book, we also come to the question: "What does it all mean?" For myself, there are lots of meanings from my association with the Robinson Crusoe Islands—both personal and professional. My personal gain from meeting, courting, and marrying Patricia Ghivarello from Santiago should be obvious. Because she comes from Italian-Chilean ancestry, I now belong to a large family, which is a new experience for me. My own family is very small; my father, deceased in 1981, was an only child. I am his only offspring. Patty's father is one of eight children, these also having married Chileans of Italian origin who came from large families, and this means tons of cousins. It feels good to belong to this large support group. Another valuable personal lesson for me has been to understand that people in all cultures are basically the same and want similar things from life—the concept of a global village.

Professionally, the Juan Fernández project in Chile was an opportunity for me to get deeply involved with a part of Latin America. This would not be a one-shot investigation in the islands, but a long-term contribution involving repeated trips and efforts to answer evolutionary questions of general biological significance. I continue to write and publish on aspects of the archipelago, and colleagues contact me for information about the islands for doing research there; others request material from herbarium collections to investigate DNA sequences. I am a resource person, but in recognition that my lifespan is limited, I have tried to publish as much as possible while time and health have permitted, which is one of the motivations for writing this book.

My hope is that you, the reader, now have a better understanding of this fascinating oceanic archipelago and the challenges of doing botanical research there, and when you pick up that old copy of *Robinson Crusoe* hidden next to more modern books on your home library shelf, you will let your imagination drift to these real islands in the southeastern Pacific. I hope also that you will be with a coffee in front of a crackling fire on some wintry evening, picking up an atlas of the world, leafing through maps, and perhaps dreaming of visiting some distant land, wanting to make it a part of your life. If so, this book has served its purpose.

Acknowledgments

Trying to acknowledge persons and institutions who have participated in the Juan Fernández Archipelago research project over more than forty years is not a simple task. All our work has built upon that of previous botanists, especially William Botting Hemsley, Carlo Bertero, Claudio Gay, Rodulfo Amando Philippi, Federico Johow, and Carl Skottsberg. These pioneers have brought us into the modern age that allowed us to focus more on evolutionary and biogeographic questions. Without this firm foundation of a basic inventory of the plants of the archipelago, we could never have made progress. This emphasizes that in science there must be a strong descriptive understanding of the world before inferences can be made about evolutionary phenomena. Description and interpretation are both needed for scientific advances to be made.

Our research in the Juan Fernández Archipelago has involved colleagues and students from universities in Argentina, Austria, Chile, Ecuador, Japan, Mexico, Sweden, and the United States. These valuable contributors have been: Gregory Anderson, Pedro Aqueveque, Jorge Arriagada, Daniela Bacho, Carlos M. Baeza, Gabriel Bernardello, Daniel J. Crawford, Michael Doyle, Leonardo Gaete, Alejandro Gatica, Fidelina González, Josef Greimler, Ana María Humaña, Gabriele Kottirsch, Thomas Lammers, Alejandro Landero, Luis Letelier, Patricio López-Sepúlveda, Oscar Matthei, Johannes Novak, Patricio Novoa, Patricia Pacheco, Oscar Parra, Patricio Peñailillo, Roberto Rodríguez, Mauricio Rondanelli, Eduardo Ruiz, Roger Sanders, Jaime Sepúlveda, Mario Silva, José Soto, Ulf Swenson, Koji Takayama, Eric Tepe, Karin Tremetsberger, Eduardo Ugarte, Hugo Valdebenito, and Delbert Wiens.

During research in the archipelago, we have been assisted by administrators and guides of the Robinson Crusoe Islands National Park. This is one of many national parks in Chile that falls under the supervision and management of the Corporación Nacional Forestal (CONAF), an organization that is charged with oversight of the natural resources of the country. Supporting our work have been the park chiefs, Bernardo Ackermann, Gastón González, and Iván Leiva. The guides and associates were fundamental for our research success, and these included Alfonso Andauer,

Jorge Angulo, Guillermo Araya, Danilo Arredondo, Danilo Arredondo, Jr., Oscar Chamorro, Esteban Charpentier, Miguel García, Alvis González, Guillermo González, José González, Michael González, Bernardo López, Guillermo López, José López, Eduardo Paredes, Aldo Recabarren, Teodoro Rivadeneira, and Ramón Schiller. It cannot be overemphasized how important these professionals and friends have been to the success of our research work. They not only showed us how to move through forests in different parts of the islands, and how to do so safely, but they also took an active part in helping with collecting activities and learning the flora.

Leaves of absence from my normal campus responsibilities to work in the Juan Fernández Archipelago were granted by The Ohio State University, the Los Angeles County Museum of Natural History, and the University of Vienna. Because of the complex logistics involved with research in the islands, especially on the remote Alejandro Selkirk Island, considerable time was needed for the expeditions, each normally four to six weeks. The Department of Botany of the University of Concepción, Chile, served as a site for many organizational aspects of the expeditions, and their personnel and students joined as collaborators on all trips and research publications.

Financial support for our Juan Fernández studies has been awarded by the U.S. National Science Foundation (from the Systematic Biology and International programs), Austrian Science Fund (FWF), Austrian National Bank, Armand Erpf Foundation, Los Angeles County Museum of Natural History, National Geographic Society, and Vicerrectoría de Investigación (University of Concepción).

Other very important contributions have come from the following individuals: Sr. Juan Ghivarello, my late father-in-law, who crafted the switchboard for the isozyme work in the islands and allowed us to use his home in Santiago for staging of expedition materials; Ann-Mari Skottsberg, the daughter of Carl Skottsberg, who welcomed me during my trip to Göteborg, Sweden, shared many memories about her parents and their trips to the islands, and translated from Swedish into English the diary kept by her mother, Inga, on their last trip in 1955; Dr. Osvaldo Jara, chief of the meteorological station (El Servicio Aeronáutico) in San Juan Bautista, who allowed us to use one of the houses under his control during two of our expeditions; Sr. Santiago Figueroa of Transportes Aéreos Robinson Crusoe, for safe flights to and from Robinson Crusoe Island; Sra. Juanita López, wife of CONAF guide Bernardo López, for her hospitality and outstanding cooking for our expedition crew on several occasions on both islands.

Most of the illustrations in this book are from my own unpublished photographs, but a few of them have been taken by other participants on the expeditions, who have given permission to include them here: Gregory Anderson (images on pp. 8, 9); Gabrielle Kottirsch (image on p. 28); Patricio López-Sepúlveda (color insert image 4, left, and images on p. 219); Josef Greimler (color insert image 4, right); and Thomas Lammers (image on p. 162). The Department of Botany of the Smithsonian Institution gave permission to reproduce a digital image of the herbarium specimen of *Robinsonia macrocephala* (p. 47). Some figures are of images published previously, the copyright holders of which have generously allowed their republication here: National Portrait Gallery (London) (image, top left, on p. 60); Royal Botanical Gardens, Kew (image, top right, on p. 63); New York Botanical Garden (image, top right, p. 68); Cambridge University Press (color insert images 4, right, 7, 8, 12; image, bottom, on p. 197, image on p. 263); University of Hawaii Press (*Pacific Science*) (color insert images 13, 17); Springer Press (images on pp. 164, 215); Ocean Treasures Memorial Library (blanket permission) (image, right, on p. 235). The maps inside the front and back covers were originally drafted by Gil Ortiz. Marcelo Baeza and Jose Becerra helped identify persons in the image on p. 27.

Heartfelt thanks go to the following people who read part or all of the final manuscript and made numerous suggestions for improvement: Daniel Crawford, Josef Greimler, Thomas Lammers, Patricio López-Sepúlveda, Jack Miller, Patricia Stuessy, and two anonymous reviewers. Michael Brooke checked information on the Juan Fernández petrels. Critical reading is never an easy job, always time-consuming, but these kind souls stepped up to offer suggestions that have greatly improved this book.

And finally, most important has been the excellent editorial work of Allison Brock of the Missouri Botanical Garden Press. She took an initial interest in the project and supported its development into what we believe is an interesting and useful book. Her critical eye caught many inconsistencies, and she made improvements in the text, especially the scientific parts, for the general reader. But even more importantly, she provided positive reinforcement for me through the many steps of the publication process. A book begins with an author, but it ends with a competent editor, and Allison played that role perfectly.

APPENDIX 1

List of dates, personnel, and collections from our Robinson Crusoe expeditions. Numbers in parentheses refer to expedition collection numbers, mostly under *Stuessy et al.*, but sometimes made by other combinations of personnel under the same number series.*

Personnel	Institution	Specialty
1980: 27 January–21 February		
(5000 to 5249)		
Jorge Arriagada	U. Concepción	flowering plants
Clodomiro Marticorena	U. Concepción	flowering plants
Oscar Parra	U. Concepción	algae
Roberto Rodríguez	U. Concepción	ferns
Roger Sanders	Ohio State U.	flowering plants
Tod Stuessy	Ohio State U.	flowering plants
Eduardo Ugarte	U. Concepción	ecology
1980: 18 November–30 November		
(5300 to 5524)		
Oscar Matthei	U. Concepción	monocots (grasses)
Roger Sanders	Ohio State U.	flowering plants
Tod Stuessy	Ohio State U.	flowering plants
Hugo Valdebenito	U. Concepción	flowering plants
1984: 15 January–13 February		
(6200 to 6680)		
Daniel Crawford	Ohio State U.	flavonoids
Alejandro Landero	U. Concepción	flowering plants
Patricia Pacheco	Ohio State U.	flavonoids
Eduardo Ruiz	U. Concepción	flavonoids
Tod Stuessy	Ohio State U.	flowering plants
Hugo Valdebenito	Ohio State U.	flavonoids

Personnel	Institution	Specialty
1986: 16 January–14 February		
(8000 to 8368; 8380 to 8518; 9000 to 9665)		
Michael Doyle	Rancho Santa Ana Bot. Gard.	cryptogams
Leonardo Gaete	U. Concepción	flowering plants
Thomas Lammers	Ohio State U.	flowering plants
Alejandro Landero	U. Concepción	flowering plants
Eduardo Ruiz	U. Concepción	flavonoids
Jaime Sepúlveda	U. Concepción	medical student
Tod Stuessy	Ohio State U.	flowering plants
Hugo Valdebenito	Ohio State U.	flowering plants
1990: 19 January–12 February		
(11030 to 11723)		
Marcelo Baeza	U. Concepción	monocots
Daniel Crawford	Ohio State U.	isozymes
Ana María Humaña	U. Valdivia	reproductive biology
Patricio López-Sepúlveda	U. Concepción	flowering plants
Patricio Peñailillo	U. Concepción	flowering plants
Mauricio Rondanelli	U. Concepción	flowering plants
Patricia Stuessy	Ohio State U.	assistant
Tod Stuessy	Ohio State U.	flowering plants
Delbert Wiens	U. Utah	*Lactoris*, reprod. biol.
1991: 13 January–3 February		
(11728 to 12124)		
Gregory Anderson	U. Connecticut	reproductive biology
Daniel Crawford	Ohio State U.	isozymes
Patricio López-Sepúlveda	U. Concepción	flowering plants
Richard Roederer	Ohio State U.	assistant
José Soto	U. Concepción	isozymes
Tod Stuessy	Ohio State U.	flowering plants
1996: 10 January–25 January		
(15000 to 15196)		
Gregory Anderson	U. Connecticut	reproductive biology
Pedro Aqueveque	U. Concepción	isozymes
Marcelo Baeza	U. Concepción	monocots
Gabriel Bernardello	U. Córdoba (Argentina)	reproductive biology
Daniel Crawford	Ohio State U.	isozymes

Personnel	Institution	Specialty
Hector Ibarra	U. Concepción	zoology
Eduardo Ruiz	U. Concepción	isozymes
Tod Stuessy	Los Angeles Nat. Hist. Mus.	flowering plants
Ulf Swenson	Los Angeles Nat. Hist. Mus.	invasive plants
Eric Tepe	Ohio State U.	isoyzmes
1997: 11 January – 26 January		
(15197 to 15407)		
Gregory Anderson	U. Connecticut	reproductive biology
Pedro Aqueveque	U. Concepción	isozymes, DNA
Marcelo Baeza	U. Concepción	monocots
Gabriel Bernardello	U. Córdoba (Argentina)	reproductive biology
Daniel Crawford	Ohio State U.	isozymes, DNA
Fidelina González	U. Concepción	isozymes
Gabi Kottirsch	U. Concepción	assistant
Patricio López-Sepúlveda	U. Concepción	flowering plants
Eduardo Ruiz	U. Concepción	isozymes, DNA
Tod Stuessy	Los Angeles Nat. Hist. Mus.	flowering plants
1999: 1–17 February		
(no collections; studies of vegetation only)		
Josef Greimler	U. Vienna	vegetation
Patricio López-Sepúlveda	U. Concepción	vegetation
Tod Stuessy	U. Vienna	flowering plants
2000: 3–18 February		
(no collections; studies of vegetation only)		
Josef Greimler	U. Vienna	vegetation
Patricio López-Sepúlveda	U. Concepción	vegetation
Alan Stuessy	U. Vienna	assistant
Tod Stuessy	U. Vienna	flowering plants
2010: 4–24 February		
(19100 to 19356)		
Daniel Crawford	U. Kansas	flowering plants
Josef Greimler	U. Vienna	vegetation
Luis Letelier	U. Talca	flowering plants
Patricio López-Sepúlveda	U. Vienna	flowering plants
Patricio Peñailillo	U. Talca	flowering plants
Tod Stuessy	U. Vienna	flowering plants

Personnel	Institution	Specialty
2011: 26 January–24 February		
(19401 to 19444; 19603 to 19683; 19800 to 19847; 20000 to 20049)		
Marcelo Baeza	U. Concepción	flowering plants
Daniela Bacho	U. San Sebastián	nurse
Alejandro Gatica	U. Talca	flowering plants
Josef Greimler	U. Vienna	vegetation
Patricio López-Sepúlveda	U. Vienna	flowering plants
Patricio Novoa	Jard. Bot. Nac. (Chile)	flowering plants
Patricio Peñailillo	U. Talca	flowering plants
Eduardo Ruiz	U. Concepción	flowering plants
Tod Stuessy	U. Vienna	flowering plants
Total of 38 participants from 11 different institutions; 4065 collection numbers		

* An additional expedition was carried out by Eduardo Ruiz in November of 1998, when he spent ten days on Robinson Crusoe Island and seven days on Alejandro Selkirk Island, as part of his doctoral research at the University of Concepción. He was accompanied by Pedro Elissetche, Fidelina González, Gabriel Mendoza, and Silvana Muñoz.

APPENDIX 2

Publications from the Robinson Crusoe Islands research group

Anderson, G. J. & G. Bernardello. 2018. Reproductive biology. Pp. 193–205 *in* T. F. Stuessy, D. J. Crawford, P. López-Sepúlveda, C. M. Baeza & E. A. Ruiz (editors), *Plants of Oceanic Islands: Evolution, Biogeography, and Conservation of the Flora of the Juan Fernández (Robinson Crusoe) Archipelago*. Cambridge University Press, Cambridge.

Anderson, G. J., G. Bernardello, P. López S., T. F. Stuessy & D. J. Crawford. 2000. Dioecy and wind pollination in *Pernettya rigida* (Ericaceae) of the Juan Fernández Islands. *Bot. J. Linn. Soc.* 132: 121–141.

Anderson, G. J., G. Bernardello, P. López S., D. J. Crawford & T. F. Stuessy. 2000. Reproductive biology of *Wahlenbergia* (Campanulaceae) endemic to Robinson Crusoe Island (Chile). *Pl. Syst. Evol.* 223: 109–123.

Anderson, G. J., G. Bernardello, T. F. Stuessy & D. J. Crawford. 2001. Breeding system and pollination of selected plants endemic to Juan Fernández Islands. *Amer. J. Bot.* 88: 230–233.

Anderson, G., G. Bernardello & M. S. Engel. 2001. Conservation implications of a newly discovered bee species on Isla Robinson Crusoe, Chile. *Conserv. Biol.* 15: 803–805.

Anderson, G., D. J. Crawford & G. Bernardello. 2013. The reproductive biology of island plants: The Juan Fernández and Canary Islands (facilitated by the work of Arnoldo Santos-Guerra). *Vieraea* 41: 109–121.

Baeza, C. M., R. Rodríguez R., M. Hoeneisen F. & T. Stuessy. 1999. Consideraciones anatómicas del leño secundario de *Santalum fernandezianum* F. Phil. (Santalaceae), especie extinta del Archipiélago de Juan Fernández, Chile. *Gayana Bot.* 56: 63–65.

Baeza, C. M., T. F. Stuessy & C. Marticorena. 2002. Notes on the Poaceae of the Robinson Crusoe (Juan Fernández) Islands, Chile. *Brittonia* 54: 154–163.

Baeza, C. M., C. Marticorena, T. Stuessy, E. Ruiz & M. Negritto. 2007. Poaceae en el archipiélago de Juan Fernández (Robinson Crusoe). *Gayana Bot.* 64: 125–174.

Baeza, C., C. Bastías, T. Stuessy, E. Ruiz, J. Greimler, P. López-Sepúlveda, P. Peñailillo, et al. 2013. Cytological investigations on populations of *Taraxacum* (Asteraceae) from the Juan Fernández Archipelago, Chile. *Gayana Bot.* 70: 401–403.

Baldwin, B. G., D. J. Crawford, J. Francisco-Ortega, S.-C. Kim, T. Sang & T. F. Stuessy. 1998. Molecular phylogenetic insights on the origin and evolution of oceanic island plants. Pp. 410–441 *in* D. E. Soltis, P. S. Soltis & J. J. Doyle (editors), *Molecular Systematics of Plants, II: DNA Sequencing*. Kluwer, New York.

Bernardello, G. & G. J. Anderson. 2018. Plant origins and dispersal. Pp. 335–354 *in* T. F. Stuessy, D. J. Crawford, P. López-Sepúlveda, C. M. Baeza & E. A. Ruiz (editors), *Plants of Oceanic Islands: Evolution, Biogeography, and Conservation of the Flora of the Juan Fernández (Robinson Crusoe) Archipelago*. Cambridge University Press, Cambridge.

Bernardello, G., G. J. Anderson, P. López S., M. A. Cleland, T. F. Stuessy & D. J. Crawford. 1999. Reproductive biology of *Lactoris fernandeziana* (Lactoridaceae). *Amer. J. Bot.* 86: 829–840.

Bernardello, G., L. Galetto & G. J. Anderson. 2000. Floral nectary structure and nectar chemical composition of some species from Robinson Crusoe Island (Chile). *Canad. J. Bot.* 78: 862–872.

Bernardello, G., G. Anderson, T. F. Stuessy & D. J. Crawford. 2002. A survey of floral traits, breeding systems, floral visitors, and pollination systems of the angiosperms of the Juan Fernandez Islands (Chile). *Bot. Rev. (Lancaster)* 67: 255–308.

Bernardello, G., R. Aguilar & G. J. Anderson. 2004. The reproductive biology of *Sophora fernandeziana* (Leguminosae), a vulnerable endemic species from Isla Robinson Crusoe. *Amer. J. Bot.* 91: 198–206.

Bernardello, G., G. J. Anderson, T. F. Stuessy & D. J. Crawford. 2006. The angiosperm flora of the Archipelago Juan Fernández (Chile): Origin and dispersal. *Canad. J. Bot.* 84: 1266–1281.

Bittner, M., M. Silva O., Z. Rozas, P. Pacheco, T. F. Stuessy, F. Bohlmann & J. Jakupović. 1988. Kaurene derivatives from *Robinsonia evenia*. *Phytochemistry* 27: 487–488.

Bittner, M., M. Silva O., Z. Rozas, J. Jakupović & T. Stuessy. 1994. Estudio químico del género *Gunnera* en Chile: II. Metabolitos secundarios de dos especies continentales y dos especies de las islas de Juan Fernández. *Bol. Soc. Chil. Quím.* 39: 79–83.

Brauner, S., D. J. Crawford & T. F. Stuessy. 1992. Ribosomal DNA and RAPD variation in the rare plant family Lactoridaceae. *Amer. J. Bot.* 79: 1436–1439.

Campos, V., J. Jakupović, M. Bittner, M. Silva O. & T. F. Stuessy. 1989. Sesquiterpene lactones from *Dendroseris neriifolia*. *Heterocycle* 28: 779–782.

Caujapé-Castells, J., A. Tye, D. J. Crawford, A. Santos-Guerra, A. Sakai, K. Beaver, W. Lobin, et al. 2010. Conservation of oceanic island floras: Present and future global challenges. *Perspect. Pl. Ecol. Evol. Syst.* 12: 107–129.

Cho, M.-S., S.-H. Kim, J. Yang, D. J. Crawford, T. F. Stuessy, P. López-Sepúlveda & S.-C. Kim. 2020. Plastid phylogenomics of *Dendroseris* (Cichorieae; Asteraceae): Insights into structural organization and molecular evolution of an endemic lineage from the Juan Fernández Islands. *Frontiers Pl. Sci.* doi: 10.3389/fpls.2020.594272.

Crawford, D. J. & T. F. Stuessy. 1997. Plant speciation on oceanic islands. Pp. 249–267 *in* K. Iwatsuki & P. H. Raven (editors), *Evolution and Diversification of Land Plants*. Springer, Tokyo.

Crawford, D. J. & T. F. Stuessy. 2016. Cryptic variation, molecular data, and the challenge of conserving plant diversity in oceanic archipelagos: The critical role of plant systematics. *Korean J. Pl. Taxon.* 46: 129–148.

Crawford, D. J., T. F. Stuessy & M. Silva O. 1986. Leaf flavonoid chemistry and the relationships of the Lactoridaceae. *Pl. Syst. Evol.* 153: 133–139.

Crawford, D. J., T. F. Stuessy & M. Silva O. 1987. Allozyme divergence and the evolution of *Dendroseris* (Compositae: Lactuceae) on the Juan Fernández Islands. *Syst. Bot.* 12: 435–443.

Crawford, D. J., R. Whitkus & T. F. Stuessy. 1987. Plant evolution and speciation on oceanic islands. Pp. 184–199 *in* K. Urbanska (editor), *Differentiation Patterns in Higher Plants*. Academic Press, London.

Crawford, D. J., T. F. Stuessy & M. Silva O. 1988. Allozyme variation in *Chenopodium sanctae-clarae*, an endemic species of the Juan Fernández Islands, Chile. *Biochem. Syst. Ecol.* 16: 279–284.

Crawford, D. J., T. F. Stuessy, T. Lammers, M. Silva O. & P. Pacheco. 1990. Allozyme variation and evolutionary relationships among three species of *Wahlenbergia* (Campanulaceae) in the Juan Fernández Islands. *Bot. Gaz.* 151: 119–124.

Crawford, D. J., T. F. Stuessy, D. W. Haines, M. B. Cosner, M. Silva O. & P. López. 1992. Allozyme diversity within and divergence among four species of *Robinsonia* (Asteraceae: Senecioneae), a genus endemic to the Juan Fernández Islands, Chile. *Amer. J. Bot.* 79: 962–966.

Crawford, D. J., T. F. Stuessy, M. B. Cosner, D. W. Haines, M. Silva O. & M. Baeza. 1992. Evolution of the genus *Dendroseris* (Asteraceae: Lactuceae) on the Juan Fernández Islands: Evidence from chloroplast and ribosomal DNA. *Syst. Bot.* 17: 676–682.

Crawford, D. J., T. F. Stuessy, R. Rodríguez & M. Rondanelli. 1993. Genetic diversity in *Rhaphithamnus venustus* (Verbenaceae), a species endemic to the Juan Fernández Islands. *Bull. Torrey Bot. Club.* 120: 23–28.

Crawford, D. J., T. F. Stuessy, M. B. Cosner, D. W. Haines & M. Silva O. 1993. Ribosomal and chloroplast DNA restriction site mutations and the radiation of *Robinsonia* (Asteraceae, Senecioneae) on the Juan Fernández Islands. *Pl. Syst. Evol.* 184: 233–239.

Crawford, D. J., S. Brauner, M. B. Cosner & T. F. Stuessy. 1993. Use of RAPD markers to document the origin of the intergeneric hybrid ×*Margyracaena skottsbergii* (Rosaceae) on the Juan Fernandez Islands. *Amer. J. Bot.* 80: 89–92.

Crawford, D. J., T. F. Stuessy, M. B. Cosner, D. W. Haines, D. Wiens & P. Peñailillo. 1994. *Lactoris fernandeziana* (Lactoridaceae) on the Juan Fernández Islands: Allozyme uniformity and field observations. *Conserv. Biol.* 8: 277–280.

Crawford, D. J., T. Sang, T. F. Stuessy, S.-C. Kim & M. Silva O. 1998. *Dendroseris* (Asteraceae: Lactuceae) and *Robinsonia* (Asteraceae: Senecioneae) on the Juan Fernández Islands: Similarities and differences in biology and phylogeny. Pp. 97–119 *in* T. F. Stuessy & M. Ono (editors), *Evolution and Speciation of Island Plants*. Cambridge University Press, Cambridge.

Crawford, D. J., E. Ruiz, T. F. Stuessy, E. Tepe, P. Aqueveque, F. González, R. J. Jensen, et al. 2001. Allozyme diversity in endemic flowering plant species of the Juan Fernández Archipelago, Chile. Ecological and historical factors with implications for conservation. *Amer. J. Bot.* 88: 2195–2203.

Crawford, D. J., M. Tago-Nakazawa, T. F. Stuessy, G. J. Anderson, G. Bernardello, E. Ruiz, R. J. Jensen, et al. 2001. Intersimple sequence repeat (ISSR) variation in *Lactoris fernandeziana* (Lactoridaceae), a rare endemic of the Juan Fernández Archipelago, Chile. *Pl. Spec. Biol.* 16: 185–192.

Crawford, D. J., T. K. Lowrey, G. J. Anderson, G. Bernardello, A. Santos-Guerra & T. F. Stuessy. 2009. Genetic diversity in the colonizing ancestors of Asteraceae endemic to oceanic islands: Baker's Law and polyploidy. Pp. 139–151 *in* V. A. Funk, A. Susanna, T. F. Stuessy & R. J. Bayer (editors), *Systematics, Evolution, and Biogeography of Compositae*. IAPT, Vienna.

Crawford, D. J., G. J. Anderson & G. Bernardello. 2011. The reproductive biology of island plants. Pp. 11–36 *in* D. Bramwell & J. Caujapé-Castells (editors), *The Biology of Island Floras*. Cambridge University Press, Cambridge.

Crawford, D. J., G. J. Anderson, G. Bernardello & A. Santos-Guerra. 2013. The reproductive biology of island plants: The Juan Fernández and Canary Islands. Pp. 166–175 *in*

J. Caujapé-Castells, G. Niet & J. M. Fernández-Palacios (editors), *Proceedings of the Amurga International Conferences on Island Biodiversity 2011: Present and Emerging Knowledge on the Evolution, Diversity and Conservation of the Canarian Flora*. Fundación Canaria Amurga-Maspalomas, Jardín Botánico Canario "Viera y Clavijo," Las Palmas de Gran Canaria.

Crawford, D. J., M. Silva O., P. Pacheco & H. Valdebenito. 2018. Flavonoid compounds. Pp. 165–187 *in* T. F. Stuessy, D. J. Crawford, P. López-Sepúlveda, C. M. Baeza & E. A. Ruiz (editors), *Plants of Oceanic Islands: Evolution, Biogeography, and Conservation of the Flora of the Juan Fernández (Robinson Crusoe) Archipelago*. Cambridge University Press, Cambridge.

Crawford, D. J., P. López-Sepúlveda, K. Takayama & E. A. Ruiz. 2018. Genetic diversity and divergence. Pp. 276–307 *in* T. F. Stuessy, D. J. Crawford, P. López-Sepúlveda, C. M. Baeza & E. A. Ruiz (editors), *Plants of Oceanic Islands: Evolution, Biogeography, and Conservation of the Flora of the Juan Fernández (Robinson Crusoe) Archipelago*. Cambridge University Press, Cambridge.

Crawford, D. J., T. F. Stuessy, K. Takayama, P. López-Sepúlveda, G. J. Anderson & G. Bernardello. 2018. Speciation. Pp. 308–331 *in* T. F. Stuessy, D. J. Crawford, P. López-Sepúlveda, C. M. Baeza & E. A. Ruiz (editors), *Plants of Oceanic Islands: Evolution, Biogeography, and Conservation of the Flora of the Juan Fernández (Robinson Crusoe) Archipelago*. Cambridge University Press, Cambridge.

Dirnböck, T., J. Greimler, P. López S. & T. F. Stuessy. 2003. Predicting future threats to the native vegetation of Robinson Crusoe Island, Juan Fernández Archipelago, Chile. *Conserv. Biol.* 17: 1650–1659.

Esselman, E. J., D. J. Crawford, S. Brauner, T. F. Stuessy, G. J. Anderson & M. Silva O. 2000. RAPD marker diversity within and divergence among species of *Dendroseris* (Asteraceae: Lactuceae). *Amer. J. Bot.* 87: 591–596.

Greimler, J., T. F. Stuessy, U. Swenson, C. M. Baeza & O. Matthei. 2002. Plant invasions on an oceanic archipelago. *Biol. Invasions* 4: 73–85.

Greimler, J., P. López S., T. F. Stuessy & T. Dirnböck. 2002. The vegetation of Robinson Crusoe Island (Isla Masatierra), Juan Fernández Archipelago, Chile. *Pacific Sci.* 56: 263–284 (plus 1 map).

Greimler, J., P. López-Sepúlveda, K. Reiter, C. Baeza, P. Peñailillo, E. Ruiz, P. Novoa, et al. 2013. Vegetation of Alejandro Selkirk Island (Isla Masafuera), Juan Fernández Archipelago, Chile. *Pacific Sci.* 67: 267–282.

Greimler, J., C. H. Schulze, P. López-Sepúlveda, P. Novoa, A. Gatica, K. Reiter, J. Wessely, et al. 2017. Strong indication of an extinction-based saturation of the flora on the Pacific Robinson Crusoe Islands. *Ecol. & Evol.* doi: 10.1002/ece3.3882.

Greimler, J., T. F. Stuessy, U. Swenson, P. López-Sepúlveda & C. M. Baeza. 2018. Invasive species. Pp. 134–148 *in* T. F. Stuessy, D. J. Crawford, P. López-Sepúlveda, C. M. Baeza & E. A. Ruiz (editors), *Plants of Oceanic Islands: Evolution, Biogeography, and Conservation of the Flora of the Juan Fernández (Robinson Crusoe) Archipelago*. Cambridge University Press, Cambridge.

Greimler, J., P. López-Sepúlveda & K. Reiter. 2018. Vegetation. Pp. 98–114 *in* T. F. Stuessy, D. J. Crawford, P. López-Sepúlveda, C. M. Baeza & E. A. Ruiz (editors), *Plants of Oceanic Islands: Evolution, Biogeography, and Conservation of the Flora of the Juan Fernández (Robinson Crusoe) Archipelago*. Cambridge University Press, Cambridge.

Hoeneisen, M., M. Silva O., M. Wink, D. J. Crawford & T. F. Stuessy. 1993. Alkaloids of *Sophora* of Juan Fernández Islands and related taxa. *Bol. Soc. Chil. Quím.* 38: 167–171.

Hoeneisen, M., J. Becerra, M. Bittner, M. Silva, C. Baeza & R. Rodríguez. 1998. Sesquiterpenes of the essential oil of *Santalum fernandezianum* F. Phil (Santalaceae), an extinct species of the Juan Fernández Islands, Chile. *Bol. Soc. Chil. Quím.* 43: 505–508.

Ingram, K., G. Bernardello, S. Cover & E. Wilson. 2006. The ants of the Juan Fernández Islands: Genesis of an invasive fauna. *Biol. Invasions* 8: 383–387.

Jensen, R. J., M. Schowzer, D. J. Crawford, T. F. Stuessy, G. J. Anderson, C. M. Baeza, M. Silva O., et al. 2002. Patterns of morphological and genetic variation among populations of *Myrceugenia fernandeziana* (Myrtaceae) on Masatierra Island: Implications for conservation. *Syst. Bot.* 27: 534–547.

Lammers, T. G. 1996. Phylogeny, biogeography, and systematics of the *Wahlenbergia fernandeziana* complex (Campanulaceae: Campanuloideae). *Syst. Bot.* 21: 397–415.

Lammers, T. & T. F. Stuessy. 1992. Lectotypification of *Lactoris fernandeziana* Philippi (Lactoridaceae). *Taxon* 41: 537–540.

Lammers, T., T. F. Stuessy & M. Silva O. 1986. Systematic relationships of the Lactoridaceae, an endemic family of the Juan Fernández Islands, Chile. *Pl. Syst. Evol.* 152: 243–266.

López-Sepúlveda, P., K. Takayama, J. Greimler, P. Peñailillo, D. J. Crawford, M. Baeza, E. Ruiz, et al. 2013. Genetic variation (AFLPs and nuclear microsatellites) in two anagenetically derived endemic species of *Myrceugenia* (Myrtaceae) on the Juan Fernández Islands, Chile. *Amer. J. Bot.* 100: 722–734.

López-Sepúlveda, P., M. Baeza, P. Peñailillo, T. Stuessy, J. Greimler, E. Ruiz, P. Novoa, et al. 2013. Introduced plant species in the Biosphere Reserve Alejandro Selkirk Island, Robinson Crusoe Archipelago, Chile, and their threat to the native and endemic vegetation. *Mem. New York Bot. Gard.* 108: 223–243.

López-Sepúlveda, P., K. Takayama, J. Greimler, D. J. Crawford, P. Peñailillo, M. Baeza, E. Ruiz, et al. 2015. Progressive migration and anagenesis in *Drimys confertifolia* of the Juan Fernández Archipelago, Chile. *J. Pl. Res.* 128: 73–90.

López-Sepúlveda, P., K. Takayama, J. Greimler, D. J. Crawford, P. Peñailillo, M. Baeza, E. Ruiz, et al. 2015. Speciation and biogeography of *Erigeron* (Asteraceae) in the Juan Fernández Archipelago, Chile, based on AFLPs and SSRs. *Syst. Bot.* 40: 888–899.

López-Sepúlveda, P., K. Takayama, D. J. Crawford, J. Greimler, P. Peñailillo, M. Baeza, E. Ruiz, et al. 2016. Genetic consequences and biogeography of anagenetic speciation of *Rhaphithamnus venustus* (Verbenaceae) in the Juan Fernández archipelago, Chile, based on AFLPs and SSRs. *Pl. Spec. Biol.* 32: 223–237.

Marticorena, C., T. F. Stuessy & C. M. Baeza. 1998. Catalogue of the vascular flora of the Robinson Crusoe or Juan Fernández Islands, Chile. *Gayana Bot.* 55: 187–211.

Matthei, O., C. Marticorena & T. F. Stuessy. 1993. La flora adventicia del archipiélago de Juan Fernández. *Gayana Bot.* 50: 69–102.

Murillo-A., J., E. Ruiz, L. R. Landrum, T. F. Stuessy & M. H. J. Barfuss. 2012. Phylogenetic relationships of *Myrceugenia* (Myrtaceae) based on plastid and nuclear DNA sequences. *Molec. Phylogen. Evol.* 62: 764–776.

Murillo-A., J., T. F. Stuessy & E. Ruiz. 2013. Phylogenetic relationships among *Myrceugenia*, *Blepharocalyx*, and *Luma* (Myrtaceae) based on paired-sites models and the secondary structures of ITS and ETS sequences. *Pl. Syst. Evol.* 299: 713–729.

Murillo-A., J., T. F. Stuessy & E. Ruiz. 2016. Explaining disjunct distributions in the flora of southern South America: Evolutionary history and biogeography of *Myrceugenia* (Myrtaceae). *J. Biogeogr.* 43: 979–990.

Murillo-Aldana, J. & E. Ruiz. 2011. Revalidación de *Nothomyrcia* (Myrtaceae), un género endémico del Archipiélago de Juan Fernández. *Gayana Bot.* 68: 129–134.

Novoa, P. 2015. *Expedición botánica a la Isla Alejandro Selkirk.* Jardín Botánico Nacional, Viña del Mar.

Pacheco, P., D. J. Crawford, T. F. Stuessy & M. Silva O. 1985. Flavonoid evolution in *Robinsonia* (Compositae) of the Juan Fernández Islands. *Amer. J. Bot.* 72: 989–998.

Pacheco, P., D. J. Crawford, T. F. Stuessy & M. Silva O. 1991. Flavonoid evolution in *Dendroseris* (Compositae) in the Juan Fernández Islands, Chile. *Amer. J. Bot.* 78: 534–543.

Pacheco, P., T. F. Stuessy & D. J. Crawford. 1991. Natural interspecific hybridization in *Gunnera* (Gunneraceae) of the Juan Fernández Islands, Chile. *Pacific Sci.* 45: 389–399.

Pacheco, P., D. J. Crawford, T. F. Stuessy & M. Silva O. 1993. Flavonoid chemistry and evolution of *Gunnera* (Gunneraceae) in the Juan Fernández Islands, Chile. *Gayana Bot.* 50: 17–30.

Peña, C. M., M. A. Negritto, E. Ruiz, C. M. Baeza & V. L. Finot. 2017. Revisión de *Megalachne* Steud. (Poaceae: Pooideae: Poeae), género endémico del Archipiélago de Juan Fernández, Chile. *Gayana Bot.* 74: 189–199.

Rodríguez-Ríos, R. 1990. *Gleichenia lepidota* n. sp. y la familia Gleicheniaceae del archipiélago de Juan Fernández, Chile. *Gayana Bot.* 47: 37–45.

Ruiz, E., J. Becerra, M. Silva O., D. J. Crawford & T. F. Stuessy. 1994. Flavonoid chemistry of the endemic species of *Myrceugenia* (Myrtaceae) of the Juan Fernández Islands and relatives in continental South America. *Brittonia* 46: 187–193.

Ruiz, E., C. Donoso, F. González, J. Becerra, C. Marticorena & M. Silva. 1999. Phenetic relationships between Juan Fernández and continental Chilean species of *Sophora* (Fabaceae) based on flavonoid patterns. *Bol. Soc. Chil. Quím.* 44: 351–356.

Ruiz, E., C. Marticorena, D. J. Crawford, T. F. Stuessy, F. González, R. Montoya, M. Silva O., et al. 2001. Morphological and ITS sequence divergence between taxa of *Cuminia* (Lamiaceae), an endemic genus of the Juan Fernández Islands, Chile. *Brittonia* 52: 341–350.

Ruiz, E., G. Fuentes, J. Becerra, F. González & M. Silva. 2002. Flavonoids as chemosystematic markers in Chilean species of *Drimys* J.F. Forst. et G. Forst. (Winteraceae). *Bol. Soc. Chil. Quím.* 47: 273–278.

Ruiz, E., D. J. Crawford, T. F. Stuessy, F. González, R. Samuel, J. Becerra & M. Silva O. 2004. Phylogenetic relationships and genetic divergence among endemic species of *Berberis*, *Gunnera*, *Myrceugenia* and *Sophora* of the Juan Fernández Islands (Chile) and their continental progenitors based on isozymes and nrITS sequences. *Taxon* 53: 321–332.

Ruiz, E., O. Toro, D. J. Crawford, T. F. Stuessy, M. A. Negritto, C. Baeza & J. Becerra. 2008. Phylogenetic relationships among Chilean species of *Drimys* (Winteraceae) based on ITS sequences and insertion/deletion events. *Gayana Bot.* 65: 220–228.

Sanders, R. W., T. F. Stuessy & C. Marticorena. 1982. Recent changes in the flora of the Juan Fernández Islands, Chile. *Taxon* 31: 284–289.

Sanders, R. W., T. F. Stuessy & R. Rodríquez. 1983. Chromosome numbers of the flora of the Juan Fernández Islands. *Amer. J. Bot.* 70: 799–810.

Sanders, R. W., T. F. Stuessy, C. Marticorena & M. Silva O. 1987. Phytogeography and evolution of *Dendroseris* and *Robinsonia*, tree-Compositae of the Juan Fernández Islands. *Opera Bot.* 92: 195–215.

Sang, T., D. J. Crawford, S.-C. Kim & T. F. Stuessy. 1994. Radiation of the endemic genus *Dendroseris* Asteraceae) on the Juan Fernández islands: Evidence from sequences of the ITS regions of nuclear ribosomal DNA. *Amer. J. Bot.* 81: 1494–1501.

Sang, T., D. J. Crawford, T. F. Stuessy & M. Silva O. 1995. ITS sequences and the phylogeny of the genus *Robinsonia* (Asteraceae). *Syst. Bot.* 20: 55–64.

Silva O., M., M. Bittner, Z. Rozas, T. F. Stuessy, D. J. Crawford & W. Steglich. 1987. Kaurene derivatives from *Robinsonia thurifera* Dcne. *Bol. Soc. Chil. Quím.* 32: 75–79.

Sontag, W. A., Jr. & T. F. Stuessy. 2018. Climate and weather. Pp. 44–53 *in* T. F. Stuessy, D. J. Crawford, P. López-Sepúlveda, C. M. Baeza & E. A. Ruiz (editors), *Plants of Oceanic Islands: Evolution, Biogeography, and Conservation of the Flora of the Juan Fernández (Robinson Crusoe) Archipelago*. Cambridge University Press, Cambridge.

Sontag, W. A., Jr. & T. F. Stuessy. 2018. Geology and soils. Pp. 31–43 *in* T. F. Stuessy, D. J. Crawford, P. López-Sepúlveda, C. M. Baeza & E. A. Ruiz (editors), *Plants of Oceanic Islands: Evolution, Biogeography, and Conservation of the Flora of the Juan Fernández (Robinson Crusoe) Archipelago*. Cambridge University Press, Cambridge.

Sontag, W. A., Jr., T. F. Stuessy & U. Swenson. 2018. Modeling species diversity. Pp. 354–366 *in* T. F. Stuessy, D. J. Crawford, P. López-Sepúlveda, C. M. Baeza & E. A. Ruiz (editors), *Plants of Oceanic Islands: Evolution, Biogeography, and Conservation of the Flora of the Juan Fernández (Robinson Crusoe) Archipelago*. Cambridge University Press, Cambridge.

Spooner, D. M., T. F. Stuessy, D. J. Crawford & M. Silva O. 1987. Chromosome numbers from the flora of the Juan Fernández Islands. II. *Rhodora* 89: 351–356.

Steinbauer, M. J., R. Field, J.-A. Grytnes, P. Trigas, C. Ah-Peng, F. Attorre, H. J. B. Birks, et al. (and 27 other authors, including T. F. Stuessy and J. Greimler). 2016. Topography-driven isolation, speciation and a global increase of endemism with elevation. *Global Ecol. Biogeogr.* 25: 1097–1107.

Stuessy, T. F. 1992. Diversidad de plantas en las Islas Robinson Crusoe. *In* J. Grau & G. Zizka (editors), Flora Silvestre de Chile. *Palmengarten Sonderh.* 19: 54–66.

Stuessy, T. F. 1994. *I Carl Skottsbergs forspår på Robinson Crusoe-öarna, Chile*. Göteborgs Botaniska Trädgård, Göteborg. [26 pp.]

Stuessy, T. F. 1995. Fernandezian Region: CPD Site PO8. Juan Fernández Islands, Chile. Pp. 565–568 *in* S. D. Davis, V. H. Heywood & A. C. Hamilton (editors), *Centres of Plant Diversity, Vol. 2*. WWF and IUCN, Cambridge.

Stuessy, T. F. 2006. Sympatric plant speciation in islands? *Nature (Brief Communications Arising)* 443: E12.

Stuessy, T. F. 2007. Evolution of specific and genetic diversity during ontogeny of island floras: The importance of understanding process for interpreting island biogeographic patterns. Pp. 117–132 *in* M. Ebach & R. S. Tangney (editors), *Biogeography in a Changing World*. CRC Press, Boca Raton.

Stuessy, T. F. 2009. Anagenesis. Pp. 8–10 *in* M. Gillespie & D. A. Clague (editors), *Encyclopedia of Islands*. University of California Press, Berkeley.

Stuessy, T. F. 2018. History of the islands. Pp. 5–14 *in* T. F. Stuessy, D. J. Crawford, P. López-Sepúlveda, C. M. Baeza & E. A. Ruiz (editors), *Plants of Oceanic Islands: Evolution, Biogeography, and Conservation of the Flora of the Juan Fernández (Robinson Crusoe) Archipelago*. Cambridge University Press, Cambridge.

Stuessy, T. F. 2020. *Environmental History of Oceanic Islands: Natural and Human Impacts on the Vegetation of the Juan Fernández (Robinson Crusoe) Archipelago*. Springer, Cham, Switzerland.

Stuessy, T. F. 2020. The importance of historical ecology for interpreting evolutionary processes in plants of oceanic islands. *J. Syst. Evol.* doi: 10.1111/jse.12673.

Stuessy, T. F. & M. Silva O. 1983. The evolution of the flora of the Juan Fernández Islands. *Chilean Univ. Life* No. 15: 3–6.

Stuessy, T. F. & C. Marticorena. 1990. Orthography of some epithets honoring Bertero in the vascular flora of Juan Fernández Islands and continental Chile. *Gayana Bot.* 47: 77–81.

Stuessy, T. F. & D. J. Crawford. 1998. Chromosomal stasis during speciation in angiosperms of oceanic islands. Pp. 307–324 *in* T. F. Stuessy & M. Ono (editors), *Evolution and Speciation of Island Plants*. Cambridge University Press, Cambridge.

Stuessy, T. F. & M. Ono (editors). 1998. *Evolution and Speciation of Island Plants.* Cambridge University Press, Cambridge.

Stuessy, T. F. & M. Ono. 1998. The current status of our knowledge and suggested research protocols in island archipelagos. Pp. 325–332 *in* T. F. Stuessy & M. Ono (editors), *Evolution and Speciation of Island Plants.* Cambridge University Press, Cambridge.

Stuessy, T. F. & C. M. Baeza. 2018. Chromosome numbers. Pp. 188–192 *in* T. F. Stuessy, D. J. Crawford, P. López-Sepúlveda, C. M. Baeza & E. A. Ruiz (editors), *Plants of Oceanic Islands: Evolution, Biogeography, and Conservation of the Flora of the Juan Fernández (Robinson Crusoe) Archipelago.* Cambridge University Press, Cambridge.

Stuessy, T. F. & C. Marticorena. 2018. Botanists in the islands. Pp. 15–28 *in* T. F. Stuessy, D. J. Crawford, P. López-Sepúlveda, C. M. Baeza & E. A. Ruiz (editors), *Plants of Oceanic Islands: Evolution, Biogeography, and Conservation of the Flora of the Juan Fernández (Robinson Crusoe) Archipelago.* Cambridge University Press, Cambridge.

Stuessy, T. F., R. W. Sanders & O. Matthei. 1983. *Juania australis* revisited in the Juan Fernández Islands, Chile. *Principes* 27: 71–74.

Stuessy, T. F., K. Foland, J. Sutter, R. W. Sanders & M. Silva O. 1984. Botanical and geological significance of potassium-argon dates from the Juan Fernández Islands. *Science* 225: 49–51.

Stuessy, T. F., R. W. Sanders & M. Silva O. 1984. Phytogeography and evolution of the flora of the Juan Fernández Islands: A progress report. Pp. 55–69 *in* F. J. Radovsky, P. H. Raven & S. H. Sohmer (editors), *Biogeography of the Tropical Pacific.* Association of Systematics Collections and B. P. Bishop Museum, Lawrence, Kansas.

Stuessy, T. F., D. J. Crawford & C. Marticorena. 1990. Patterns of phylogeny in the endemic vascular flora of the Juan Fernández Islands, Chile. *Syst. Bot.* 15: 338–346.

Stuessy, T. F., D. J. Crawford & M. Silva O. 1991. Evolution of the endemic vascular flora of the Juan Fernández Islands. *In* E. C. Dudley (editor), *The Unity of Evolutionary Biology, Proceedings of the 4th International Congress of Systematic and Evolutionary Biology.* Dioscorides Press, Portland. [4 pp.]

Stuessy, T. F., C. Marticorena, R. Rodríguez, D. J. Crawford & M. Silva O. 1992. Endemism in the vascular flora of the Juan Fernández Islands. *Aliso* 13: 297–307.

Stuessy, T. F., U. Swenson, D. J. Crawford, G. Anderson & M. Silva O. 1997. Plant conservation in the Juan Fernández archipelago, Chile. *Aliso* 16: 89–102.

Stuessy, T. F., D. J. Crawford, C. Marticorena & R. Rodríguez. 1998. Island biogeography of angiosperms of the Juan Fernández archipelago. Pp. 121–138 *in* T. F. Stuessy & M. Ono (editors), *Evolution and Speciation of Island Plants.* Cambridge University Press, Cambridge.

Stuessy, T. F., D. J. Crawford, C. Marticorena & M. Silva O. 1998. Isolating mechanisms and modes of speciation in endemic angiosperms of the Juan Fernández Islands. Pp. 79–96 *in* T. F. Stuessy & M. Ono (editors), *Evolution and Speciation of Island Plants.* Cambridge University Press, Cambridge.

Stuessy, T. F., U. Swenson, C. Marticorena, O. Matthei & D. J. Crawford. 1998. Loss of plant diversity and extinction on Robinson Crusoe Islands, Chile. Pp. 243–257 *in* C.-I. Peng & P. P. Lowry II (editors), *Rare, Threatened, and Endangered Floras of Asia and the Pacific Rim.* Institute of Botany, Academia Sinica, Taipei. [*Monograph Series* No. 16.]

Stuessy, T. F., D. J. Crawford, G. Anderson & R. Jensen. 1999. Systematics, biogeography and conservation of Lactoridaceae. *Perspect. Pl. Ecol. Evol. Syst.* 1: 267–290.

Stuessy, T. F., J. Greimler & T. Dirnböck. 2005. Landscape modification and impact on specific and genetic diversity in oceanic islands. Pp. 89–101 *in* H. Balslev & I. Friis

(editors), *Plant Diversity and Complexity Patterns: Local, Regional and Global Dimensions*. Danish Academy of Sciences, Copenhagen. [*Biol. Skr.* 55.]
Stuessy, T. F., E. Ruiz, K. Tremetsberger & D. J. Crawford. 2005. Testing degrees of genetic divergence and populational variation in oceanic island archipelagos: Juan Fernández as a model system. *Nova Acta Leopoldina* 92: 147–165.
Stuessy, T. F., G. Jakubowsky, R. Salguero Gómez, M. Pfosser, P. M. Schlüter, T. Fer, B.-Y. Sun, et al. 2006. Anagenetic evolution in island plants. *J. Biogeogr.* 33: 1259–1265.
Stuessy, T. F., K. Takayama & P. López-Sepúlveda. 2012. Founder effects are invisible in endemic species of oceanic islands. *J. Biogeogr.* 39: 1565–1566.
Stuessy, T. F., K. Takayama, P. López-Sepúlveda & D. J. Crawford. 2014. Interpretation of patterns of genetic variation in endemic plant species of oceanic islands. *Bot. J. Linn. Soc.* 174: 276–288.
Stuessy, T. F., C. Koenig & P. López Sepúlveda. 2014. Paraphyly and endemic genera of oceanic islands: Implications for conservation. *Ann. Missouri Bot. Gard.* 100: 50–78.
Stuessy, T. F., D. J. Crawford, C. M. Baeza, P. López-Sepúlveda, J. Greimler, E. A. Ruiz, K. Takayama, et al. 2018. Conclusions. Pp. 367–376 *in* T. F. Stuessy, D. J. Crawford, P. López-Sepúlveda, C. M. Baeza & E. A. Ruiz (editors), *Plants of Oceanic Islands: Evolution, Biogeography, and Conservation of the Flora of the Juan Fernández (Robinson Crusoe) Archipelago*. Cambridge University Press, Cambridge.
Stuessy, T. F., C. M. Baeza, P. López-Sepúlveda, R. Rodríguez & U. Swenson. 2018. Conservation of native and endemic species. Pp. 149–161 *in* T. F. Stuessy, D. J. Crawford, P. López-Sepúlveda, C. M. Baeza & E. A. Ruiz (editors), *Plants of Oceanic Islands: Evolution, Biogeography, and Conservation of the Flora of the Juan Fernández (Robinson Crusoe) Archipelago*. Cambridge University Press, Cambridge.
Stuessy, T. F., C. Marticorena, U. Swenson, J. Greimler & P. López-Sepúlveda. 2018. Impacts on the vegetation. Pp. 117–133 *in* T. F. Stuessy, D. J. Crawford, P. López-Sepúlveda, C. M. Baeza & E. A. Ruiz (editors), *Plants of Oceanic Islands: Evolution, Biogeography, and Conservation of the Flora of the Juan Fernández (Robinson Crusoe) Archipelago*. Cambridge University Press, Cambridge.
Stuessy, T. F., D. J. Crawford & E. A. Ruiz. 2018. Patterns of phylogeny. Pp. 209–275 *in* T. F. Stuessy, D. J. Crawford, P. López-Sepúlveda, C. M. Baeza & E. A. Ruiz (editors), *Plants of Oceanic Islands: Evolution, Biogeography, and Conservation of the Flora of the Juan Fernández (Robinson Crusoe) Archipelago*. Cambridge University Press, Cambridge.
Stuessy, T. F., D. J. Crawford, P. López-Sepúlveda, C. M. Baeza & E. A. Ruiz (editors). 2018. *Plants of Oceanic Islands: Evolution, Biogeography, and Conservation of the Flora of the Juan Fernández (Robinson Crusoe) Archipelago*. Cambridge University Press, Cambridge.
Stuessy, T. F., R. Rodríguez, C. M. Baeza & P. López-Sepúlveda. 2018. Taxonomic inventory. Pp. 57–97 *in* T. F. Stuessy, D. J. Crawford, P. López-Sepúlveda, C. M. Baeza & E. A. Ruiz (editors), *Plants of Oceanic Islands: Evolution, Biogeography, and Conservation of the Flora of the Juan Fernández (Robinson Crusoe) Archipelago*. Cambridge University Press, Cambridge.
Sun, B.-Y., T. F. Stuessy & D. J. Crawford. 1990. Chromosome counts from the flora of the Juan Fernández Islands, Chile. III. *Pacific Sci.* 44: 258–264.
Sun, B.-Y., T. F. Stuessy, A. M. Humaña, M. Riveros G. & D. J. Crawford. 1996. Evolution of *Rhaphithamnus venustus* (Verbenaceae), a gynodioecious hummingbird-pollinated endemic of the Juan Fernández Islands, Chile. *Pacific Sci.* 50: 55–65.
Swenson, U., T. F. Stuessy, M. Baeza & D. J. Crawford. 1997. New and historical plant introductions, and potential pests in the Juan Fernández Islands, Chile. *Pacific Sci.* 51: 233–253.

Takayama, K., P. López-Sepúlveda, G. Kohl, J. Novak & T. F. Stuessy. 2012. Development of microsatellite markers in species of *Erigeron* (Asteraceae) endemic to the Juan Fernández Archipelago, Chile. *Amer. J. Bot. Primer Notes* e1–e3. doi:10.3732/ajb.1200218.

Takayama, K., P. López-Sepúlveda, J. Greimler, D. J. Crawford, P. Peñailillo, M. Baeza, E. Ruiz, et al. 2015. Genetic consequences of cladogenetic vs. anagenetic speciation in endemic plants of oceanic islands. *AoB Pl.* 7: plv102. doi:10.1093/aobpla/plv102.

Takayama, K., P. López-Sepúlveda, J. Greimler, D. J. Crawford, P. Peñailillo, M. Baeza, E. Ruiz, et al. 2015. Relationships and genetic consequences of contrasting modes of speciation among endemic species of *Robinsonia* (Asteraceae, Senecioneae) of the Juan Fernández Archipelago, Chile, based on AFLPs and SSRs. *New Phytol.* 205: 415–428.

Takayama, K., D. J. Crawford, P. López-Sepúlveda, J. Greimler & T. F. Stuessy. 2018. Factors driving adaptive radiation in plants of oceanic islands: A case study from the Juan Fernández Archipelago. *J. Pl. Res.* 131: 469–485.

Tobe, H., T. F. Stuessy, P. H. Raven & K. Oginuma. 1993. Embryology and karyomorphology of Lactoridaceae. *Amer. J. Bot.* 80: 933–946.

Valdebenito, H., T. F. Stuessy & D. J. Crawford. 1990. A new biogeographic connection between islands in the Atlantic and Pacific Oceans. *Nature* 347: 549–550.

Valdebenito, H., T. F. Stuessy & D. J. Crawford. 1990. Synonymy in *Peperomia berteroana* (Piperaceae) results in biological disjunction between Pacific and Atlantic Oceans. *Brittonia* 42: 121–124.

Valdebenito, H., T. F. Stuessy, D. J. Crawford & M. Silva O. 1992. Evolution of *Erigeron* (Compositae) in the Juan Fernández Islands, Chile. *Syst. Bot.* 17: 470–480.

Valdebenito, H., T. F. Stuessy, D. J. Crawford & M. Silva O. 1992. Evolution of *Peperomia* (Piperaceae) in the Juan Fernández Islands, Chile. *Pl. Syst. Evol.* 182: 107–119.

Notes

Generally available data, such as population of a city, have been taken from Wikipedia or other online sources and are not specifically referenced in the text or listed here.

Chapter 1. Corrales de Molina

1. Ricci, M. 2001. Evaluation of conservation status of *Lactoris fernandeziana* Philippi (Lactoridaceae) in Chile. *Biodivers. & Conserv.* 10: 2129–2138.
2. Wanke, S., M. A. Jaramillo, T. Borsch, M.-S. Samain, D. Quandt & C. Neinhuis. 2007. Evolution of Piperales—*matK* gene and *trnK* intron sequence data reveal lineage specific resolution contrast. *Molec. Phylogen. Evol.* 42: 477–497; Soltis, D. E., S. A. Smith, N. Cellinese, K. J. Wurdack, D. C. Tank, S. F. Brockington, N. F. Refulio-Rodríguez, et al. 2011. Angiosperm phylogeny: 17 genes, 640 taxa. *Amer. J. Bot.* 98: 704–730; Naumann, J., K. Salomo, J. P. Der, E. K. Wafula, J. F. Bolin, E. Maass, L. Frenzke, et al. 2013. Single-copy nuclear genes place haustorial Hydnoraceae within Piperales and reveal a Cretaceous origin of multiple parasitic angiosperm lineages. *PLOS One* 8(11): 1–12. E79204.
3. Zavada, M. S. & J. M. Benson. 1987. First fossil evidence for the primitive angiosperm family Lactoridaceae. *Amer. J. Bot.* 74: 1590–1594; MacPhail, M. K., A. D. Partridge & E. M. Truswell. 1999. Fossil pollen records of the problematical primitive angiosperm family Lactoridaceae in Australia. *Pl. Syst. Evol.* 214: 199–210; Gamerro, J. C. & V. Barreda. 2008. New fossil record of Lactoridaceae in southern South America: A palaeobiogeographical approach. *Bot. J. Linn. Soc.* 158: 41–50.
4. The usual dating for the appearance of our human species has been about 200,000 years from deposits in the Oma Valley, Ethiopia (McDougall, I., F. H. Brown & J. G. Fleagle. 2005. Stratigraphic placement and age of modern humans from Kibish, Ethiopia. *Nature* 433: 733–736), but recent studies on fossils in Morocco have suggested extending this to ca. 300,000 years (Hublin, J. J., A. Ben-Ncer, S. E. Bailey, S. E. Freidline, M. M. Skinner, I. Bergmann, A. Le Cabec, et al. 2017. New fossils from Jebel Irhoud, Morocco and the pan-African origin of *Homo sapiens*. *Nature* 546: 289–292).
5. Dates of photographs in figures: color insert image 1, 2010; pp. 3, 4, 6, 8, 9, 1996.

Chapter 2. Meeting Mario

1. Carlquist, S. 1965. *Island Life: A Natural History of the Islands of the World*. Natural History Press, Garden City, New York.
2. The historical name for these islands is the Juan Fernández Archipelago, named after the Spanish navigator who first discovered them in 1574. The island closest to the continent was called Más a Tierra ("Closer to the Land") and the far island Más a Fuera ("Farther Away"). To stimulate tourism, the Chilean government in 1962 officially renamed them the Robinson Crusoe Archipelago, a connection with the popular novel by Daniel Defoe. The near island is now officially designated Robinson Crusoe Island (Isla Piloto Robinson Crusoe) and the one farther away Alejandro Selkirk Island (Isla Marinero Alejandro Selkirk). There is a legitimate tie between these islands

and Defoe's famous book. In 1704 a Scottish sailor, Alexander Selkirk, had a dispute with the captain of his ship and demanded to be put off on the nearest land, which happened to be Robinson Crusoe Island. He stayed there in isolation for more than four years, finally returning on an English ship to London, where he created quite a sensation. He was interviewed by many journalists, and information written about him served as material for Defoe to write *Robinson Crusoe* in 1719, an enduring tale of solitude, perseverance, and religious conversion.

3. The sunflower family has two names officially accepted by the International Code of Nomenclature for algae, fungi and plants (ICN). The older and more descriptive name is Compositae, which refers to having flowers aggregated into tight clusters called heads. The other name is Asteraceae, bearing the mandated "-aceae" ending for names of plant families, and based on the well-known genus *Aster*. Following the rule of priority of use of scientific names of plants, the earliest name, Compositae, should be used, but it does not have the proper ending. Because of the long usage of this name for the family, the ICN has *sanctioned* it for use. Asteraceae, although it has been formed properly, is not the earliest name (hence illegitimate), but the ICN has officially *conserved* it for usage. Some botanists feel very strongly about using one or the other name.
4. Such isolation, in fact, happened on 13 October 1972 when a plane carrying an amateur Uruguayan rugby team and other passengers crashed in the interior of the Andes Mountains, killing many of the crew and passengers. As the months passed and food ran out, survivors resorted to eating flesh from the cadavers, which created much media attention when they were finally rescued. A book by Piers Paul Read appeared in 1974 (*Alive: The Story of the Andes Survivors*, Lippincott) and a film in 1993 (*Alive: The Miracle of the Andes*, United International Pictures), among numerous other journalistic and cinematic essays.
5. Sater, W. F. 2007. *Andean Tragedy: Fighting the War of the Pacific, 1879–1884*. University of Nebraska Press, Lincoln.
6. An exception to this respect for the Santiago Metro system occurred in October of 2019. Due to a 3.5 percent rise in cost of subway tickets, massive protests erupted throughout the city, combined with anger against income inequality and lack of adequate pension and health facilities. The protests became violent, eventually causing the death of nineteen people, and many Metro stations were vandalized and some were burned. See Alarcón, D. 2020. Chile at the barricades: Can a once-in-a-generation political revolution survive the pandemic? *The New Yorker*, 12 October.
7. The most recent taxonomic treatment of the genus *Chaetanthera* is Davies, A. M. R. 2010. The Systematic Revision of *Chaetanthera* Ruiz & Pav., and the Reinstatement of the Genus *Oriastrum* Poepp. & Endl. (Asteraceae; Mutisieae). Ph.D. Dissertation, Ludwig-Maximilians-Universität, München.
8. Marticorena, C. 1992. *Bibliografía Botánica Taxonómica de Chile*. Monogr. Syst. Bot. Missouri Bot. Gard. 41: 1–587.
9. Skottsberg, C. (editor). 1920–1956. *The Natural History of Juan Fernandez and Easter Island, Vol. 1, Geography, Geology, Origin of Island Life, Vol. 2, Botany*. Almqvist & Wiksells, Uppsala; Skottsberg, C. (editor). 1921–1940. *The Natural History of Juan Fernandez and Easter Island, Vol. 3, Zoology*. Almqvist & Wiksells, Uppsala.
10. Skottsberg, C. 1921. The phanerogams of the Juan Fernandez Islands. Pp. 95–240 (plus plates 10–20) *in* C. Skottsberg (editor), *The Natural History of Juan Fernandez and Easter Island, Vol. 2, Botany*. Almqvist & Wiksells, Uppsala.
11. Bailey, I. W. & A. C. Smith. 1942. Degeneriaceae, a new family of flowering plants from Fiji. *J. Arnold Arbor.* 23: 356–365; Miller, J. M. 1988. A new species of *Degen-*

eria (Degeneriaceae) from the Fiji archipelago. *J. Arnold Arbor.* 69: 275–280; Miller, J. M. 1989. The archaic flowering plant family Degeneriaceae: Its bearing on an old enigma. *Natl. Geogr. Res.* 512: 218–231.

12. Johow, F. 1896. *Estudios sobre la Flora de las Islas de Juan Fernández*. Gobierno de Chile, Santiago.
13. Carine, M. & M. Menezes de Sequeira. 2020. Sir Joseph Hooker on *Insular Floras*: Human impact and the natural laboratory paradigm. *Sci. Insularum* 3: 73–88.
14. Dates of photographs in figures: p. 15, 1985; p. 18, 1991; p. 20, 1986; p. 25, 1980; pp. 26, 27, 29, 1977.

Chapter 3. Expedition Preparations

1. Lots of biographies of Frank Sinatra exist, but my favorite is Zehme, B. 1997. *The Way You Wear Your Hat: Frank Sinatra and the Lost Art of Livin'*. HarperCollins, New York.
2. Culberson, W. L. 1976. Water purification for field botanists. *Syst. Bot.* 1: 194; Curtis, R. 1998. OA [Outdoor Action] guide to water purification. Chapter 4 in *The Backpacker's Field Manual: A Comprehensive Guide to Mastering Backcountry Skills*. Three Rivers Press (Random House), New York.
3. Bilz is an artificially flavored red soft drink produced by the Compañía de Cervecerías Unidas in Chile. It had a sparkle that refreshed, and with some alcoholic addition at the end of the day, it earned its place in the expeditions.
4. This famous piece of folk wisdom is attributed to Benjamin Franklin, who published it (under the authorship of "A. A.") in his own newspaper, the *Pennsylvania Gazette*, in 1735. See Brands, H. W. 2000. *The First American. The Life and Times of Benjamin Franklin*. Anchor Books (Random House), New York. [p. 135.]
5. Notes taken with pen, if with permanent ink, are fine, but many inks from cheap office pens run when immersed in water. Some inks also fade over time, but graphite pencil does not. This mode of documentation even works well for writing numbers on chits of paper (high-quality cardstock) that are then immersed in preservative solutions.
6. Properly prepared herbarium specimens can last centuries. The oldest known specimens reside in collections in Italy (e.g., Florence), products of the Renaissance, now some 500 years old. See Thiers, B. M. 2020. *Herbarium: The Quest to Preserve & Classify the World's Plants*. Timber Press, Portland.
7. For a good review of the value of herbarium (voucher) specimens, see Funk, V. A., P. C. Hoch, L. A. Prather, W. L. Wagner. 2005. The importance of vouchers. *Taxon* 54: 127–129.
8. Skottsberg, C. 1921. The phanerogams of the Juan Fernandez Islands. Pp. 95–240 (plus plates 10–20) *in* C. Skottsberg (editor), *The Natural History of Juan Fernandez and Easter Island, Vol. 2, Botany*. Almqvist & Wiksells, Uppsala; Christensen, C. & C. Skottsberg. 1920. The Pteridophyta of the Juan Fernandez Islands. Pp. 1–46 *in* C. Skottsberg (editor), *The Natural History of Juan Fernandez and Easter Island, Vol. 2, Botany*. Almqvist & Wiksells, Uppsala.
9. Skottsberg, C. 1953. The vegetation of Juan Fernandez and Easter Islands. Pp. 793–960 *in* C. Skottsberg (editor), *The Natural History of Juan Fernandez and Easter Island, Vol. 2, Botany*. Almqvist & Wiksells, Uppsala; Skottsberg, C. 1928. Pollinationsbiologie und Samenverbreitung auf den Juan Fernandez-Inseln. Pp. 503–547 *in* C. Skottsberg (editor), *The Natural History of Juan Fernandez and Easter Island, Vol. 2, Botany*. Almqvist & Wiksells, Uppsala.
10. Dates of photographs in figures: p. 44, 1990; p. 46, top, 1980; p. 46, bottom, 2011.

Chapter 4. Previous Visitors

1. Santayana, G. 1905. *The Life of Reason, or the Phases of Human Progress, Vol. 1: Reason in Common Sense*. Charles Scribner's Sons, New York. [See chapter 12, Flux and constancy in human nature, p. 284.]
2. Defoe, D. 1719. *The Life and Strange Surprising Adventures of Robinson Crusoe, of York, Mariner, Who Lived Eight and Twenty Years All Alone in an Uninhabited Island on the Coast of America, Near the Mouth of the Great River of Oronoque; Having Been Cast on Shore by Shipwreck, Wherein All the Men Perished but Himself. With an Account How He Was at Last as Strangely Delivered by Pyrates*. London.
3. Watt, I. 1957. *The Rise of the Novel: Studies in Defoe, Richardson and Fielding*. University of California Press, Berkeley; Skilton, D. 1977. *Defoe to the Victorians: Two Centuries of the English Novel*. David & Charles, Newton Abbot, United Kingdom; Hasan, M. N. 2015. The eighteenth century and the rise of the English novel. *Int. J. Lit. & Arts* 3(2): 18–21.
4. Mégroz, R. L. 1939. *The Real Robinson Crusoe, Being the Life and Strange Surprising Adventures of Alexander Selkirk of Largo, Fife, Mariner*. The Cresset Press, London.
5. For more details of Selkirk and his stay on the island, see Howell, J. 1829. *The Life and Adventures of Alexander Selkirk; Containing the Real Incidents upon Which the Romance of Robinson Crusoe Is Founded*. . . . Oliver & Boyd, Edinburgh; Woodward, R. L., Jr. 1969. *Robinson Crusoe's Island*. University of North Carolina Press, Chapel Hill.
6. For detailed information on Daniel Defoe, see Richetti, J. 2005. *The Life of Daniel Defoe: A Critical Biography*. Blackwell, Oxford.
7. Medina, J. T. 1974. *El Piloto Juan Fernández, Descubridor de las Islas Que Lleven Su Nombre, y Juan Jufre, Armador de la Expedición Que Hizo en Busca de Otras en el Mar del Sur*, ed. 2. Editora Nacional Gabriela Mistral, Santiago.
8. Radiometric dates for the islands of the archipelago were provided by Stuessy, T. F., K. A. Foland, J. F. Sutter, R. W. Sanders & M. Silva O. 1984. Botanical and geological significance of potassium-argon dates from the Juan Fernandez Islands. *Science* 255: 49–51.
9. For many historical details of this early period, see Stuessy, T. F. 2020. *Environmental History of Oceanic Islands: Natural and Human Impacts on the Vegetation of the Juan Fernández (Robinson Crusoe) Archipelago*. Springer, Cham, Switzerland.
10. Schouten, W. C. 1619. *The Relation of a Wonderfull Voiage Made by William Cornelison Schouten of Horne. Shewing How South from the Straights of Magelan, in Terra Del-fuogo: He Found and Discovered a Newe Passage through the Great South Sea, and That Way Sayled Round about the World. Describing What Islands, Countries, People, and Strange Adventures He Found in His Saide Passage*. London. [Translated from the Dutch by "W. P.," William Philip *fide* Dougan, R. O. 1966. Bibliographical notes, Pp. 16–24 *in* [a booklet] *The World Encompassed by Sir Francis Drake, 1628, and The Relation of a Wonderfull Voyage by Cornelison Schouten*, 1619. 24 pp. Together with a combined facsimile reprint of the two works. The World Publishing Co., Cleveland.]
11. The drawing by Le Hermite was published in Gottfried, J. L. & M. Merian. 1631. *Newe Welt vnd Americanische Historien, Jnhaltende Warhaffige vnd volkommene Beschreibungen aller West-Jndianischen Landschafften*. Frankfurt.
12. These historical data were derived from Woodward, R. L., Jr. 1969. *Robinson Crusoe's Island*. University of North Carolina Press, Chapel Hill.
13. For details of the fur seal harvesting, see Torres, D. 1987. Antecedentes sobre el lobo fino de Juan Fernández *Arctocephalus philippii* y proyecciones para su estudio. Pp. 289–317 *in* J. C. Castilla (editor), *Islas Oceánicas Chilenas: Conocimiento Científico y*

Necesidades de Investigaciones. Universidad Católica de Chile, Santiago. [See Apéndice 1, pp. 312–313, for numbers of fur seals harvested by crews from each of the ships.]

14. See details of this time period classification in Stuessy, T. F. 2020. *Environmental History of Oceanic Islands: Natural and Human Impacts on the Vegetation of the Juan Fernández (Robinson Crusoe) Archipelago*. Springer, Cham, Switzerland.
15. Middleton, R. M. 1909. The first Fuegian collection. *J. Bot.* 47: 207–212; Gunckel L., H. 1971. Las primeras plantas herborizadas en Chile en 1690. *Anales Inst. Patagonia* 2 (1–2): 134–141; Carine, M. & T. Stuessy. 2020. George Handisyd. Pp. 192–197 *in* M. Carine (editor), *The Collectors. Creating Hans Sloane's Extraordinary Herbarium*. The Natural History Museum, London.
16. Cordingly, D. 2007. *Cochrane the Dauntless. The Life and Adventures of Admiral Thomas Cochrane, 1775–1860*. Bloomsbury, London.
17. For more information on her trip to Chile, see Graham, M. 1824. *Journal of a Residence in Chile, during the Year 1822. And a Voyage from Chile to Brazil in 1823*. Longman, Hurst, Rees, Orme, Brown, and Green, and John Murray, London.
18. Delprete, P. G., G. Forneris & A. Pistarino. 2002. Carlo Bertero (1789–1831) in the New World. *Sida* 20: 621–644.
19. Bertero, C. G. L. 1830. Notice sur l'histoire naturelle de l'ile Juan Fernandez [extraite d'une lettre de M. Bertero]. *Ann. Sci. Nat.* 21: 344–351.
20. Fuenzalida V., H. 1944. Don Claudio Gay y el Museo Nacional de Historia Natural. *Bol. Mus. Nac. Hist. Nat.* 22: 5–16.
21. Gay, C. 1845–1854. *Historia Física y Política de Chile Según Documentos Adquiridos en Esta República Durante Doce Años de Residencia en Ella y Publicada Bajo los Auspicios del Supremo Gobierno. Botánica*. Published by the author, Paris.
22. Gay, C. 1832. Exploración de las islas de Juan Fernández. *El Araucano* No. 77, 3 March 1832. [Reprinted pp. 124–130 *in* C. S. Ortiz & G. F. Cruz, *Vida de Claudio Gay, Vol. 2*, Editorial Nascimento, Santiago. Also reprinted as La Isla de Juan Fernández, 1974, *Mus. Nac. Hist. Nat. Ser. Educativa* no. 9, pp. 6–16.]
23. Sutcliffe, T. ["T. S."]. 1839. *The Earthquake of Juan Fernandez, as It Occurred in the Year 1835, Authenticated by the Retired Governor of That Island, to Which Is Added, a Refutation of Several Misstatements That Have Been Published in the "Nautical Magazine" of 1837, and the Public Papers*. Published by the author, Manchester. For a geological analysis of this tsunami (and rejection of the idea of an earthquake), see Lara, L. E., R. Moreno, V. Valdivia, R. Aránguiz & M. Lagos. 2020. The AD1835 eruption at Robinson Crusoe Island discredited: Geological and historical evidence. *Progr. Phys. Geogr.* 2020: 1–20. doi: 10.1177/0309133320937998.
24. Sutcliffe, T. 1841. *Sixteen Years in Chile and Peru from 1822 to 1839 by the Retired Governor of Juan Fernandez*. Fisher, Son, and Co., London. [No name is given as author, but the dates of residency in Chile and Peru of the former governor in the island allow for no other than Thomas Sutcliffe; no date is provided on the title page (or reverse) either, but on p. x at the end of the preface, a date of 1 June 1841 is given.]
25. Gotschlich, B. 1904. *Biografía del Dr. Rodulfo Amando Philippi (1803–1904)*. Imprenta Central J. Lampert, Valdivia.
26. Philippi, R. A. 1860. *Viaje al Desierto de Atacama Hecho por Orden del Gobierno de Chile en el Verano 1853–54*. Halle; Philippi, R. A. 1860. *Florula Atacamesis seu Enumeratio Plantarum, quas in Itinere per Desertum Atacamense Observavit Dr. R. Philippi*. Halle; Muñoz-Schick, M. 2008. Viajes de los Philippi a la Región de Atacama. Pp. 357–370 *in* F. A. Squeo, G. Arancio & J. R. Gutiérrez (editors), *Libro Rojo de la Flora Nativa y de los Sitios Prioritarios para Su Conservación: Región de Atacama*. Ediciones Universidad de La Serena, La Serena, Chile.

27. Philippi, R. 1856. Observaciones sobre la flora de Juan Fernández. *Anal. Univ. Chile* 13: 157–169; Philippi, R. A. 1856. Bemerkungen über die Flora der Insel Juan Fernandez. *Bot. Zeitung (Berlin)* 14: 625–636 [5 September]; 641–650 [12 September].
28. Philippi, R. A. 1865. Ueber zwei neue Pflanzen-Gattungen. Ein Schreiben an Hrn. Prof u. Director Dr. Eduard Fenzl. *Verh. K. K. Zool.-Bot. Ges. Wien* 15: 517–524.
29. Linklater, E. 1972. *The Voyage of the* Challenger. John Murray, London. [A Spanish edition was published in 1982 under the title *El Viaje del Challenger (1872–1876)*, Ediciones del Serbal, Barcelona.]
30. Moseley, H. N. 1892. *Notes by a Naturalist. An Account of Observations Made during the Voyage of H.M.S.* "Challenger" *Round the World in the Years 1872–1876, under the Command of Capt. Sir G. S. Nares, R.N., K.C.B., F.R.S., and Capt. F. T. Thomson, R.N.* John Murray, London.
31. Hemsley, W. B. 1884. Report on the botany of Juan Fernández, the south-eastern Moluccas, and the Admiralty Islands. Pp. 1–96 *in* C. W. Thomson & J. Murray (superintendents). *Report on the Scientific Results of the Voyage of H.M.S.* Challenger *during the Years 1873–76, Vol. 1, Botany*. H. M. Stationary Office, London.
32. Muñoz Pizarro, C. 1974. *Recuerdos del Profesor Federico Adalberto Johow 1859–1933*. Museo Nacional de Historia Natural, Ser. Educativa 10, Santiago; Aguirre de Maino, I. M. (editor). 1981. *Chile a Color: Biografías, Vols. 1 y 3*. Editorial Antártica, Santiago.
33. Johow, F. 1896. *Estudios sobre la Flora de las Islas de Juan Fernández*. Gobierno de Chile, Santiago.
34. Peterson, B. 1964. Carl Skottsberg 1880–1963. *Taxon* 13: 1–7.
35. Salisbury, E. J. 1964. Carl Johan Fredrik Skottsberg 1880–1963. *Biogr. Mem. Fellows Roy. Soc.* 10: 244–256.
36. Skottsberg, C. 1909. Swedish Magellanic Expedition, 1907–1909. The excursion to the Juan-Fernandez-group: Biological remarks. *Geogr. J.* 33: 292–294; Skottsberg, C. 1911. *The Wilds of Patagonia. A Narrative of the Swedish Expedition to Patagonia, Tierra del Fuego and the Falkland Islands in 1907–1909*. The Macmillan Co., New York; Skottsberg, C. 1914. Botanische Ergebnisse der Schwedischen Expedition nach Patagonien und dem Feuerlande 1907–1909. IV. Studien über die Vegetation der Juan Fernandez-Inseln. *Kungl. Svenska Vetenskapsakad. Handl.* 51: 1–73.
37. Skottsberg, C. 1917. La expedición científica sueca a las islas esporádicas de Chile 1916–17. Informe preliminar sobre los resultados botánicos. *Revista Univ. Nac. Córdoba* 4(8): 110–115.
38. Skottsberg, C. (editor). 1920–1956. *The Natural History of Juan Fernandez and Easter Island, Vol. 1, Geography, Geology, Origin of Island Life, Vol. 2, Botany*. Almqvist & Wiksells, Uppsala; Skottsberg, C. (editor). 1921–1940. *The Natural History of Juan Fernandez and Easter Island, Vol. 3, Zoology*. Almqvist & Wiksells, Uppsala.
39. Christensen, C. & C. Skottsberg. 1920. The Pteridophyta of the Juan Fernandez Islands. Pp. 1–46 *in* C. Skottsberg (editor), *The Natural History of Juan Fernandez and Easter Island, Vol. 2, Botany*. Almqvist & Wiksells, Uppsala; Skottsberg, C. 1921. The phanerogams of the Juan Fernandez Islands. Pp. 95–240 (plus plates 10–20) *in* C. Skottsberg (editor), *The Natural History of Juan Fernandez and Easter Island, Vol. 2, Botany*. Almqvist & Wiksells, Uppsala; Skottsberg, C. 1928. Pollinationsbiologie und Samenverbreitung auf den Juan Fernandez-Inseln. Pp. 503–547 *in* C. Skottsberg (editor), *The Natural History of Juan Fernandez and Easter Island, Vol. 2, Botany*. Almqvist & Wiksells, Uppsala; Skottsberg, C. 1953. The vegetation of Juan Fernandez and Easter Islands. Pp. 793–960 *in* C. Skottsberg (editor), *The Natural History of Juan Fernandez and Easter Island, Vol. 2, Botany*. Almqvist and Wiksells, Uppsala;

Skottsberg, C. 1953. A supplement to the phanerogams of the Juan Fernandez Islands. Pp. 763–792 *in* C. Skottsberg (editor), *The Natural History of Juan Fernandez and Easter Island, Vol. 2, Botany*. Almqvist & Wiksells, Uppsala; Skottsberg, C. 1954. A geographical sketch of the Juan Fernandez Islands. Pp. 89–192 *in* C. Skottsberg (editor), *The Natural History of Juan Fernandez and Easter Island, Vol. 1, Geography, Geology, Origin of Island Life*. Almqvist & Wiksells, Uppsala.

40. Skottsberg, C. 1918. *Til Robinson-ön och Världens ände*. Albert Bonniers, Stockholm. Ann-Marie Skottsberg was a lovely person—intelligent, curious, warm, and with an appreciation of what her father had accomplished in the Robinson Crusoe Islands. She died peacefully on 14 July 2021 at the age of 100.
41. Looser, G. 1927. Sobre algunos objetos que venden los habitantes de las islas de Juan Fernández—Apuntes folklóricos. *Revista Chilena Hist. Nat.* 31: 240–244.
42. Meyer, F. G. 1966. Chile-United States botanical expedition to Juan Fernandez Islands, 1965. *Antarc. J.* 1: 238–242.
43. Moore, H. E., Jr. 1969. The genus *Juania* (Palmae-Arecoideae). *Gentes Herb.* 10: 385–393; Tomlinson, P. B. 1969. The anatomy of the vegetative organs of *Juania australis* (Palmae). *Gentes Herb.* 10: 412–424; Uhl, N. W. 1969. Floral anatomy of *Juania, Ravenea*, and *Ceroxylon* (Palmae-Arecoideae). *Gentes Herb.* 10: 394–411.
44. For a convenient summary of our cooperative research, see Stuessy, T. F., D. J. Crawford, P. López-Sepúlveda, C. M. Baeza & E. A. Ruiz (editors). 2018. *Plants of Oceanic Islands: Evolution, Biogeography, and Conservation of the Flora of the Juan Fernández (Robinson Crusoe) Archipelago*. Cambridge University Press, Cambridge.
45. Danton, P., E. Breteau & M. Baffray. 1999. *Les Îles de Robinson: Trésor Vivant des Mers du Sud. Entre Légende et Realité*. Nathan, Paris; Danton, P. 2004. *Plantas Silvestres de la Isla Robinson Crusoe*. CONAF, Viña del Mar, Chile; Danton, P., C. Perrier & G. Martínez R. 2006. Nouveau catalogue de la flore vasculaire de l'archipel Juan Fernández (Chili) [Nuevo catálogo de la flora vascular del Archipiélago Juan Fernández (Chile)]. *Acta Bot. Gallica* 153: 399–587; Danton, P. & C. Perrier. 2011. *Estudio para una Monografía de la Flora Vascular del Archipiélago Juan Fernández: Joyas Verdes en el Océano*. Published by the authors, Santiago.
46. Danton, P. & C. Perrier. 2020. *Monographie de la Fore Vasculaire de l'Archipel Juan Fernández (Chili): Essai de Valorisation pour su Préservation*. Editions Privat, Paris.
47. Dates of photographs in figures: p. 56, 1990; p. 67, bottom left and right, 1994; p. 68, top left, 1965; p. 68, top right, 1976; p. 68, middle left, bottom left, and right, 2008; p. 68, middle right, 2010.

Chapter 5. Bizarre, Unusual, and Interesting Plants

1. Carlquist, S. 1974. *Island Biology*. Columbia University Press, New York; König, C., P. Weigelt, A. Taylor, A. Stein, W. Dawson, F. Essl, J. Pergl, et al. 2020. Source pools and disharmony of the world's island floras. *Ecography* 43: 1–12.
2. Funk, V. A., A. Susanna, T. F. Stuessy & R. J. Bayer (editors). 2009. *Systematics, Evolution, and Biogeography of Compositae*. IAPT, Vienna.
3. Kim, S.-C., D. J. Crawford & R. K. Jansen. 1996. Phylogenetic relationships among the genera of the subtribe Sonchinae (Asteraceae): Evidence from ITS sequences. *Syst. Bot.* 21: 417–432.
4. Colwell, R. K. 1989. Hummingbirds of the Juan Fernández Islands: Natural history, evolution and population status. *Ibis* 131: 548–566.
5. Crawford, D. J., T. F. Stuessy & M. Silva O. 1987. Allozyme divergence and the evolution of *Dendroseris* (Compositae: Lactuceae) on the Juan Fernández Islands. *Syst. Bot.* 12: 435–443; Crawford, D. J., T. F. Stuessy, M. B. Cosner, D. W. Haines,

M. Silva O. & M. Baeza. 1992. Evolution of the genus *Dendroseris* (Asteraceae: Lactuceae) on the Juan Fernández Islands: Evidence from chloroplast and ribosomal DNA. *Syst. Bot.* 17: 676–682; Sang, T., D. J. Crawford, S.-C. Kim & T. F. Stuessy. 1994. Radiation of the endemic genus *Dendroseris* (Asteraceae) on the Juan Fernández Islands: Evidence from sequences of the ITS regions of nuclear ribosomal DNA. *Amer. J. Bot.* 81: 1494–1501; Crawford, D. J., T. Sang, T. F. Stuessy, S.-C. Kim & M. Silva O. 1998. *Dendroseris* (Asteraceae: Lactuceae) and *Robinsonia* (Asteraceae: Senecioneae) on the Juan Fernández Islands: Similarities and differences in biology and phylogeny. Pp. 97–119 *in* T. F. Stuessy & M. Ono (editors), *Evolution and Speciation of Island Plants*. Cambridge University Press, Cambridge.

6. Pelser et al. (2010) have submerged *Robinsonia* taxonomically into the large genus *Senecio*, but based on morphological and breeding system distinctions, I do not accept this disposition. Pelser, P. B., E. J. Tepe, A. H. Kennedy & L. E. Watson. 2010. The fate of *Robinsonia* (Asteraceae): Sunk in *Senecio*, but still monophyletic? *Phytotaxa* 5: 31–46.
7. CONAF. 2016. *Robinsonia berteroi* (DC.) Sanders, Stuessy & Martic. http://www.mma.gob.cl/clasificacionespecies/fichas13proceso/pdf. Accessed 22 February 2020.
8. Crawford, D. J., T. F. Stuessy, D. W. Haines, M. B. Cosner, M. Silva O. & P. López. 1992. Allozyme diversity within and divergence among four species of *Robinsonia* (Asteraceae: Senecioneae), a genus endemic to the Juan Fernández Islands, Chile. *Amer. J. Bot.* 79: 962–966; Crawford, D. J., T. F. Stuessy, M. B. Cosner, D. W. Haines & M. Silva O. 1993. Ribosomal and chloroplast DNA restriction site mutations and the radiation of *Robinsonia* (Asteraceae, Senecioneae) on the Juan Fernández Islands. *Pl. Syst. Evol.* 184: 233–239; Sang, T., D. J. Crawford, T. F. Stuessy & M. Silva O. 1995. ITS sequences and the phylogeny of the genus *Robinsonia* (Asteraceae). *Syst. Bot.* 20: 55–64.
9. Sanders, R. W., T. F Stuessy, C. Marticorena & M. Silva O. 1987. Phytogeography and evolution of *Dendroseris* and *Robinsonia*, tree-Compositae of the Juan Fernández Islands. *Opera Bot.* 92: 195–215.
10. Anagenesis is a form of transformational speciation whereby a plant immigrant arrives to an island and over thousands or millions of years changes into a different species, as judged by morphological and genetic distinctions from the progenitor. See Stuessy, T. F., G. Jakubowsky, R. Salguero Gómez, M. Pfosser, P. M. Schluter, T. Fer, B.-Y. Sun, et al. 2006. Anagenetic evolution in island plants. *J. Biogeogr.* 33: 1259–1265.
11. Recent molecular phylogenetic studies on *Blechnum* s.l. have resulted in the creation of several smaller genera: Gasper, A. L. de, V. A. de Oliveira Dittrich, A. R. Smith & A. Salino. 2016. A classification for Blechnaceae (Polypodiales: Polypodiopsida): New genera, resurrected names, and combinations. *Phytotaxa* 275: 191–227; Gasper, A. L. de, T. E. Almeida, V. A. de Dittrich, A. R. Smith & A. Salino. 2017. Molecular phylogeny of the fern family Blechnaceae (Polypodiales) with a revised genus-level treatment. *Cladistics* 33: 429–446.
12. The relationships of *Lactoris* based on morphology have long been controversial, but recent DNA sequences have helped to suggest affinities. The genus clearly belongs in Piperales near family Aristolochiaceae (the Dutchman's pipe family), but different molecular markers support specific alternative placements. Some workers recommend placing *Lactoris* within Aristolochiaceae, but others suggest recognition as a closely related neighboring family. Tucker, S. C. & A. W. Douglas. 1996. Floral structure, development, and relationships of paleoherbs: *Saruma*, *Cabomba*, *Lactoris*, and selected Piperales. Pp. 141–175 *in* D. W. Taylor & L. J. Hickey (editors), *Flowering*

Plant Origin, Evolution & Phylogeny. Chapman & Hall, New York; Stuessy, T. F., D. J. Crawford, G. J. Anderson & R. J. Jensen. 1998. Systematics, biogeography, and conservation of Lactoridaceae. *Perspect. Pl. Ecol. Evol. Syst.* 1: 267–290; González, F. & P. Rudall. 2001. The questionable affinities of *Lactoris*: Evidence from branching pattern, inflorescence morphology, and stipule development. *Amer. J. Bot.* 88: 2143–2150; Wanke, S., M. A. Jaramillo, T. Borsch, M.-S. Samain, D. Quandt & C. Neinhuis. 2007. Evolution of Piperales—*matK* gene and *trnK* intron sequence data reveal lineage-specific resolution contrast. *Molec. Phylogen. Evol.* 42: 477–497; Soltis, D. E., S. A. Smith, N. Cellinese, K. J. Wurdack, D. C. Tank, S. F. Brockington, N. F. Refulio-Rodríguez, et al. 2011. Angiosperm phylogeny: 17 genes, 640 taxa. *Amer. J. Bot.* 98: 704–730; Naumann, J., K. Salomo, J. P. Der, E. K. Wafula, J. F. Bolin, E. Maass, L. Frenzke, et al. 2013. Single-copy nuclear genes place haustorial Hydnoraceae within Piperales and reveal a Cretaceous origin of multiple parasitic angiosperm lineages. *PLOS One* 8(11): 1–12. E79204.

13. Ricci, M. 2001. Evaluation of conservation status of *Lactoris fernandeziana* Philippi (Lactoridaceae) in Chile. *Biodivers. & Conserv.* 10: 2129–2138.
14. Philippi, R. A. 1865. Ueber zwei neue Pflanzen-Gattungen. Ein Schreiben an Hrn. Prof u. Director Dr. Eduard Fenzl. *Verh. K. K. Zool.-Bot. Ges. Wien* 15: 517–524.
15. Bernardello, G., G. J. Anderson, P. López S., M. A. Cleland, T. F. Stuessy & D. J. Crawford. 1999. Reproductive biology of *Lactoris fernandeziana* (Lactoridaceae). *Amer. J. Bot.* 86: 829–840.
16. Zavada, M. S. & T. N. Taylor. 1986. Pollen morphology of Lactoridaceae. *Pl. Syst. Evol.* 154: 31–39; Sampson, F. G. 1995. Pollen morphology of Lactoridaceae: A re-examination. *Grana* 34: 100–107.
17. Zavada, M. S. & J. M. Benson. 1987. First fossil evidence for the primitive angiosperm family Lactoridaceae. *Amer. J. Bot.* 74: 1590–1594.
18. MacPhail, M. K., A. D. Partridge & E. M. Truswell. 1999. Fossil pollen records of the problematical primitive angiosperm family Lactoridaceae in Australia. *Pl. Syst. Evol.* 214: 199–210.
19. Gamerro, J. C. & V. Barreda. 2008. New fossil record of Lactoridaceae in southern South America: A palaeobiogeographical approach. *Bot. J. Linn. Soc.* 158: 41–50.
20. Smith, A. R., K. M. Pryer, E. Schuettpelz, P. Korall, H. Schneider & P. G. Wolf. 2006. A classification for extant ferns. *Taxon* 55: 705–731.
21. Menéndez, C. A. 1966. La presencia de *Thyrsopteris* en el Cretácico superior de Cerro Guido, Chile. *Ameghiniana* 4: 299–302.
22. Li, C., R. C. Moran, J. Ma, B. Wang & J. Hao. 2020. A mid-Cretaceous tree fern of Thyrsopteridaceae (Cyatheales) preserved in Myanmar amber. *Cretaceous Res.* 105: 104050.
23. Nishida, H. & M. Nishida. 1979. *Thyrsopterorachis*, gen. nov., a tree fern rachis from the Upper Cretaceous of Hokkaido, Japan. *Bot. Mag. Tokyo* 92: 187–195.
24. Stuessy, T. F., D. J. Crawford & E. A. Ruiz. 2018. Patterns of phylogeny. Pp. 209–275 *in* T. F. Stuessy, D. J. Crawford, P. López-Sepúlveda, C. M. Baeza & E. A. Ruiz (editors), *Plants of Oceanic Islands: Evolution, Biogeography, and Conservation of the Flora of the Juan Fernández (Robinson Crusoe) Archipelago*. Cambridge University Press, Cambridge.
25. The estimates of total numbers of species on Earth vary widely. Mora et al. (Mora, C., D. P. Tittensor, S. Adl, A. G. B. Simpson & B. Worm. 2011. How many species are there on Earth and in the ocean? *PLOS Biol.* doi.org/10.1371/journal.pbio.1001127) estimated more than 8.7 million eucaryotic species, but Larsen et al. (Larsen, B. B., E. C. Miller, M. K. Rhodes & J. J. Wiens. 2017. Inordinate fondness multiplied and

redistributed: The number of species on Earth and the new Pie of Life. *Quart. Rev. Biol.* 92: doi.org/10.1086/693564) suggest as many as one to six billion species by including estimates for procarykotes (e.g., bacteria) and other lesser-known groups.
26. Dates of photographs in figures: color insert image 2, 1980; color insert image 6, left, 2011; color insert image 8, right, 1980.

Chapter 6. Arrivals and Departures

1. Systematic and evolutionary biologists make particularly good international ambassadors, because they normally work in the countryside together with host colleagues. As one example: Stuessy, T. F. 1992. International collaboration in natural history museums: Organisms as international ambassadors. *Korean J. Pl. Taxon.* 21: 217–227.
2. Darwin, C. 1839. *Journal of Researches into the Geology and Natural History of the Various Countries Visited by H.M.S.* Beagle, *under the Command of Captain Fitzroy, R.N. from 1832 to 1836.* Henry Colburn, London.
3. The children, in particular, loved to roll their eyes and comment that it was difficult living with "Chevy Chase" for a father, referring to the *Vacation* movies in which he was forever overly enthusiastic about seeing some new sight and waxing poetically about it (e.g., *National Lampoon's Vacation*, 1983, and *European Vacation*, 1985).
4. Verne, J. 1871. *Vingt mille lieues sous les mers: Tour du monde sous marin.* P.-J. Hetzel, Paris. [First serialized in 1869–1870 in the *Magasin d'éducation et de recreation*; first English translation in 1873 by L. P. Mercier as *Twenty Thousand Leagues Under the Seas: A World Tour Under Water.* Sampson Low, London]
5. *Cholgas* are a small mussel (*Aulacomya ater*), available fresh or canned, and often used in Chilean seafood dishes.
6. A tragic accident occurred during an intended landing on Robinson Crusoe Island on 2 September 2011. A much larger plane (C-212 Aviocar) of the Chilean Air Force with twenty-one persons attempted a landing in strong winds. Their first attempt was unsuccessful, so they continued toward Santa Clara Island to make a circle and try a second attempt, but unpredictable winds apparently caused the plane to crash in the ocean. There were no survivors.
7. *Alfajor* cookies, or *alfajores*, consist of two sugar cookie halves held together with *dulce de leche* (caramelized sugar), and sometimes with a spot of pisco or brandy. These originated with the Moors, who brought them to Spain when they invaded the Iberian Peninsula beginning in the eighth century, and Spain later introduced them to her New World colonies. An *alfajor*, with coffee, is a delectable snack!
8. Dates of photographs in figures: p. 90, 1997; p. 91, 1986; p. 94, 1984; p. 95, 1980; p. 96, 1980; p. 99, 1990; p. 103, 1996; p. 104, top, 1980; p. 104, bottom, 1996.

Chapter 7. Island Living

1. For good overviews of the changing population of Robinson Crusoe Island, see Woodward, R. L., Jr. 1969. *Robinson Crusoe's Island.* University of North Carolina Press, Chapel Hill; Stuessy, T. F. 2020. *Environmental History of Oceanic Islands: Natural and Human Impacts on the Vegetation of the Juan Fernández (Robinson Crusoe) Archipelago.* Springer, Cham, Switzerland.
2. From the discovery of Robinson Crusoe Island by Juan Fernández in 1574 to the middle of the eighteenth century, Spain controlled the island, but they did not bother to fortify it. As a result, numerous privateers from other European nations used the island for recuperation of tired and sick crew members, obtaining fresh water, repairing ships, and as a lair from which to attack Spanish colonies. This constant harassment led to the building in 1750 of Fort Santa Bárbara on the slope above Bahía

Cumberland, which was well fortified with eighteen cannons and 7,000 cannon balls. From this point onward, Spain controlled access to the bay, but this did nothing to prohibit ships from visiting Alejandro Selkirk Island.

3. All systematic and evolutionary studies, involving chromosome numbers, secondary products, or DNA sequences, require that herbarium specimens be made from these same individual plants (or at least from the same natural population), and these must be deposited in a reputable institution. In this manner one can always check the identity of the plants that have been analyzed. These are called "vouchers" because they permanently document the identity of the organisms.
4. Culberson, W. L. 1976. Water purification for field botanists. *Syst. Bot.* 1: 194.
5. The cumbia is a type of rhythmic music that originated in Colombia in the 1940s, and from there it expanded to other Latin American countries. It has a lively, infectious beat that is suitable for rapid dancing, and it is hard to sit still while listening to it. We preferred the Chilean group Pachuco (Roberto Fonseco) y la Cubanacán, who first achieved acclaim at the 1980 music festival in Viña del Mar, Chile. We enjoyed many of their tunes, but my personal favorite was always their upbeat "Oye." Pachuco died in 2001 at age sixty-four, but his music and happy spirit live on.
6. Dates of photographs in figures: p. 107, 1980; pp. 108, 109, 1986; p. 110, 1984; p. 112, 1990; p. 114, 2010; p. 115, left, 1984; p. 115, right, 2011; p. 117, 2010; p. 118, left, 2011; p. 118, right, p. 119, 1980; p. 121, 1984.

Chapter 8. Real Isolation

1. Stuessy, T. F., K. A. Foland, J. F. Sutter, R. W. Sanders & M. Silva O. 1984. Botanical and geological significance of potassium-argon dates from the Juan Fernández Islands. *Science* 255: 49–51.
2. Sterno, originally invented about 1918 by Sternau & Co. in Brooklyn, New York, is a pinkish gel infused with methanol that comes in a small, round can. Once the lid is removed, a match can light the contents, which gives off a low blue flame suitable for camp cooking. Any remaining gel can be saved by closing the lid tight and carrying it until the next meal. It is often used by backpackers, in food buffets to keep metal serving trays warm, and by soldiers during military campaigns.
3. Skottsberg (1921) originally described *Abrotanella crassipes* as an endemic species, restricted to the top of Los Inocentes on Alejandro Selkirk Island. Further comparative studies by Ulf Swenson have revealed it to be the same as *A. linearifolia*, distributed on the Chilean mainland mostly in southern Magellanic provinces. See Skottsberg, C. 1921. The phanerogams of the Juan Fernandez Islands. Pp. 95–240 (plus plates 10–20) *in* C. Skottsberg (editor), *The Natural History of Juan Fernandez and Easter Island, Vol. 2, Botany*. Almqvist & Wiksells, Uppsala; Swenson, U. 1995. *Systematics of the Blennospermatinae (Asteraceae, Senecioneae)*. Doctoral Thesis [*Acta Universitatis Upsaliensis*], University of Uppsala, Uppsala; Swenson, U. 1995. Systematics of *Abrotanella*, an amphi-pacific genus of Asteraceae (Senecioneae). *Pl. Syst. Evol.* 197: 149–193; Swenson, U. & K. Bremer. 1997. Pacific biogeography of the Asteraceae genus *Abrotanella* (Senecioneae, Blennospermatinae). *Syst. Bot.* 22: 493–508; Swenson, U. & K. Bremer. 1999. On the circumscription of the Blennospermatinae (Asteraceae, Senecioneae) based on *ndh*F sequence data. *Taxon* 48: 7–14; Wagstaff, S. J., I. Breitwieser & U. Swenson. 2006. Origin and relationships of the austral genus *Abrotanella* (Asteraceae) inferred from DNA sequences. *Taxon* 55: 95–106.
4. Skottsberg, C. 1953. The vegetation of the Juan Fernandez Islands. Pp. 793–960 *in* C. Skottsberg (editor), *The Natural History of Juan Fernandez and Easter Island, Vol. 2, Botany*. Almqvist & Wiksells, Uppsala. Greimler et al. (2013) refer to this zone

as the *Lophosoria quadripinnata* fern community. Greimler, J., P. López-Sepúlveda, K. Reiter, C. Baeza, P. Peñailillo, E. Ruiz, P. Novoa, et al. 2013. Vegetation of Alejandro Selkirk Island (Isla Masafuera), Juan Fernández Archipelago, Chile. *Pacific Sci.* 67: 267–282.

5. Stuessy, T. F. 2020. *Environmental History of Oceanic Islands: Natural and Human Impacts on the Vegetation of the Juan Fernández (Robinson Crusoe) Archipelago.* Springer, Cham, Switzerland. [p. 285.]
6. Dates of photographs in figures: pp. 124, 127, 128, 133, 135, 136, 138, 139, 1986; color insert image 9, 1984; p. 141, 1986; p. 143, left, 2011; p. 143, right, pp. 144, 145, 149, 151, 155, color insert image 10, 1986.

Chapter 9. An Improbable Encounter

1. The *Dresden* was a German cruiser that was sunk on 8 March 1915 in Bahía Cumberland during WWI. As Chile had declared itself neutral during the conflict, the German ship entered the bay seeking fresh water and fuel. Tracking the *Dresden* were three English warships, *Kent, Glasgow*, and *Orama*, and they waited patiently nearby. At some point it was believed that the *Dresden* was making a run out of the bay, which caused the English ships to open fire. This resulted in serious damage to the *Dresden*, leading to its scuttling. The captain and most of the crew escaped to land, after which explosions destroyed records and sent the ship to the bottom of the bay, where it still resides at 60 meters. A monument to this event was placed in the cemetery in 1922 by the German community in Valparaíso. See Woodward, R. L., Jr. 1969. *Robinson Crusoe's Island.* University of North Carolina Press, Chapel Hill; Parker de Bassi, M. T. 1987. *Tras la Estela del Dresden.* Ediciones Tusitala, Santiago.
2. Amazingly, nearly 7,000 languages are spoken, and many of them are now endangered. At first, I was indifferent about such a catastrophic loss of human culture, thinking that this would simply make global communications easier. But after reading the book by K. D. Harrison (2007, *When Languages Die: The Extinction of the World's Languages and the Erosion of Human Knowledge*, Oxford University Press, New York), I realized that each language is one human vision of the reality of our world. Each language describes things, events, and emotions differently, and to lose even one of these is to lose part of our human heritage. It is inevitable that some languages will disappear as our population increases and globalization continues, but we must place a priority on capturing these languages before they are gone. Losing a language also makes it more difficult to understand the evolution of extinct languages, because part of the puzzle will be lost. This is analogous to attempting to reconstruct evolutionary histories (phylogenies) of organisms when some lineages have already gone extinct.
3. The term *once* literally means "eleven" in Spanish. It is presumed to have originated in Spain to refer to a late morning snack at about 11:00. Some Latin American countries have adopted this general idea, using the same label, but having the snack at different times of the day. Chile adopted 4:00 to 5:00 in the afternoon because it helps to tide people over to their typical dinner after 8:00.
4. Written by Jerry Fuller, the song became a hit in 1961 with Ricky Nelson and Imperial Records.
5. Salvador Allende of the Popular Unity Party was elected President of Chile and took office on 3 November 1970. He was openly Marxist in orientation and sought help from communist countries, especially Russia and Cuba. Land reform, the taking of farmland and giving it to the poor, plus general economic decline (exacerbated by pressure from the United States) eventually led to civil unrest. The military intervened in a coup on 11 September 1973 and created a governing junta of the heads of the

armed services. This military government established order in the country, markedly improved the economy, and focused on infrastructure projects, but the cost was restriction of civil liberties and massive persecution of political opponents.

6. The cocktail, pisco sour, is popular in Peru and Chile. There are many recipes, but the usual ingredients are pisco, lemon juice, sugar, and egg white. The Peruvians sometimes add a dash of Angostura bitters.
7. This was the situation in Chile until finally in 1988 a plebiscite was held to determine if Pinochet should continue in power. He lost this vote, and a new election was held the following year, won by Patricio Aylwin of the Christian Democratic Party. Democratic institutions were restored to the country in 1990.
8. A code-breaking group was established in Bletchley Park by the British government, under leadership of the brilliant, but eccentric, mathematician Alan Turing. Their principal objective was to break the Enigma Code used by the Germans to communicate within their armed forces. The effort was successful, but there were tensions within the research group as well as between them and the government. The movie *The Imitation Game* (2014, Black Bear Pictures) provides an instructive and entertaining overview of how it all happened.
9. Dates of photographs in figures: pp. 161, 162, 163, 1986; pp. 164, 165, 167, 1980; p. 179, 1988.

Chapter 10. The Call

1. Greimler, J. 1997. Pflanzengesellschaften und Vegetationsstruktur in den südlichen Gesäusebergen (nordöstliche Kalkalpen, Steiermark). *Mitt. Abteil. Bot. Landesmus. "Joanneum" Graz* 25–26: 1–238.
2. Johow, F. 1896. *Estudios sobre la Flora de las Islas de Juan Fernández*. Gobierno de Chile, Santiago.
3. Skottsberg, C. 1953. The vegetation of the Juan Fernandez Islands. Pp. 793–960 *in* C. Skottsberg (editor), *The Natural History of Juan Fernandez and Easter Island, Vol. 2, Botany*. Almqvist & Wiksells, Uppsala.
4. Kunkel, G. 1957. Beobachtungen über die Vegetation auf dem Yunque-Massiv. Beiträge zur flora der Juan Fernández-Inseln. *Bot. Jahrb. Syst.* 77: 149–157; Schwaar, J. 1979. Feuchtwälder auf Juan Fernandez. *Phytocoenologia* 6: 514–523; Ortiz R., A. (editor). 1982. *Estudio Recursos Físicos: Archipiélago Juan Fernández*. Instituto Nacional de Investigación de Recursos Naturales – CORFO, Santiago.
5. Grant no. 6399-99 entitled "Vegetation of Robinson Crusoe Island."
6. Greimler, J., P. López-Sepúlveda, K. Reiter, C. Baeza, P. Peñailillo, E. Ruiz, P. Novoa, et al. 2013. Vegetation of Alejandro Selkirk Island (Isla Masafuera), Juan Fernández Archipelago, Chile. *Pacific Sci.* 67: 267–282.
7. The appellation *Don* attached to men's first names (*Doña* for women) is an informal title, showing more respect than the commonly used *Señor* (or *Señora*), which accompanies the family name. Although he was my father-in-law, I usually referred to him in this more formal way.
8. Two planes have crashed into the ocean in recent decades on flights from the Chilean continent to Robinson Crusoe Island. The first occurred in 1972, when a small bimotor Piper PA-31 Navajo crashed halfway toward the island, killing all eight persons aboard. No trace of the plane or passengers was ever found. The second was in 2011, when strong and unpredictable winds caused a plane of the Chilean Air Force to crash into the ocean upon attempting to land at the airstrip.
9. The Nomad airplane was produced in Melbourne, Australia, by the Government Aircraft Factories, and 172 were manufactured from 1975 to 1985. It was designed as a

twin-engine utility aircraft with short takeoff and landing requirements and saw considerable service in Australia, but safety concerns in the 1990s led to its lack of popularity. The airplane acquired by Transportes Aéreos Robinson Crusoe was obviously a used airplane, but it gave good service to and from the islands for a number of years. It had good storage capacity, which was useful for dispatch of supplies to the island and for sending fish and lobsters back to the continent.

10. The Juan Fernández fur seal (*Arctocephalus philippii*) was incredibly abundant on both islands of the archipelago at the time of European discovery in the sixteenth century. Historical documents indicate that there were so many animals in Bahía Cumberland that a person could not easily arrive to shore. Due to a profitable market for fur pelts in Canton (China), millions of animals were slaughtered at the end of the eighteenth and beginning of the nineteenth centuries. This stopped only when the number of seals was so low as to make the enterprise no longer profitable. For more details, see Stuessy, T. F. 2020. *Environmental History of Oceanic Islands: Natural and Human Impacts on the Vegetation of the Juan Fernández (Robinson Crusoe) Archipelago*. Springer, Cham, Switzerland.
11. Danton, P. & C. Perrier. 2020. *Monographie de la Flore Vasculaire de l'Archipel Juan Fernández (Chili): Essai de Valorisation pour Su Préservation*. Editions Privat, Paris.
12. *Machas* are saltwater clams, *Mesodesma donacium*, native to Chile. They are smaller than *cholgas* (see chapter 6, note 5), but equally good eating either straight from the can (already cooked) or in *machas a la parmasana*, a very tasty Italian dish.
13. Braun-Blanquet, J. 1928. *Pflanzensoziologie, Grundzüge der Vegetationskunde*. Springer, Berlin. (Ed. 3, 1964, Springer, Wien.)
14. The data from the relevés were run with TWINSPAN of the program JUICE 7. Hill, M. O. 1979. *Twinspan - A Fortran Program for Arranging Multivariate Data in an Ordered Two-way Table by Classification of the Individuals and Attributes*. Ithaca, New York; Tichy, L. 2002. JUICE, software for vegetation classification. *J. Veg. Sci.* 13: 451–453.
15. It is remarkable that Skottsberg dared to describe (1929) a new genus and species based only on leaves. Because of his broad experience with the flora of the island (and elsewhere in South America), he was able to infer familial affinities from vegetative material only. An additional collection showing old flowering heads later confirmed his initial judgment (Skottsberg, 1958). Susanna and Garcia-Jacas (2001) have placed *Yunquea* in synonymy with the genus *Centaurodendron*, also endemic to the island and in Asteraceae, but more material is needed to test this hypothesis. Skottsberg, C. 1929. Notes on some recent collections made in the islands of Juan Fernandez. *Acta Horti Gothob.* 4: 155–171; Skottsberg, C. 1958. Über *Yunquea tenziii* Skottsb. *Ber. Deutsch. Bot. Ges.* 71: 45–52; Susanna, A. & N. Garcia-Jacas. 2001. Tribe Cardueae. Pp. 123–147 *in* A. Anderberg, et al. (editors), Compositae. *In* J. W. Kadereit & C. Jeffrey (editors), *Flowering Plants, Vol. VIII: Eudicots*, Asterales. *In* K. Kubitzki (series editor), *The Families and Genera of Vascular Plants*. Springer, Berlin.
16. Danton, P. 2000. Une ascension au Mont Yunque dans l'île de Robinson Crusoe (Chili). *Bull. Mens. Soc. Linn. Lyon* 69: 205–216.
17. Alarcón, T. 2015. Expedición a Cerro el Yunque, en la Isla de Robinson Crusoe encuentra especie considerara extinta. *Identidad y Futuro* (24 marzo). https://identidadyfuturo.cl/2015/03/24/expedicion-a-cerro-el-yunque-en-la-isla-de-robinson-crusoe-encuentra-especie-considerara-extinta/. Accessed 10 February 2020.
18. Holland, E. 1983. OSU researcher dies in Peruvian avalanche. *OSU on Campus* 13(4): 1, 5.

19. The lobster is *Jasus frontalis*, endemic to the Robinson Crusoe Islands as well as to the Desventuradas Islands located 824 kilometers to the north, also belonging to Chile.
20. Figueroa, G. J. 2005. La maldición del tesoro. *El Sábado* (supplement in *El Mercurio*), 9 October, pp. 22–29.
21. Cabrera, M. 2019. Polémica en archipiélago de Juan Fernández por uso de retroexcavadora en búsqueda de tesoro. *BiobioChile*. http//:www.biobiochile.cl/noticias/nacional/región-de-valparaiso/2019/12/16/polemica-en-archipielago-juan-fernandez. Accessed 1 February 2020.
22. A parallel and remarkable story occurred with the French journalist, Jean Dominique Bauby, who suffered a similar stroke in 1995 at age forty-five. After coming out of a three-week coma, he was paralyzed from the neck down, able only to blink his eyelids and move his eyebrows. As a journalist, he wished to write about his condition, his frustration of not being able to hug or kiss his two children, and his inability to converse with anyone. He decided to write a book about these experiences, and this he did with the help of an assistant. A system was devised where she would list letters of the alphabet and he would blink for the one he wanted. This was extremely tedious, obviously, but as the text progressed, she was able to anticipate words from only one initial letter, which made progress slightly more efficient. It took 200,000 blinks to yield a small book of more than one hundred pages. It was published in 1997 (Robert Laffont, Paris) to acclaim with the title *Le Scaphandre et le Papillon* (*The Diving Bell and the Butterfly*), which referred to Bauby's active mind trapped within his rigid body. He died shortly after the book was published. See Dahlberg, J.-T. 1997. A writer's spirit soared with a blink of his eye. *Los Angeles Times*, 15 March.
23. Dates of photographs in figures: p. 182, left, 2003; p. 182, right, 1999; p. 183, ca. 1980; p. 186, 1999; p. 187, 2010; p. 188, 1980; p. 189, 2010; p. 192, 1986; pp. 193, 194, color insert image 11, 1997; color insert image 12, 1980; p. 197, top, 1991; p. 197, bottom, 1980.

Chapter 11. Isolation Revisited

1. Jonathan Franzen was featured on *Time*'s cover of 23 August 2010, under the heading "Great American Novelist."
2. Franzen, J. 2001. *The Corrections*. Farrar, Straus & Giroux, New York.
3. Bird watchers can be extremely serious about their hobby, maintaining life lists that they add to with great enthusiasm. For a somewhat humorous take on the competitive nature of some of these hobbyists, see the film *The Big Year* (2011, Twentieth Century-Fox Film Corporation), starring Steve Martin, Owen Wilson, and Jack Black.
4. The *fardelas*, *Pterodroma externa* and *P. longirostris*, are petrels that burrow in the soil and breed only on Alejandro Selkirk Island. They are currently under conservation watch due to predation pressure from feral cats and disturbance of habitat from feral goats. See Brooke, M. de L. 1987. Population estimates and breeding biology of the petrels *Pterodroma externa* and *P. longirostris* on Isla Alejandro Selkirk, Juan Fernandez Archipelago. *The Condor* 89: 581–586; Brooke, M. 2004. *Albatrosses and Petrels across the World*. Oxford University Press, Oxford. For an excellent general overview of seabird biology, see Brooke, M. 2018. *Far from Land. The Mysterious Lives of Seabirds*. Princeton University Press, Princeton.
5. This article, "Farther Away," appeared in the *New Yorker* on 18 April 2011.
6. David Foster Wallace was an author of novels, short stories, and essays, with his book *Infinite Jest* (1996) receiving critical acclaim. He taught English at several American

universities, but after suffering from bouts of depression, he died of suicide in 2008 at the age of 46.

7. Franzen, J. 2012. *Farther Away*. Fourth Estate, London.
8. Greimler, J., P. López S., T. F. Stuessy & T. Dirnböck. 2002. The vegetation of Robinson Crusoe Island (Isla Masatierra), Juan Fernández Archipelago, Chile. *Pacific Sci.* 56: 263–284 + 1 map.
9. Crawford, D. J., T. F. Stuessy & M. Silva O. 1987. Allozyme divergence and the evolution of *Dendroseris* (Compositae: Lactuceae) on the Juan Fernández Islands. *Syst. Bot.* 12: 435–443; Crawford, D. J., T. F. Stuessy, D. W. Haines, M. B. Cosner, M. Silva O. & P. López. 1992. Allozyme diversity within and divergence among four species of *Robinsonia* (Asteraceae: Senecioneae), a genus endemic to the Juan Fernandez Islands, Chile. *Amer. J. Bot.* 79: 962–966; Crawford, D. J., E. Ruiz, T. F. Stuessy, E. Tepe, P. Aqueveque, F. González, R. J. Jensen, et al. 2001. Allozyme diversity in the endemic flowering plant species of the Juan Fernández Archipelago, Chile: Ecological and historical factors with implications for conservation. *Amer. J. Bot.* 88: 2195–2203; Ruiz, E., D. J. Crawford, T. F. Stuessy, F. González, R. Samuel, J. Becerra & M. Silva. 2004. Phylogenetic relationships and genetic divergence among endemic species of *Berberis*, *Gunnera*, *Myrceugenia* and *Sophora* of the Juan Fernández Islands (Chile) and their continental progenitors based on isozymes and nrITS sequences. *Taxon* 53: 321–332; Stuessy, T. F., E. Ruiz, D. J. Crawford & K. Tremetsberger. 2005. Testing degrees of genetic divergence and populational variation in oceanic island archipelagos: Juan Fernández as a model system. *Nova Acta Leopoldina* 92: 147–165.
10. The topographic map of Alejandro Selkirk Island, scale 1:10,000 and 20 meter contour interval, was prepared by Henry H. Brecher in 1990 at the Byrd Polar Research Center, The Ohio State University, from 1980 Chilean Air Force aerial photographs. Linear scale was established by laser measurements taken by Marcelo Baeza and Patricio Peñailillo on their short visit to the island on 5 February 1990. Production of this map and its use on the expedition of 2011 allowed precise plotting of relevés and other collections, as well as landscape orientation in the field.
11. Dates of photographs in figures: p. 209, 2011; color insert image 14, 1997; pp. 211, 212, 214, 215, 216, 219, 220, 221, 223, 226.

Chapter 12. The Heavy Hand of Humans

1. For the most detailed chronicle of these reports (from seventy persons), see Stuessy, T. F. 2020. *Environmental History of Oceanic Islands: Natural and Human Impacts on the Vegetation of the Juan Fernández (Robinson Crusoe) Archipelago*. Springer, Cham, Switzerland.
2. Stuessy, T. F. 2020. The importance of historical ecology for interpreting evolutionary processes in plants of oceanic islands. *J. Syst. Evol.* doi: 10.1111/jse.12673.
3. Stuessy, T. F., D. J. Crawford, C. Marticorena & R. Rodríguez. 1998b. Island biogeography of angiosperms of the Juan Fernández archipelago. Pp. 121–138 *in* T. F. Stuessy & M. Ono (editors), *Evolution and Speciation of Island Plants*. Cambridge University Press, Cambridge. See also Becerril, L., L. E. Lara & V. I. Astudillo. 2021. The strong competition between growth and erosive processes on the Juan Fernández Archipelago (SE Pacific, Chile). *Geomorphology* 373: 107513.
4. Dalrymple, G. B. 1991. *The Age of the Earth*. Stanford University Press, Stanford; Dalrymple, G. B. 2001. The age of the Earth in the twentieth century: A problem (mostly) solved. *Spec. Publ., Geol. Soc. London* 190: 205–221.

5. For many cultural events in the history of the archipelago, information has been taken from Woodward, R. L., Jr. 1969. *Robinson Crusoe's Island*. University of North Carolina Press, Chapel Hill.
6. Hubbs, C. L. & K. S. Norris. 1971. Original teeming abundance, supposed extinction, and survival of the Juan Fernández fur seal. *Antarctic Res. Ser.* 18: 35–52; Torres, D. 1987. Antecedentes sobre el lobo fino de Juan Fernández *Arctocephalus philippii* y proyecciones para su estudio. Pp. 289–317 *in* J. C. Castilla (editor), *Islas Oceánicas Chilenas: Conocimiento Científico y Necesidades de Investigaciones*. Universidad Católica de Chile, Santiago. [See Apéndice 1, pp. 312–313, for numbers of fur seals harvested by crews from each of the ships.]
7. Sutcliffe, T. ["T. S."]. 1839. *The Earthquake of Juan Fernandez, as It Occurred in the Year 1835, Authenticated by the Retired Governor of That Island, to Which Is Added, a Refutation of Several Misstatements That Have Been Published in the "Nautical Magazine" of 1837, and the Public Papers*. Published by the author, Manchester.
8. Arana, P. M. 2010. *La Isla de Robinson Crusoe*. Ediciones Universitarias de Valparaíso, Chile, Valparaíso.
9. Anson, G., R. Walter & B. Robins. 1748. *A Voyage Round the World in the Years MDCCXL, I, II, III, IV by George Anson. London*. [Edited with an introduction by Glyndwr Williams, 1974, Oxford University Press, Oxford.]
10. Gay, C. 1854. *Atlas de la Historia Física y Política de Chile*. E. Thunot, Paris. [Reprint edition 2010, LOM Ediciones, Santiago.]
11. Barría, A. 1996. Incendio penetró bosque nativo en Juan Fernández. *El Mercurio*, February 22.
12. Prebble, M. & J. L. Dowe. 2008. The late Quaternary decline and extinction of palms on oceanic Pacific islands. *Quatern. Sci. Rev.* 27: 2546–2567.
13. Stuessy, T. F., R. W. Sanders & O. R. Matthei. 1983. *Juania australis* revisited in the Juan Fernández Islands, Chile. *Principes* 27: 71–74; Ricci, M. 2006. Conservation status and *ex situ* cultivation efforts of endemic flora of the Juan Fernández Archipelago. *Biodivers. & Conserv.* 15: 3111–3130.
14. Trénel, P., M. H. G. Gustafsson, W. J. Baker, C. B. Asmussen-Lange, J. Dransfield & F. Borchsenius. 2007. Mid-Tertiary dispersal, not Gondwanan vicariance explains distribution patterns in the wax palm subfamily (Ceroxyloideae: Arecaceae). *Molec. Phylogen. Evol.* 45: 272–288.
15. Shineberg, D. 1967. *They Came for Sandalwood: A Study of the Sandalwood Trade in the South-west Pacific, 1830–1865*. Melbourne University Press, Carlton. [Facsimile in 2014 (UQ ePress), University of Queensland Press, St. Lucia.]. The drawing of the sandalwood by Bernardo Krüssel (p. 234, right) was published in Johow, F. 1896. *Estudios sobre la Flora de las Islas de Juan Fernández*. Gobierno de Chile, Santiago.
16. St. John, H. 1947. The history, present distribution, and abundance of sandalwood on Oahu, Hawaiian Islands: Hawaiian Plant Studies 14. *Pacific Sci.* 1: 5–20.
17. Skottsberg, C. 1910. Juan Fernandez-öarnas sandeltrad. *Svensk Bot. Tidskr.* 4: 167–173; Skottsberg, C. 1911. *The Wilds of Patagonia. A Narrative of the Swedish Expedition to Patagonia, Tierra del Fuego and the Falkland Islands in 1907–1909*. The Macmillan Co., New York.
18. Dyer, B. S. & M. W. Westneat. 2010. Taxonomy and biogeography of the coastal fishes of Juan Fernández Archipelago and Desventuradas Islands, Chile. *Rev. Biol. Mar. Oceanograf.* 45: 589–617.
19. See data on lobster catches in Guzmán P., J. Circa 1950. *Cumbres Oceánicas. Las Islas Juan Fernández*. Published by the author, Santiago. [No date of publication exists in

this work; the latest date cited in the "Bibliografía" is 1945. Some booksellers have picked 1950 or 1951 as the year of publication, but this appears to be guesswork.]; Arana E., P. 1985. Análisis y recomendaciones sobre medidas de regulación en la pesquería de la langosta de Juan Fernández (*Jasus frontalis*). Pp. 291–300 *in* P. Arana E. (editor), *Investigaciones Marinas en el Archipiélago de Juan Fernández*. Escuela de Ciencias del Mar, Universidad Católica de Valparaíso, Valparaíso.

20. Bahamonde, N. 1966. Capitulo VIII. El mar y sus recursos. Pp. 81–89 *in* Anonymous (editor), *Geografía Económica de Chile*, primer apéndice. Corporación de Fomento de la Producción (CORFO), Santiago.
21. Biodiversa. 2009. *Archipiélago Juan Fernández. Sitio Prioritario para la Conservación de la Biodiversidad Global. Sistematización del Estado Actual del Conocimiento.* DIPROREN-SAG, Santiago.
22. Camus P., S. Castro & F. Jaksic. 2008. El conejo europeo en Chile: Historia de una invasión biológica. *Historia* 41: 305–339.
23. Saiz, F. & P. Ojeda. 1988. *Oryctolagus cuniculus* L. en Juan Fernández. Problema y control. *Anal. Mus. Hist. Nat. Valparaíso* 19: 91–98.
24. Looser, G. 1927. La zarzamora (*Rubus ulmifolius* Schott) en Juan Fernández. *Rev. Chil. Hist. Nat.* 31: 84–85.
25. Smith-Ramírez, C., G. Arellano, E. Hagin, R. Vargas, J. Castillo & A. Miranda. 2013. El rol de *Turdus falcklandii* (Aves: Passeriforme) como dispensor de plantas invasoras en el archipiélago de Juan Fernández. *Rev. Chil. Hist. Nat.* 86: 33–48.
26. Davis, M. A., M. K. Chew, R. J. Hobbs, A. E. Lugo, J. J. Ewel, G. J. Vermeij, J. H. Brown, et al. 2011. Don't judge species on their origins. *Nature* 474: 153–154; Vince, G. 2011. Embracing invasives. *Science* 331: 1383–1384.
27. Alarcón, D., P. López-Sepúlveda, G. Fuentes, H. Montoya, P. Peñailillo & P. Carrasco. 2019. Parches invadidos por *Ugni molinae* en isla Robinson Crusoe: ¿Hay plantas nativas y endémicas capaces de vivir en ellos? *Gayana Bot.* 76: 126–131.
28. Stuessy, T. F., R. Rodríguez, C. M. Baeza & P. López-Sepúlveda. 2018. Taxonomic inventory. Pp. 57–97 *in* T. F. Stuessy, D. J. Crawford, P. López-Sepúlveda, C. M. Baeza & E. A. Ruiz (editors), *Plants of Oceanic Islands: Evolution, Biogeography, and Conservation of the Flora of the Juan Fernández (Robinson Crusoe) Archipelago*. Cambridge University Press, Cambridge.
29. Date of photograph in figure: p. 235, left, 1980.

Chapter 13. Scientific Accomplishments

1. Rodríguez-Ríos, R. 1990. *Gleichenia lepidota* n. sp. y la familia Gleicheniaceae del archipiélago de Juan Fernández, Chile. *Gayana Bot.* 47: 37–45; Danton, P. 2006. Contribution à la flore de l'archipel Juan Fernández (Chili). Description de deux taxons nouveaux: *Nicotiana cordifolia* subsp. *sanctaclarae* subsp. nov. (Solanaceae), *Robinsonia saxatilis* sp. nov. (Asteraceae). *Acta Bot. Gallica* 153: 249–255; Danton, P. 2014. Contribution à la flore de l'archipel Juan Fernández (Chili). Description de 4 taxons nouveaux: Angiospermae – *Erigeron corrales-molinensis* sp. nov. (Asteraceae), *Euphrasia formossissima* Skottsb. subsp. *cucharensis* subsp. nov. (Orobanchaceae), *Haloragis masatierrana* var. *applanata* var. nov. et var. *scabrida* var. nov. (Haloragidaceae). *Acta Bot. Gallica* 161: 355–371; Danton, P., M. Boudrie, A. Bizot & R. L. L. Viane. 2015. *Pleopeltis* ×*cerro-altoensis* (Polypodiaceae), a new fern hybrid from Robinson Crusoe Island (Juan Fernández Archipelago, Chile). *Fern Gaz.* 20: 65–78; Valdebenito, H. 2018. *Erigeron stuessyi* sp. nov. Pp. 381–382 *in* T. F. Stuessy, D. J. Crawford, P. López-Sepúlveda, C. M. Baeza & E. A. Ruiz (editors), *Plants of Oceanic*

Islands: Evolution, Biogeography, and Conservation of the Flora of the Juan Fernández (Robinson Crusoe) Archipelago. Cambridge University Press, Cambridge.

2. Stuessy, T. F., K. A. Foland, J. F. Sutter, R. W. Sanders & M. Silva O. 1984. Botanical and geological significance of potassium-argon dates from the Juan Fernandez Islands. *Science* 255: 49–51; Stuessy, T. F. 2020. *Environmental History of Oceanic Islands: Natural and Human Impacts on the Vegetation of the Juan Fernández (Robinson Crusoe) Archipelago*. Springer, Cham, Switzerland. [See p. 14, table 2.1.]
3. For good discussions of the efficacy of different types of data in plant systematics and evolution, see Stuessy, T. F. 2009. *Plant Taxonomy: The Systematic Evaluation of Comparative Data*, ed. 2. Columbia University Press, New York; Stuessy, T. F., D. J. Crawford, D. E. Soltis & P. S. Soltis. 2014. *Plant Systematics: The Origin, Interpretation, and Ordering of Plant Biodiversity*. Koeltz Scientific Books, Königstein.
4. Bernardello, G., G. J. Anderson, T. F. Stuessy & D. J. Crawford. 2006. The angiosperm flora of the Archipelago Juan Fernández (Chile): Origin and dispersal. *Canad. J. Bot.* 84: 1266–1281.
5. We have estimated that as much as 95 percent of the surface area of Robinson Crusoe Island has disappeared over the four million years of its existence; Alejandro Selkirk Island would have suffered only 28 percent reduction. See Sanders, R. W., T. F. Stuessy, C. Marticorena & M. Silva O. 1987. Phytogeography and evolution of *Dendroseris* and *Robinsonia*, tree-Compositae of the Juan Fernández Islands. *Opera Bot.* 92: 195–215; Stuessy, T. F., D. J. Crawford, C. Marticorena & R. Rodríguez. 1998. Island biogeography of angiosperms of the Juan Fernández archipelago. Pp. 121–138 *in* T. F. Stuessy & M. Ono (editors), *Evolution and Speciation of Island Plants*. Cambridge University Press, Cambridge.
6. Stuessy, T. F., J. Greimler & T. Dirnböck. 2005. Landscape modification and impact on specific and genetic diversity in oceanic islands. Pp. 89–101 *in* H. Balslev & I. Friis (editors), *Plant Diversity and Complexity Patterns: Local, Regional and Global Dimensions*. Danish Academy of Sciences, Copenhagen. [*Biol. Skr.* 55: 89–101.]; Stuessy, T. F. 2020. The importance of historical ecology for interpreting evolutionary processes in plants of oceanic islands. *J. Syst. Evol.* doi: 10.1111/jse.12673.
7. For much more detail of the human impacts on the vegetation and marine life of the Robinson Crusoe Islands, see Stuessy, T. F. 2020. *Environmental History of Oceanic Islands: Natural and Human Impacts on the Vegetation of the Juan Fernández (Robinson Crusoe) Archipelago*. Springer, Cham, Switzerland.
8. This concept apparently originated in the nineteenth century from the British botanist, Joseph Hooker. See Carine, M. & M. Menezes de Sequeira. 2020. Sir Joseph Hooker on *Insular Floras*: Human impact and the natural laboratory paradigm. *Sci. Insularum* 3: 73–88.
9. Darwin, C. 1859. *On the Origin of Species by Means of Natural Selection, or the Preservation of Favoured Races in the Struggle for Life*. John Murray, London.
10. Behnke, H.-D. 1988. Sieve-element plastids, phloem protein, and evolution of flowering plants: III. Magnoliidae. *Taxon* 37: 699–732.
11. Weiss-Schneeweiss, H., T. F. Stuessy & J. L. Villaseñor. 2009. Chromosome numbers, karyotypes, and evolution in *Melampodium* (Asteraceae). *Int. J. Pl. Sci.* 170: 1168–1182.
12. Stuessy, T. F. & D. J. Crawford. 1998. Chromosomal stasis during speciation in angiosperms of oceanic islands. Pp. 307–324 *in* T. F. Stuessy & M. Ono (editors), *Evolution and Speciation of Island Plants*. Cambridge University Press, Cambridge.
13. Crawford, D. J., G. J. Anderson & G. Bernardello. 2011. The reproductive biology of island plants. Pp. 11–36 *in* D. Bramwell & J. Caujapé-Castells (editors), *The Biology*

of Island Floras. Cambridge University Press, Cambridge; Anderson, G., D. J. Crawford & G. Bernardello. 2013. The reproductive biology of island plants: The Juan Fernández and Canary Islands (facilitated by the work of Arnoldo Santos-Guerra). *Vieraea* 41: 109–121; Crawford, D. J., G. J. Anderson, G. Bernardello & A. Santos-Guerra. 2013. The reproductive biology of island plants: The Juan Fernández and Canary Islands. Pp. 166–175 *in* J. Caujapé-Castells, G. Niet & J. M. Fernández-Palacios (editors), *Proceedings of the Amurga International Conferences on Island Biodiversity 2011: Present and Emerging Knowledge on the Evolution, Diversity and Conservation of the Canarian Flora*. Fundación Canaria Amurga-Maspalomas, Jardín Botánico Canario "Viera y Clavijo," Las Palmas de Gran Canaria.

14. Bernardello, G., L. Galetto & G. J. Anderson. 2000. Floral nectary structure and nectar chemical composition of some species from Robinson Crusoe Island (Chile). *Canad. J. Bot.* 78: 862–872; Anderson, G., G. Bernardello, T. Stuessy & D. J. Crawford. 2001. Breeding system and pollination of selected plants endemic to Juan Fernández Islands. *Amer. J. Bot.* 88: 220–233; Bernardello, G., G. J. Anderson, T. F. Stuessy & D. J. Crawford. 2001. A survey of floral traits, breeding systems, floral visitors, and pollination systems of the angiosperms of the Juan Fernández Islands (Chile). *Bot. Rev. (Lancaster)* 67: 255–308.
15. Bernardello, G., G. Anderson, P. López, M. A. Cleland, T. F. Stuessy & D. J. Crawford. 1999. Reproductive biology of *Lactoris fernandeziana* (Lactoridaceae). *Amer. J. Bot.* 86: 829–840.
16. Anderson, G. & G. Bernardello. 2012. Robinson Crusoe principle: The pollination biology of the flora of the Juan Fernández Archipelago and how the species "made do" with what they had [abstract]. Columbus, Ohio: Botany 2012.
17. For excellent overviews of these compounds, see Crawford, D. J. 1979. Flavonoid chemistry and angiosperm evolution. *Bot. Rev.* 44: 431–456; Bohm, B. A. 1998. *Introduction to Flavonoids*. Harwood, Amsterdam.
18. Crawford, D. J., M. Silva O., P. Pacheco & H. Valdebenito. 2018. Flavonoid compounds. Pp. 165–187 *in* T. F. Stuessy, D. J. Crawford, P. López-Sepúlveda, C. M. Baeza & E. A. Ruiz (editors), *Plants of Oceanic Islands: Evolution, Biogeography, and Conservation of the Flora of the Juan Fernández (Robinson Crusoe) Archipelago*. Cambridge University Press, Cambridge.
19. Bohm, B. A. 1998. Secondary compounds and evolutionary relationships of island plants. Pp. 233–306 *in* T. F. Stuessy & M. Ono (editors), *Evolution and Speciation of Island Plants.* Cambridge Univeresity Press, Cambridge.
20. Lewontin, R. C. & J. L. Hubby. 1966. A molecular approach to the study of genetic heterozygosity in natural populations. II. Amount of variation and degree of heterozygosity in natural populations of *Drosophila pseudoobscura*. *Genetics* 54: 595–609; Crawford, D. J. 1985. Electrophoretic data and plant speciation. *Syst. Bot.* 10: 405–416; Crawford, D. J. 1989. Enzyme electrophoresis and plant systematics. Pp. 146–164 *in* D. E. Soltis & P. S. Soltis (editors), *Isozymes in Plant Biology*. Dioscorides, Portland; Crawford, D. J. 1990. *Plant Molecular Systematics: Macromolecular Approaches*. Wiley, New York; Crawford, D. J., E. Ruiz, T. F. Stuessy, E. Tepe, P. Aqueveque, F. González, R. J. Jensen, et al. 2001. Allozyme diversity in the endemic flowering plant species of the Juan Fernández Archipelago, Chile: Ecological and historical factors with implications for conservation. *Amer. J. Bot.* 88: 2195–2203.
21. Palmer, J. D. 1986. Chloroplast DNA and phylogenetic relationships. Pp. 63–80 *in* S. K. Dutta (editor), *DNA Systematics, Vol. 2: Plants*. CRC Press, Boca Raton.
22. Chase, M. W. & H. H. Hills. 1991. Silica gel: An ideal material for field preservation of leaf samples for DNA studies. *Taxon* 40: 215–220.

23. Crawford, D. J., T. F. Stuessy, M. B. Cosner, D. W. Haines & M. Silva O. 1993. Ribosomal and chloroplast DNA restriction site mutations and the radiation of *Robinsonia* (Asteraceae, Senecioneae) on the Juan Fernández Islands. *Pl. Syst. Evol.* 184: 233–239; Crawford, D. J., T. F. Stuessy, M. B. Cosner, D. W. Haines, M. Silva O. & M. Baeza. 1992. Evolution of the genus *Dendroseris* (Asteraceae: Lactuceae) on the Juan Fernández Islands: Evidence from chloroplast and ribosomal DNA. *Syst. Bot.* 17: 676–682; Crawford, D. J., T. Sang, T. F. Stuessy, S.-C. Kim & M. Silva O. 1998. *Dendroseris* (Asteraceae: Lactuceae) and *Robinsonia* (Asteraceae: Senecioneae) on the Juan Fernandez Islands: Similarities and differences in biology and phylogeny. Pp. 97–119 *in* T. F. Stuessy & M. Ono (editors), *Evolution and Speciation of Island Plants*. Cambridge University Press, Cambridge. [p. 106.]
24. Sang, T., D. J. Crawford, S.-C. Kim & T. F. Stuessy. 1994. Radiation of the endemic genus *Dendroseris* (Asteraceae) on the Juan Fernández Islands: Evidence from sequences of the ITS regions of nuclear ribosomal DNA. *Amer. J. Bot.* 81: 1494–1501; Sang, T., D. J. Crawford, T. F. Stuessy & M. Silva O. 1995. ITS sequences and the phylogeny of the genus *Robinsonia* (Asteraceae). *Syst. Bot.* 20: 55–64.
25. Vos, P., R. Hogers, M. Bleeker, M. Reijans, T. van de Lee, M. Hornes, A. Friters, et al. 1995. AFLP: A new technique for DNA fingerprinting. *Nucleic Acids Res.* 23: 4407–4414.
26. Takayama, K., P. López-Sepúlveda, J. Greimler, D. J. Crawford, P. Peñailillo, M. Baeza, E. Ruiz, et al. 2015. Relationships and genetic consequences of contrasting modes of speciation among endemic species of *Robinsonia* (Asteraceae, Senecioneae) of the Juan Fernández Archipelago, Chile, based on AFLPs and SSRs. *New Phytol.* 205: 415–428.
27. Cho, M.-S., S.-H. Kim, J. Yang, D. J. Crawford, T. F. Stuessy, P. López-Sepúlveda & S.-C. Kim. 2020. Plastid phylogenomics of *Dendroseris* (Cichorieae; Asteraceae): Insights into structural organization and molecular evolution of an endemic lineage from the Juan Fernández Islands. *Frontiers Pl. Sci.* doi: 10.3389/fpls.2020.594272
28. Greimler, J. 1997. Pflanzengesellschaften und Vegetationsstruktur in den südlichen Gesäusebergen (nordöstliche Kalkalpen, Steiermark). *Mitt. Abteil. Bot. Landesmus. "Joanneum" Graz* 25–26: 1–238.
29. Braun-Blanquet, J. 1928. *Pflanzensoziologie, Grundzüge der Vegetationskunde*. Springer, Berlin. [Ed. 3, 1964, Springer, Wien.]
30. Greimler, J., P. López S., T. F. Stuessy & T. Dirnböck. 2002. The vegetation of Robinson Crusoe Island (Isla Masatierra), Juan Fernández Archipelago, Chile. *Pacific Sci.* 56: 263–284 + 1 map; Greimler, J., P. López-Sepúlveda, K. Reiter, C. Baeza, P. Peñailillo, E. Ruiz, P. Novoa, et al. 2013. Vegetation of Alejandro Selkirk Island (Isla Masafuera), Juan Fernández Archipelago, Chile. *Pacific Sci.* 67: 267–282.
31. Stuessy, T. F., D. J. Crawford, C. Marticorena & M. Silva O. 1998e. Isolating mechanisms and modes of speciation in endemic angiosperms of the Juan Fernández Islands. Pp. 79–96 *in* T. F. Stuessy & M. Ono (editors), *Evolution and Speciation of Island Plants*. Cambridge University Press, Cambridge; Crawford, D. J., T. F. Stuessy, K. Takayama, P. López-Sepúlveda, G. J. Anderson & G. Bernardello. 2018. Speciation. Pp. 308–331 *in* T. F. Stuessy, D. J. Crawford, P. López-Sepúlveda, C. M. Baeza & E. A. Ruiz (editors), *Plants of Oceanic Islands: Evolution, Biogeography, and Conservation of the Flora of the Juan Fernández (Robinson Crusoe) Archipelago*. Cambridge University Press, Cambridge.
32. Levin, D. A. 1978. The origin of isolating mechanisms in flowering plants. *Evol. Biol.* 11: 185–317.
33. This was illustrated in studies of elevation, soils, and associated species among taxa of *Dendroseris* and *Robinsonia*. See Sanders, R. W., T. F. Stuessy, C. Marticorena &

M. Silva O. 1987. Phytogeography and evolution of *Dendroseris* and *Robinsonia*, tree-Compositae of the Juan Fernández Islands. *Opera Bot.* 92: 195–215.

34. For a discussion of the concepts of cladogenesis and anagenesis, see Stuessy, T. F., D. J. Crawford & C. Marticorena. 1990. Patterns of phylogeny in the endemic vascular flora of the Juan Fernández Islands, Chile. *Syst. Bot.* 15: 338–346.
35. Stuessy, T. F., D. J. Crawford & E. A. Ruiz. 2018. Patterns of phylogeny. Pp. 209–275 *in* T. F. Stuessy, D. J. Crawford, P. López-Sepúlveda, C. M. Baeza & E. A. Ruiz (editors), *Plants of Oceanic Islands: Evolution, Biogeography, and Conservation of the Flora of the Juan Fernández (Robinson Crusoe) Archipelago*. Cambridge University Press, Cambridge.
36. Stuessy, T. F., G. Jakubowsky, R. Salguero Gómez, M. Pfosser, P. M. Schluter, T. Fer, B.-Y. Sun, et al. 2006. Anagenetic evolution in island plants. *J. Biogeogr.* 33: 1259–1265.
37. Sun, B.-Y. & T. F. Stuessy. 1998. Preliminary observations on the evolution of endemic angiosperms of Ullung Island, Korea. Pp. 181–202 *in* T. F. Stuessy & M. Ono (editors), *Evolution and Speciation of Island Plants*. Cambridge University Press, Cambridge; Takayama, K., B.-Y. Sun & T. F. Stuessy. 2012. Genetic consequences of anagenetic speciation in *Acer okamotoanum* (Sapindaceae) on Ullung Island, Korea. *Ann. Bot.* 109: 321–330; Takayama, K., B.-Y. Sun & T. Stuessy. 2013. Anagenetic speciation in Ullung Island, Korea: Genetic diversity and structure in the island endemic species, *Acer takesimense* (Sapindaceae). *J. Pl. Res.* 126: 323–333.
38. Takayama, K., P. López-Sepúlveda, J. Greimler, D. J. Crawford, P. Peñailillo, M. Baeza, E. Ruiz, et al. 2015. Genetic consequences of cladogenetic vs. anagenetic speciation in endemic plants of oceanic islands. *AoB PLANTS* 7: plv102; doi:10.1093/aobpla/plv102
39. Francisco-Ortega, J., R. K. Jansen & A. Santos-Guerra. 1996. Chloroplast DNA evidence of colonization, adaptive radiation, and hybridization in the evolution of the Macaronesian flora. *Proc. Natl. Acad. Sci. U.S.A.* 93: 4085–4090; Fjellheim, S., M. H. Jørgensen, M. Kjos & L. Borgen. 2009. A molecular study of hybridization and homoploid hybrid speciation in *Argyranthemum* (Asteraceae) on Tenerife, the Canary Islands. *Bot. J. Linn. Soc.* 159: 19–31; Van Hengstum, T., S. Lachmuth, J. G. B. Oostermeijer, J. C. M. den Nijs, P. G. Meirmans & P. H. van Tienderen. 2012. Human-induced hybridization among congeneric endemic plants on Tenerife, Canary Islands. *Pl. Syst. Evol.* 298: 1119–1131; White, O. W., J. A. Reyes-Betancort, M. A. Chapman & M. A. Carine. 2020. Geographical isolation, habitat shifts and hybridization in the diversification of the Macaronesian endemic genus *Argyranthemum* (Asteraceae). *New Phytol.* 228: 1953–1971.
40. Carr, G. D. & D. W. Kyhos. 1981. Adaptive radiation in the Hawaiian silversword alliance (Compositae-Madiinae). I. Cytogenetics of spontaneous hybrids. *Evolution* 35: 543–556; Caraway, V., G. D. Carr & C. W. Morden. 2001. Assessment of hybridization and introgression in lava-colonizing Hawaiian *Dubautia* (Asteraceae: Madiinae) using RAPD markers. *Amer. J. Bot.* 88: 1688–1694; García-Verdugo, C., E. Friar & L. S. Santiago. 2013. Ecological role of hybridization in adaptive radiations: A case study in the *Dubautia arborea–Dubautia ciliolata* (Asteraceae) complex. *Int. J. Pl. Sci.* 174: 749–759.
41. Pacheco, P., T. F. Stuessy & D. J. Crawford. 1991. Natural interspecific hybridization in *Gunnera* (Gunneraceae) of the Juan Fernández Islands, Chile. *Pacific Sci.* 45: 389–399.
42. Crawford, D. J., S. Brauner, M. B. Cosner & T. F. Stuessy. 1993. Use of RAPD markers to document the origin of the intergeneric hybrid ×*Margyracaena skottsbergii* (Rosaceae) on the Juan Fernández Islands. *Amer. J. Bot.* 80: 89–92.

43. Danton, P., M. Boudrie, A. Bizot & R. L. L. Viane. 2015. *Pleopeltis ×cerro-altoensis* (Polypodiaceae), a new fern hybrid from Robinson Crusoe Island (Juan Fernández Archipelago, Chile). *Fern Gaz.* 20: 65–78.
44. Sanders, R., T. F. Stuessy & C. Marticorena. 1982. Recent changes in the flora of the Juan Fernández Islands, Chile. *Taxon* 31: 284–289; Swenson, U., T. F. Stuessy, M. Baeza & D. J. Crawford. 1997. New and historical plant introductions, and potential pests, in the Juan Fernández Islands, Chile. *Pacific Sci.* 51: 233–253; Stuessy, T. F., U. Swenson, D. J. Crawford, G. Anderson & M. Silva O. 1998. Plant conservation in the Juan Fernández archipelago, Chile. *Aliso* 16: 89–101; Stuessy, T. F., U. Swenson, C. Marticorena, O. Matthei & D. J. Crawford. 1998. Loss of plant diversity and extinction on Robinson Crusoe Islands, Chile. Pp. 243–257 *in* C.-I. Peng & P. P. Lowry II (editors), *Rare, Threatened, and Endangered Floras of Asia and the Pacific Rim.* Institute of Botany, Academica Sinica, Taipei. [Monograph Series No. 16.]; Stuessy, T. F., C. M. Baeza, P. López-Sepúlveda, R. Rodríguez & U. Swenson. 2018. Conservation of native and endemic species. Pp. 149–161 *in* T. F. Stuessy, D. J. Crawford, P. López-Sepúlveda, C. M. Baeza & E. A. Ruiz (editors), *Plants of Oceanic Islands: Evolution, Biogeography, and Conservation of the Flora of the Juan Fernández (Robinson Crusoe) Archipelago.* Cambridge University Press, Cambridge.
45. Sanders, R., T. F. Stuessy & C. Marticorena. 1982. Recent changes in the flora of the Juan Fernández Islands, Chile. *Taxon* 31: 284–289; Stuessy, T. F., R. Rodríguez, C. M. Baeza & P. López-Sepúlveda. 2018. Taxonomic inventory. Pp. 57–97 *in* T. F. Stuessy, D. J. Crawford, P. López-Sepúlveda, C. M. Baeza & E. A. Ruiz (editors), *Plants of Oceanic Islands: Evolution, Biogeography, and Conservation of the Flora of the Juan Fernández (Robinson Crusoe) Archipelago.* Cambridge University Press, Cambridge.
46. Dates of photographs in figures: p. 243, top left, 1991; p. 243, top middle left, 2011; p. 243, top middle right, 1984; p. 243, top right and bottom left, 2011; p. 243, bottom middle left, 2010; p. 243, bottom middle right, 2011; p. 243, bottom right, 1980; p. 252, 1984; p. 254, top, 1991; p. 254, middle, 1990; p. 254, bottom, 1991; p. 261, bottom, 1996.

Chapter 14. Perspectives

1. Estimates suggest that the number of eukaryotic species on earth might be somewhere near 8.7 million. Mora, C., D. P. Tittensor, S. Adl, A. G. B. Simpson & B. Worm. 2011. How many species are there on earth and in the ocean? *PLOS Biol.* 9(8): e1001127. doi.org/10.1371/journal.pbio.1001127.
2. For a readable insight to the human need to inventory all other life forms, see Dunn, R. R. 2009. *Every Living Thing: Man's Obsessive Quest to Catalog Life, from Nanobacteria to New Monkeys.* HarperCollins, New York.
3. Ruiz, E., D. J. Crawford, T. F. Stuessy, F. González, R. Samuel, J. Becerra & M. Silva O. 2004. Phylogenetic relationships and genetic divergence among endemic species of *Berberis*, *Gunnera*, *Myrceugenia* and *Sophora* of the Juan Fernández Islands (Chile) and their continental progenitors based on isozymes and nrITS sequences. *Taxon* 5: 321–332; Stuessy, T. F., E. Ruiz, D. J. Crawford & K. Tremetsberger. 2005. Testing degrees of genetic divergence and populational variation in oceanic island archipelagos: Juan Fernández as a model system. *Nova Acta Leopoldina* 92: 147–165.
4. The Emperor Seamounts are a series of submarine volcanoes up to eighty-one million years of age that stretch 5,800 kilometers northwest of the present Hawaiian Archipelago. See Tarduno, J. A. 2003. The Emperor Seamounts: Southward motion of the Hawaiian hotspot plume in Earth's mantle. *Science* 301: 1064–1069.

5. For a readable account of these famous islands, see Perry, R. 1972. *The Galápagos Islands.* Dodd, Mead & Company, New York.
6. Hoernle, K. & J.-C. Carracedo. 2009. *Canary Islands, Geology*. Pp. 133–143 *in* R. G. Gillespie & D. A. Clague (editors), *Encyclopedia of Islands.* University of California Press, Berkeley.
7. For a detailed history of the development of cladistics within botany, see Stuessy, T. F. 2009. *Plant Taxonomy: The Systematic Evaluation of Comparative Data*, ed. 2. Columbia University Press, New York.
8. Wagner, W. L., D. R. Herbst & S. H. Sohmer. 1999. *Manual of the Flowering Plants of Hawai'i*, ed. 2. University of Hawai'i and Bishop Museum Presses, Honolulu.
9. W. Preston Adams passed away on 19 August 2017 at the age of eighty-seven. He was a fundamental part of my academic career development.
10. Buero Vallejo, A. 1954. *En la Ardiente Oscuridad: Drama en Tres Actos*. Charles Scribner's Sons, New York. [Edited by S. A. Wotsy.]
11. Stuessy, T. F., K. A. Foland, J. F. Sutter, R. W. Sanders & M. Silva O. 1984. Botanical and geological significance of potassium-argon dates from the Juan Fernandez Islands. *Science* 255: 49–51. It was in this paper that we showed Robinson Crusoe Island to be about four million years old and Alejandro Selkirk Island about one million.
12. Sanders, R. W., T. F. Stuessy, C. Marticorena & M. Silva O. 1987. Phytogeography and evolution of *Dendroseris* and *Robinsonia*, tree-Compositae of the Juan Fernández Islands. *Opera Bot.* 92: 195–215.
13. For good discussions of evolution versus creationism, see Berra, T. M. 1990. *Evolution and the Myth of Creationism*. Stanford University Press, Stanford; Nye, B. [edited by C. S. Powell]. 2014. *Undeniable: Evolution and the Science of Creation*. St. Martin's Press, New York.
14. The Botanical Society of America met jointly with the Canadian Botanical Society at Vancouver, British Columbia, on the University of Vancouver campus from 26 to 30 July 2008.
15. The literature on hominin evolution is vast; a good start for an overview is Wikipedia. Ancient DNA from fossils is now providing more insights, especially routes of human migration out of Africa and our relationship to Neandertals and Denisovans, e.g., Green, R. E., A.-S. Malaspinas, J. Krause, A. W. Briggs, P. L. F. Johnson, C. Uhler, M. Meyer, et al. 2008. A complete Neandertal mitochondrial genome sequence determined by high-throughput sequencing. *Cell* 134: 416–426; Sankararaman, S., N. Patterson, H. Li, S. Pääbo & D. Reich. 2012. The date of interbreeding between Neandertals and modern humans. *PLOS Genet.* 8(10): e1002947.
16. I learned sadly on 15 January 2021 that Roger Sanders passed away in November 2018 from Alzheimer's disease. He was a warm and decent person who made significant contributions to the Robinson Crusoe research project.

Index

Location of names in legends of the unpaginated color (C) insert are indicated by C1, C2, C3, etc.